Third Edition

MANAGING THE LABORATORY ANIMAL FACILITY

Third Edition

MANAGING THE LABORATORY ANIMAL FACILITY

Jerald Silverman

CRC Press
Taylor & Francis Group
Boca Raton London New York

CRC Press is an imprint of the
Taylor & Francis Group, an **informa** business

CRC Press
Taylor & Francis Group
6000 Broken Sound Parkway NW, Suite 300
Boca Raton, FL 33487-2742

Printed on acid-free paper
Version Date: 20160427

International Standard Book Number-13: 978-1-4987-4278-8 (Hardback)

Library of Congress Cataloging-in-Publication Data

Names: Silverman, Jerald, author.
Title: Managing the laboratory animal facility / Jerald Silverman.
Description: Third edition. | Boca Raton : Taylor & Francis, 2017. | Includes bibliographical references and index.
Identifiers: LCCN 2016016334 | ISBN 9781498742788
Subjects: | MESH: Laboratories--organization & administration | Laboratory Animal Science--organization & administration
Classification: LCC SF406 | NLM QY 23 | DDC 636.088/5--dc23
LC record available at https://lccn.loc.gov/2016016334

Visit the Taylor & Francis Web site at
http://www.taylorandfrancis.com

and the CRC Press Web site at
http://www.crcpress.com

The first edition of this book was dedicated to my family, and the second was dedicated to my colleagues in laboratory animal science. I am very pleased to dedicate this third edition to the untold number of animals who have been used on behalf of scientific inquiries intended to help many other animals. Only time and the shifting sands of conscience will judge if that goal, no matter how noble it may seem today, will have justified the means. For now, this book is a small contribution that I hope will enhance our efforts as a profession to do the best we can for those animals who, day after day, depend on us for their care and comfort.

Contents

Preface to the Third Edition

During the time the first edition of *Managing the Laboratory Animal Facility* was being written, there was no working title; it was just called "the management book." But books have titles, and the first title considered was *Managing Laboratory Animal Resources*. We eventually decided that title was too formal and some potential readers might not understand what laboratory animal resources implied; thus, the title of *Managing the Laboratory Animal Facility* was chosen. Nonetheless, this third edition of *Managing the Laboratory Animal Facility* continues to focus on the management of the resources, such as people, money, time, and information, that managers use on a daily basis. The proper use of resources is the key to successful animal facility management.

Just as plants and animals evolve, management also evolves. Theories come and go, with new thinking replacing old concepts that no longer integrate with the needs of a changing world. In this book, we focus on the basics of modern animal facility management. It is primarily a text to help you maximize your management skills in order to provide high-quality service to your customers while minimizing the daily stressors that often occur in work environments. This focus on providing a strong general management background helps prevent problems from arising in the first place. It almost goes without saying that it is more important for a manager to prevent a problem from occurring than to have to correct that problem. This prevention perspective extends beyond working with people. Managers are expected to manage resources, and people are just one of those resources. You still have to manage money, information, and even time. A chapter is devoted to each of these subjects, yet because working with people is such an important part of many managers' jobs, an additional chapter is devoted to leadership. Managers often must take the role of leaders, and conversely, it's the rare leader who has no managerial responsibilities.

Not all management needs are going to be the same. Even though the basics of modern management are the central theme of this book, their implementation may require some adaptation when working in different environments. The animal facility manager in an academic environment may have challenges that differ from those of managers in the pharmaceutical industry. Managers in different academic institutions have to adapt to their own institution's culture, which may be substantially different from the culture of a nearby college or university. What works for you in New York City may not work in Atlanta, Georgia. Another distinction is that managing and leading an animal facility is unlike traditional for-profit corporate management, as developing strategies to gain, retain, and expand a market share is usually not a primary concern of vivarium managers, and vivarium managers have a relatively limited need to market their services. Nevertheless, in line with managers in other industries, we do have strategic planning needs, and on occasion, we do choose to market our services.

Management isn't easy. It takes a lot of work, practice, and adaptability. I want to emphasize the need to keep learning and practicing your management skills, and then practice some more. You will only be fooling yourself if you try to improve yourself as a manager by reading this book, going to management seminars, talking about good management, and then reverting right back to old, comfortable, yet outdated management methods when you get to work. We are all very talented at convincing ourselves that we don't have to improve; it's the rest of the world that needs to change. You probably know that's just not true and it's not going to happen. The truth is that management is a learned skill, and the more you practice, the better you become. But you have to practice what you learn and not stuff new knowledge into your back pocket, never again to see the light of day. Great scientists work hard to find answers to important questions. Discovery is their passion. Great athletes continually hone their skills to become even greater athletes. Being the best at what they do is their passion. Great managers do the same. They continually practice and improve their management skills so performing every day at a high level becomes routine, not an exception, and then they strive for an even higher level of performance. Don't be afraid to make some mistakes. The goal is to keep improving and not let the usual setbacks that every manager experiences eat away at your determination. Every setback is an opportunity to learn and to avoid making the same mistake in the future.

The third edition of *Managing the Laboratory Animal Facility* provides updated and expanded information in every chapter and appendix. Most of the original references remain, as they set the stage for supplementation by more recent management thinking. New sections include discussions of Lean management, change management, negotiations, mentoring, and performance reviews. As in past editions, the appendices focus more on the "how to do it" aspects of certain topics than does the main text, although in this third edition, some additional "how to do it" information has been incorporated into the main text.

This book was written by me, and I will take full responsibility for its content and, in particular, for my tendency to repeat important concepts over and over, in different chapters and using slightly different phrasing. I hope that this redundancy will be helpful to you. At the same time, I have listened carefully to the comments made by my colleagues who donated their time to review the text and provide me with much welcomed feedback. I am grateful to these fine people: Ms. Maryellen Berry, Ms. Pamela Belanger, Ms. Ellen Bridge, Mr. Ethan Hallet, Dr. Samer Jaber, Mr. Edward Jaskolski, Ms. Eva Miele, Mr. Corey Nunes, Ms. Amy Rivera, and Ms. Holly Such.

Jerald Silverman

Author

Jerald Silverman received his degree in nonprofit organization management from the Graduate School of Management and Urban Professions of the New School for Social Research in New York City. Previously, he received degrees in vertebrate zoology and veterinary medicine from Cornell University. After a decade of private veterinary practice, he moved to biomedical research, in which he has more than 40 years of experience managing laboratory animal facilities in private and academic research institutions. Dr. Silverman is the director of the Department of Animal Medicine and professor of pathology at the University of Massachusetts Medical School in Worcester and adjunct professor at the Cummings School of Veterinary Medicine at Tufts University, North Grafton, Massachusetts.

Dr. Silverman is a diplomate of the American College of Laboratory Animal Medicine and past president of the American Society of Laboratory Animal Practitioners. He has published more than 75 peer-reviewed research papers, book chapters, and books, and has coordinated the "Protocol Review" column for *Lab Animal* since 1988. His professional interests include management processes, Institutional Animal Care and Use Committee functioning, and the assessment of pain in laboratory animals, while his leisure interests focus on nature photography, stamp collecting, and watching his daughter mature into a fine, caring young woman.

Chapter 1

The Basics of Managing a Laboratory Animal Facility

> Most of what we call management consists of making it difficult for people to get their work done.
>
> **Peter Drucker**

I grew up in Brooklyn's Borough Park neighborhood. We lived in a one-bedroom apartment on the fourth floor of an apartment house surrounded by other apartment houses. When we moved down the hallway to an apartment with two bedrooms, I thought we had struck it rich. In retrospect, my experiences in Brooklyn were typical. We were all kids from middle-class families, and like most of the neighborhood kids, I was pretty sure that I knew a lot about life because I equated being streetwise with knowing everything there was to know. In reality, I knew very little. None of us living in Borough Park knew much about what happened in Bensonhurst, Bay Ridge, Canarsie, or Flatbush. That was because Brooklyn was, and still is, a series of neighborhoods. You lived in your neighborhood, not Brooklyn. Yet, there were some areas of commonality among Brooklyn's 4 million residents. The rides and food at Coney Island, the Prospect Park zoo, kids diving for pennies in the murky waters of Sheepshead Bay, and even the gangs in some of the schools were shared experiences for many of us. But one of the strongest threads that held us together was the Dodgers. The Brooklyn Dodgers, or the Bums, as they were affectionately known, were an integral part of our world. The Dodgers were the good guys. The Yankees were the bad guys. The Giants weren't worth mentioning. I left Brooklyn many years ago, but I still remember the day—actually the moment—that Bobby Thompson hit the home run that gave the Giants the 1951 pennant over Brooklyn. I was just a kid, but I remember feeling sick and dejected, as if somebody squeezed all the air out of me. The only salvation was that the Yankees beat the Giants in the World Series.

We would usually take the bus to Ebbets Field to watch the Dodgers play. The manager of the team was Chuck Dressen, Jake Pitler coached first base, and Cookie Lavagetto was the third base coach. As a kid, I could never understand if there was any difference between a manager and a coach. I thought that the words meant the same thing. In fact, I was sure that Lavagetto and Pitler, who you always saw, did most of the work, and Dressen, the manager who stayed in the dugout, did very little. The only time I saw him was at the beginning of the game or when he went out to the mound to change the pitcher. At the time, that didn't seem particularly important to me. Now, many years later, I've figured things out. Dressen was responsible for overseeing the entire field operation; he made the final decisions on the batting order, the playing positions, and the game strategy. He had final responsibility for molding the individual players into a team. No matter what the quality of his players, it was his job to get the most out of them. Lavagetto, Pitler, and the other coaches were his helpers, consultants, and administrators. Dressen delegated certain responsibilities to them, but he was ultimately responsible for the game and the season. He was also responsible to the team's upper management. If the Dodgers didn't win, Dressen and his coaches would be looking for jobs.

Chuck Dressen wasn't much different from today's laboratory animal facility manager. On the field, Dressen was top management, but in the scheme of the entire organization, he was middle management. He had to make the best use of the resources he had, and even if resources were scarce, he was expected to win the game. The same analogy holds for most business executives, head chefs, orchestra conductors, and fire chiefs. Even though they have different duties on a day-to-day basis, as managers, they have similar responsibilities and problems.

As you might expect, the word *manager* can mean different things to different people. In large part, it depends on whether you are the manager, the one who is being managed, or somebody in between. If we were to ask a hundred people on the street what it means to be a manager, I'm willing to wager that the majority would provide an answer such as "being a manager is the same as being a boss. You're in charge of supervising other people." They would say that because many people believe that telling other people what to do, where to do it, and when to do it is what management is all about. Actually, I used to think the same thing. That's not to say that managers never do those things, but supervising people is only one part of a manager's responsibilities. There is more to management than meets the eye, and one of the goals of this book is to help you understand what a manager is and what you should be doing as a manager. This is more important than it sounds because almost nobody can do a job well unless they know what they're supposed to do. Now, the cynic who has been in laboratory animal science for 20 or more years might say that there aren't too many things he hasn't seen or handled, so there really isn't much more to know about management in laboratory animal science. That, as you might guess, is nonsense. Management changes just as fast as laboratory animal science changes. Bad or outdated techniques with animals and bad or outdated techniques in management are inevitable. I vividly remember the admonition given to me and my classmates by Dr. George Poppensiek, dean of the veterinary school at Cornell, on the occasion of our first day of class. Poppensiek said that in order to be considered an expert in performing a surgical procedure, it was generally accepted that you had to do it at least 500 times. And even then, you could do it wrong 500 times. I've never forgotten that caution. Over the years, I've seen lots of people who became an expert at doing things wrong for many, many years. They survived in laboratory animal science because of reasonably good technical skills, a modicum of basic management skills, a lot of luck, and more job openings than there were people to fill them. I want you to be better than those people. I want you to know what a manager is supposed to do. I want you to have, at the least, the theoretical knowledge needed to be a competent manager. Then, as I wrote in the preface, you have to put this knowledge into action every day you are at work. It doesn't mean much to read a book and then fall back into tried-and-true bad habits when you are on the animal facility floor. That's just a shortcut to becoming an expert at doing things wrong.

In some ways, this book will give you some very traditional thoughts about management, and in other ways, I hope that it will help you with the unique aspects of managing a laboratory animal facility. Overall, I like to think that this book is a guide, not a compilation of specifications that must be rigidly followed. More often than not, a good manager is an adaptable person, and if you want to do a good job managing an animal facility, you will have to be very adaptable.

A Definition of Management

We all know how to manage. If you can get out of bed in the morning, brush your teeth, and go to work, you have managed to do something. In this case, you managed to reach one small goal—getting to work. For those of us whose hearts don't start to beat until noontime, that's no small accomplishment. On a small scale, that is (in part) what managers do. There is a goal (getting to work on time), a specific plan to reach that goal (getting out of bed on time and getting washed up in a set amount of time), and a means of organizing the goal (using strategies such as setting an alarm clock the night before and making sure clothing is laid out for the morning), and you even made a decision that doing all this was the best way to proceed. On a more serious note, the concept of setting goals and developing plans to accomplish those goals are two of the cornerstones of successful management.

> **Setting goals and developing plans to accomplish those goals are two cornerstones of successful management.**

Since most people can manage (or at least manage a little), for most of you this book will take some of the things that you intuitively know and place them in a logical order to help you become a better manager. I'll talk to you about the things expected of good managers by other good managers. I'll try to point out problems that have confronted many managers of laboratory animal facilities and give you suggestions on how to approach them. In many instances, you many think, "That's just common sense." That's probably true, but the trick is to recognize that there are some basic actions that good managers use in almost all situations in order to reach a goal.

Let me begin by providing a working definition of management. *Management* is the art and science of using resources efficiently and effectively in order to accomplish a goal. This definition should hold whether you are managing an animal facility, a baseball team, or a multinational corporation. It's not perfect; some people would add or subtract a little, and some would disagree, but in general, it will do.

> **Management is the art and science of using resources efficiently and effectively in order to accomplish a goal.**

To put this into plain language, if you have a job that has to be completed or a problem that has to be resolved (both are goals), you should develop a plan and use all the resources at your disposal to reach that goal. Don't waste time or money, but make sure it gets done right. If that sounds like common sense, you're right—it is. I've used the phrase "art and science" to emphasize that although there is a good deal of textbook knowledge that is part of the science of management, there still is a large amount of finesse that is needed, and that only comes with experience. That is the "art" of management, and it is often more important than the science of management. This definition of management raises further questions that are important to managers. What is meant by *resources, efficient, effective,* and *goal?* As you read through this book, you will start to find the answers to these questions. First, however, let's see if you are a manager.

Are You a Manager?

There is no universally accepted definition of what it is about work responsibilities that makes a person a manager. One well-known management theorist defined a manager as "that person in charge of an organization or subunit" [1]. I don't subscribe to that definition because "in charge of" is far too vague a statement and says nothing about goals. It can include every person who has any control, no matter how limited, over a specific resource, such as a cage washer. My preferred definition of a manager, which is shown below, is derived from the definition of management I just presented.

The most important criterion for being an animal facility manager (or any other manager) is that you are expected to establish goals that can significantly affect your animal facility's operations and you have the authority to use all or some of the resources of your organization (such as money or people) to reach those goals. If you can do that, you're a manager. Everyone has goals, but not everyone has a responsibility to establish goals that can significantly affect the functioning of his or her business (in this case, the animal facility). Managers have that responsibility. On the other hand, if the only work you are responsible for is your own, and if you really don't have the authority to use resources in order to make significant decisions toward reaching a goal, then you are not yet a manager. The fact that you are not a manager has nothing to do with the importance of your work. For example, we all know that laboratory animal facilities could hardly function without animal care technicians.

Don't worry about titles. You may be called a facility director, operations manager, supervisor of animal care, administrative assistant, or any such designation. The question is not one of titles; rather, it is whether you have the authority to use resources to reach goals that can significantly affect your organization. If you have authority to hire a new person, you have controlled at least one resource (people) that can have a significant impact on organizational operations. If your authority includes making purchasing decisions for the animal facility, you have control over another resource (money). In either case, you have fulfilled the basic criterion for being categorized as a manager.

Even though a manager has the authority to use an organization's resources, it does not follow that a manager has unlimited authority. I manage my department and I have the authority to approve a vacation request from a direct report of mine, but I have no authority to close the school due to a snowstorm. The chancellor of the medical school has the authority to close the school in a snowstorm, but does not have the authority to dictate how often a mouse is to be given a pain-relieving drug. In short, almost all managers have limitations to their authority.

A manager is a person who has the authority to use an organization's resources to establish goals and make decisions that can significantly affect the organization's operations.

In a laboratory animal facility, most animal care technicians are not managers. They may have the authority to separate animals if they are fighting, to perform certain diagnostic tests when they feel it is appropriate, and to meet with vendors or order supplies when needed. But they probably do not have the authority to purchase animals, cancel a contract with a vendor, hire another technician, or order a new cage washer. They are responsible for their own actions, and they have very limited ability to make decisions that significantly affect the organization's overall operations.

Unfortunately, not all people with the title of manager are truly managers. Consider the "manager" whose primary responsibility is to make sure that animal care technicians are getting their assigned work done in an efficient and effective manner. If that person requires the approval of a higher authority within the animal facility to reassign a person to a different area, initiate a salary raise, send a letter of reprimand, approve time off, initiate a purchase order, cancel an order, and so forth, then we have to wonder about that so-called manager's ability to use resources to make significant decisions that can affect the animal facility's operations. That person may be a very talented supervisor, but at this time he doesn't fit the definition of a manager.

A manager does not have to use all of her available resources all of the time. Consider an accountant in a small company. There may be no people working for this person (people are a resource), but because the accountant may have the authority to set goals and make decisions about another resource (money) that can significantly affect the company's operations, that person is considered to be a manager.

Are You a New Manager?

So far, I've given you a definition of management and told you a few things about management in general. In a little while, we'll begin a discussion about the resources managers have at their disposal, and then we'll delve into what managers actually do on a day-to-day basis. But for right now, let me say a few words about becoming a new manager because it's easy to talk or read about management until the time you actually have to do it. I want to try to put your mind at ease (a little) by reassuring you that whatever you are feeling as a new manager, others before you have had that same feeling. There are very few people who are natural managers. In fact, it's been stated that only about 1 in 10 people have the inborn ability to be a manager [2]. The good news is that management can be learned. The bad news is that you have to be diligent about learning to be a good manager.

As I will repeat in more than one place, *managers have to manage*. Not all the time, but most of the time. If you don't like the concept of being a manager you should not strive to become one because you have to enjoy your work in order to succeed. The extra money that often comes with a promotion into a managerial position isn't worth much if you are unhappy with your responsibilities. Management isn't a "step up" from your previous job, such as being a lead technician. It's an entirely new job, and you will have to work hard to be a good manager. If you continue to do most of your old job just to prove that you're still one of the good guys, you're going to be in for a rude awakening because chances are that your former coworkers aren't going to be as close to you as they used to be. You have to cut that umbilical cord. There may be times when you have to chip in so that the job gets done, but you have a new job now, and that's what you have to do.

Most new managers have a degree of trepidation when entering their new world. Some are out-and-out afraid that they're going to fail (that was me). Some are afraid that if they succeed, their success may lead to personal problems. A few are overconfident, perhaps to their detriment. One way or the other, most new managers will have some rude awakenings. It's far beyond the scope of this book (and my expertise) to write about all of the trials and tribulations of new managers, but consider reading books such as *Becoming a Manager* [3] or *Managing People* [4]. To veer just a little from the subject, one of my pet peeves about postveterinary training programs in laboratory animal medicine is that they provide minimal formal management training. The residents learn by watching and doing what others do, but do the "others" really know what to do? The American Association for Laboratory Animal Science (AALAS) tries to help its members improve their management skills through its certification process and various training programs, but there is no substitute for on-the-job experience. You can use this book as a resource, but at some point you're going to have to put the information I'm giving you into a real-world context.

One thing that I can say for sure is that none of us will be able to learn and implement everything that there is to know about laboratory animal science or animal facility management. Likewise, laboratory animal veterinarians cannot know everything there is to know about laboratory animal medicine. Therefore, as a new manager you should continually try to improve your management skills, but it is better for you to focus on improving a limited number of skills (such as communicating during difficult conversations, taking initiatives, setting goals, or understanding your institution's budgeting processes), rather than trying to be an expert at everything. In fact, Zenger et al. [5] claim that trying to develop strength in more than five areas ceases to make much of a difference. As I stated in the preface to this book, great athletes practice their sport when others have gone home for the day. It's no different with management; if you want to be a star, you have to practice. You have to associate with other ambitious people, but don't try to surpass everybody at everything. You don't want to waste a lot of time practicing what you consider to be your strengths (e.g., developing a budget). Rather, practice improving the areas of your weaknesses, perhaps, as I just mentioned, difficult conversations or clearly expressing your thoughts in writing [6]. As time goes by, not only will you become a better manager, but your exposure to different management and life challenges will help you develop the mental toughness that will ease any trepidation that you may have now about your managerial career [7].

Let's return to the main subject. To succeed as a manager, you will have to earn the respect of the people who report to you. This discussion is developed much more in Chapter 3, but the take-home message is that whatever authority your position gives you will be diluted unless people respect and trust you. People will *not* respect you if you try to do their job, always hang out with them when they take a break, complain about your job, talk negatively about your own boss, and so on. Respect and trust, not raw power, lead to the kind of managerial authority you are seeking and you will probably need.

Gaining the respect and trust of coworkers leads to managerial authority.

Even with the importance of respect and trust, you have to be competent in your new role. You may have been the best veterinary technician of the group, but now, as the supervisor of veterinary services, your old skills count—but far less so than your new skills as a manager. Can you convince the director to purchase a new endoscope? Can you deal fairly and firmly with two technicians who are fighting with each other? How will you react to the weak link in your department who threatens to quit unless she gets a raise? Will you make sure that there are adequate continuing educational opportunities? Can you put together a budget? Your managerial competency will be tested in many ways. Don't fear this; just understand that it's going to happen.

Throughout this book, I will advise you to listen to your direct reports, give them honest feedback, and when possible, have them participate in decision making. In general, that's good advice. However, there is an indication that some new managers may need to take exception to this rule and become more assertive in their management style, at least initially. Based on his research, Sauer [8] suggested that a new manager should try to recognize where he stands within the existing status structure of his new position. If he has relatively low status (e.g., if he's quite young or brand new to the vivarium, having come from another institution), it may be better for this person to display self-confidence and use his authority to take charge and tell people what has to be done. On the other hand, if the new manager is well known to his direct reports and has a reputation for competency, this new manager would likely be better off relying on a more participative management style and attempt to influence his team members by asking for their opinions, delegating responsibilities, and coaching them more than directing them.

Perhaps the biggest mistake I made early in my career was to never ask for help. In 1975, when I left my private veterinary practice and entered laboratory animal medicine, I really didn't know anything about the field. When I interviewed for my first lab animal job, I went so far as to tell my future employer that I knew nothing about laboratory animal medicine. When I began work, I knew essentially nothing about running a department with almost 30 employees. In both instances, I didn't (at first) turn to anybody for help and I made some whopping errors. Part of the problem was that I didn't have any peers (other vivarium directors) to turn to for help. I was alone, or so I thought I was. I finally figured out that I could have gone to my managers (my immediate subordinates) for some technical help on how to run an animal facility, and interestingly, I learned that I could *not* go to my boss for general help on management (he was worse than I was and invariably gave me bad advice). Eventually, and before I went back to school to earn my master's degree in not-for-profit management, I began networking with a few other animal facility directors. I learned a lot from them because they had experienced the same problems I was experiencing and they were willing to share their experiences with me. I listened, learned, and implemented their advice.

Now that you're a manager, and once you get a little time and skill under your belt, you can't just sit back and relax (do you remember "sophomore slump" from high school or college?). You have to keep up your responsibilities and activities. Think about new managers and how you can mentor them. They need help getting away from the work that they did in the past and getting into a managerial frame of mine. They also may need help seeing the big picture and learning the management culture of the organization. Take them to the people they will be dealing with (e.g., investigators and administrators), and make sure they know the key people with whom they need to interact to get their jobs done. In other words, put a face with a name. In a like manner, remind new managers that it can only help them to have individual "get-to-know-you" meetings with each person they will be supervising. And, most likely, they'll need general help learning how to be a manager. Appendix 4, which discusses training and mentoring, will give you more information on this subject.

Support new managers by introducing them to investigators and those they have to know to accomplish their jobs.

Basically, I learned, and you will learn, that you have to communicate openly and honestly, value people for their contributions, ask for help when it's needed, develop social networks within and outside of the animal facility, and look and listen to people who you respect *even if they aren't among your favorites*. And one more thing: read, read, read your laboratory animal science and management journals. You have to be able to talk with veterinarians, scientists, technicians, administrators, vendors, and so forth. Part of your present value and future growth potential as a manager and leader is based on what you know about many different but related topics, and you will have trouble grabbing for opportunities and parrying off threats unless you can recognize them as such.

"Levels" of Management

I hope you are starting to form a mental picture of who is a manager. You probably realize by now that managers are not just people who sit in large offices and have big desks, assistants, and little concern about the day-to-day operations of their organization. There are all kinds of managers, and most of them do not sit in large offices with those big desks and assistants.

This book is directed to those of you who are (or would like to be) involved in the day-to-day management and leadership of a laboratory animal facility. Some of you may be first-line managers with a major responsibility for ensuring that the work of nonmanagers is performed efficiently and effectively. Nevertheless, you should be able to manage whatever resources you have at your disposal using the same concepts used by the president of a major corporation.

There are many levels of management, from *first-level managers* (sometimes called front-level, frontline, or first-line managers), who make sure that animals and their environment are properly cared for, up to that person in the large office with the big desk who has the title of president. As a first-level manager, you typically are assumed to have the same technical skill sets as the people you manage, even if you rarely are called on to use those skills. If you don't have those skills, you may have trouble gaining the respect of those you manage. First-level (and middle-level) management may be the toughest managerial positions, because you are a supervisor to some and, in turn, are supervised by others. You are the person who is expected to carry out decisions that were made with or without your input. Even though you may have some resources at your disposal, they may be relatively limited and your sphere of influence may also be relatively limited. On a daily basis, your main focus is on making sure that procedures already in place are being performed efficiently and effectively. Nevertheless, your own supervisor may ask you to fulfill vaguely stated goals that he or she has no idea how to accomplish (perhaps this can force some managers to become more creative in how they accomplish goals [9]). You may be asked to develop specific goals, but because you are usually not involved in major organizational decisions, you may have to change your goals to fit organizational goals. If that isn't enough of a problem, as a first- or middle-level manager you have to supervise your staff as well as represent their interests to upper management. This can be a problem, since many employees believe that upper-level managers live in an ivory tower and have neither knowledge nor interest in their problems.

Above first-level management is that great nebulous area called *middle management*. There is nothing specific that defines middle-level management. It can include anybody from the animal care supervisor to a veterinarian who directs the animal facility. It includes all the managers who are not first-level managers but aren't quite the top executives of their organization. Generally speaking, first-level and middle-level managers are closely involved in the day-to-day operations of the animal facility. Some people call this *operational management*. Of course, first- and middle-level managers have to establish long-range goals and plans at times, but for the most part, middle-level managers focus on what's happening in the short run.

The next step is *upper-level (or senior) management.* In general, upper-level managers make the decisions that establish the major programs and policies that the entire organization will follow. Senior managers should have a long-range outlook for their business, whether it is a university or a pharmaceutical corporation. They have the most resources at their disposal. These managers usually don't get involved in day-to-day operations. For animal facilities, upper management can include the person the facility director reports to, such as the vice president for research, the dean, the provost, or the president. As many of you already know, the vice president for research (or a similarly titled person) is frequently the individual who has the greatest direct influence on laboratory animal facilities. The vice president reports to the president, and the president reports to the board of trustees (or a board of directors). So as you can see, everybody reports to somebody else.

I have to be honest and tell you that I loathe the term *senior management* when it is used by certain managers to distinguish themselves from all other employees. For example, you may hear a university president thank a school's senior management team for a job well done, or tell us that at the senior management level, certain decisions were made. To my fragile ego, this is ostentatious rhetoric and suggests a class system that has no place in any organization. We all know that there are managers who are in higher positions than we are, and we also know that many decisions are made without our input, but to emphasize that there is a special group of people who have the power to do this detracts from the goal of working together as a smoothly functioning unit—a goal (stated or unstated) that essentially every organization strives for. There is no "us" and "them" or "senior" and "junior" management. We are all coworkers. Words are powerful tools, and we have to choose them carefully when we want to engender a culture of working together. Maybe in your animal facility you are a senior manager, but never hold that over the head of the other managers. We just have different levels of responsibility.

From my ranting in the last paragraph, you have probably deduced the obvious. That is, most laboratory animal facilities have a classical pyramid structure. There's a hierarchy with a director at the top of the pyramid, one or more levels of animal facility managers under the director, and technicians supporting the base of the pyramid. An advantage of hierarchies is that there are clear reporting lines, clear levels of responsibility, and many helpful management books and articles that stem from a mindset that assumes a typical hierarchical organizational structure. On the negative side, hierarchical organizations are not particularly nimble. It may take forever to get a request fulfilled or a plan approved due to the number of people that have to give their approval, sometimes right up to the very top of the pyramid. Nevertheless, animal facilities are parts of larger organizations, and typically they have an organization structure that mimics that of their parent organization.

Although theories about organizational structure may come and go, the basic pyramid structure has been with us for a long time and seems to work fairly well. A few years ago, I attended a lecture by a colleague who said that he didn't like a pyramid structure because once you start dividing your employees into discrete working groups (such as animal care and veterinary services groups), they stop talking to each other and become compartmentalized (also known as working in silos). Rather, he said that his department was more of a flat matrix where everybody readily interacted with everybody else. He then went on to thank his associate directors and assistant directors for making it all possible. As I sat there, I almost laughed out loud when he made that last comment because he just described a traditional pyramidal structure.

Not every hierarchy has to have one person at the top. There are numerous examples in the corporate world where two or more people coequally lead an organization, with the investment firm of Goldman Sachs being a well-known example [10]. As another example, where I work, the veterinary services division was formerly led by two people, one with primary responsibility for surgery and large animals and the other with primary responsibility for veterinary technical activities and rodents. They had to work together because there was only one budget for all of veterinary services, and they both reported to the director. There are some distinct advantages to such an arrangement: The division of responsibility allows each of the two people to focus their managerial skills on areas of their interest, they can back up each other when necessary, it is easier for them to develop relationships and expertise in their areas of interest, and they have a colleague with whom to discuss ideas and concerns about the division. On the negative side, if you have the wrong people in those roles, there can be jealousy, backstabbing, lack of coordination, resource hoarding, and a slew of other problems. Choosing the right people for shared leadership is an art in itself.

Perhaps, for some laboratory animal facilities, a pyramid isn't necessary and other forms of managerial leadership will work, but for many others, the pyramid seems to work just fine. As I wrote a little earlier, at least with a pyramid there are clear lines of responsibility; it helps in making decisions, allocating resources, and managing people who need it [11]. On the negative side, it can isolate managers from other employees, be dictatorial, stifle creativity, and (as you now know) take forever to get something approved. Whether we agree or disagree about the use of a classical hierarchy, we should be able to agree that if there is a pyramid, then it is up to us, as managers, to make sure that we can talk to just about anybody without needing permission from someone else to do so. This is part of an "authoritarian democracy" in which managers use their authority to create a people-oriented management system. These managers know when to socialize and when not to socialize. They know when to push and when not to push. In other words, they've become both managers and leaders [12]. I'll have more to say about communicating in Chapter 3 and leadership in Chapter 7.

No matter how good a pyramid may look on paper, no matter how much money is planned for animal-related work, anybody who has worked in animal facilities for any length of time knows that the animal facility pyramid is supported by the animal care and veterinary technicians (Figure 1.1). If technicians are not qualified and motivated, the entire facility will suffer.

A pyramid will not long stand with a top of gold but a base of sand

Figure 1.1   An animal facility is only as strong as its technicians.

Think about the basic truth of this statement. We have to work hard every day to be quality managers; otherwise, morale will be low, turnover will be high, and the quality of animal care will suffer.

Resources Managers Have

I've already emphasized that managers use their available resources efficiently and effectively to reach their goals. Soon, you will learn that while using their resources, they plan, make decisions, organize, direct, and establish control systems. Now it's time to be somewhat more specific about what is meant by *resources*. Then, in later chapters, I'll go into much more detail about the management of specific resources.

All organizations have resources; therefore, before you read on, stop for a few moments at the end of this sentence to think about what resources your own organization has.

Chances are that you came up with money as a resource. Did you realize, however, that you are a resource, a surgeon is a resource, an x-ray machine is a resource, medical records are a resource, and even the building you work in is a resource? Let's divide resources into five broad categories: human, fiscal, capital, information, and time.

Resources Available to Managers

- **Human resources (people)**
- **Fiscal resources (money)**
- **Capital resources (major equipment and physical plant)**
- **Information resources**
- **Time**

These categories, like the definition of management itself, are somewhat arbitrary. They are simply convenient pigeonholes. Still, I think that most professional managers would not argue strongly about them. More important, you can use them to help organize your thinking about resource management.

Human Resources

Human resources are the heart and soul of a business. In laboratory animal management, human resources are *all* of the people whose work affects the functioning of a laboratory animal facility, not just the veterinarians and technicians. This includes bookkeepers, receptionists, chemists, food suppliers, and anyone else, inside or outside of your organization, who you may call upon to help you reach a specific goal. Sometimes, we get so wound up in our own organization that we forget that there are many outsiders whom we routinely use as resources. Food, bedding, and animal suppliers are obvious, but there are also other managers, technicians, a friend of a friend who can help you, and anybody else who forms part of that great network of people that we call human resources. In fact, I wrote part of this chapter while on the way to a national meeting of AALAS. One of my responsibilities as the director of my animal facility is to network with my colleagues, at all levels of lab animal science, to find better ways of doing things and to make friends that I might be able to call on in the future. I also try to establish networking opportunities for others in my department to help them better understand the workings of the school and the research we do. In other words, I am establishing, maintaining, and using human resources.

Fiscal Resources

Fiscal resources are the funds that your organization has or can get. In laboratory animal facilities, money may be acquired by charging investigators for the maintenance of their animals, by direct support from the parent organization, by charging for miscellaneous services such as the use of a surgery suite, by ordering animals, and by many other procedures that are common in laboratory animal facilities. Your inventory of cages and supplies can also be considered fiscal resources since, in theory, you can sell them to make money. Your parent organization may acquire some of its fiscal resources from government grants, from the sale of products, by investments, by borrowing, and through many other ways.

Capital Resources

Some of your organization's capital resources are the building in which you work, the grounds on which the building is located, and the major pieces of equipment that support your activities (such as a digital x-ray machine or a cage washer). When you hear people talking about "capital equipment," they are referring to the expensive, more or less permanent equipment items. Different organizations have different ways of categorizing what is and isn't a capital resource, but many define a capital resource as a building or piece of equipment worth at least $5000 and with an expected life span of five or more years.

I will discuss the use of capital resources in Chapter 4 on financial management, but not to any great degree. This is not because they are unimportant, but because animal facility managers are usually more involved in the *use* of these resources, as opposed to their management.

Information Resources

When I suggested that you think about the resources your organization has, chances were that you did not consider information as a resource. But information is a real and valuable resource that all of us use on a daily basis. Some examples of information resources are animal health records, animal censuses, electronic mail, the World Wide Web, records needed to comply with regulatory agencies, your library, professional meetings, and vendors' representatives. In the last example (vendor representatives), you can readily appreciate that there is no clear-cut line as to where one resource ends and another begins since these people can be considered both human and information resources.

Time

As with information, most of us don't think of time as a resource. Not only is time a resource, but also it is a limiting resource because there are only 24 hours in a day. Part of your job is to use time in the most efficient and effective way you can. You can use time to schedule the optimum number of surgeries, to begin new experiments, to have staff meetings, and to have on-the-job training. There are many first-level managers who believe they need a 25-hour day, and there are some upper-level managers who want their staff to give 110% of their time. Time (as well as these people) has to be managed.

Now that you know a little more about these five resources, what use will you make of them? What do they have to do with being a manager? It's a fair question, and here is the answer: a manager's job is to use these five resources in an efficient and effective manner in order to reach a goal. This should sound familiar to you by now. Resources are the basic tools with which you work. In the vast majority of instances, you will use more than one resource at a time. Indeed, you cannot effectively manage an animal facility using only one resource any more than you can build a house using only one tool. As a simple example, consider a laboratory animal facility where an employee spends five minutes cleaning each rat cage. If your goal is to have each cage cleaned in one minute in order to complete the day's work, both human and time resources must be considered. Of course, the manager realizes that "time is money"; thus, indirectly, fiscal resources are considered as well. You will probably use a cage washer, and that machine is a capital resource. In fact, if the cage washer can't handle the number of cages you would like to put into it in one load, then you will have to decide if the cage washer is the problem or if the way people are using it is the problem. In this simple example, you can see that managers use their resources every day as part of the goal attainment process.

Let me give you another thought on how resources are used in day-to-day management. There are three questions that you will hear over and over in the management world (or at least in management textbooks):

- Where are we now?
- Where do we want to be?
- How are we going to get there?

These three questions simply help us focus our planning efforts and resource utilization. There is usually a presumption that we know what our goal is (and reaching that goal is where we want to be), but that's not always true. In fact, sometimes we have to do some hard thinking about where we want to be (i.e., what our goals should be), often involving many people from many different parts of the organization and perhaps outside of our own organization. As an example, suppose an airline has two flights a day between Boston and Los Angeles. It sells about 80% of the seats on one of the flights and about 50% on the other flight. The latter flight actually loses money. Should the airline's goal be to phase out the less full flight, or should the goal be to find out why one flight sells better than the other? Only thoughtful discussion will provide an answer, but it is very important that your goal, whatever it is, be clear in your own mind and certainly clear enough that you can readily communicate it to others. Let's assume that one or more goals have been established, so at least one of the questions (Where do we want to be?) has been answered. Very often, you know where you are now, so another question has been answered, although sometimes managers need to check and verify where they are now. For example, if you want to wash more cages per day, you have to know how many cages per day are currently being washed. Therefore, in many instances the key remaining question for the management team is, how are we going to get there? This is where the plans and strategies come into play. This is the point where managers start using all of their resources and also try to figure out what specific items are needed to make things happen. You gather a group of people together for some directed brainstorming. Maybe money is needed for a study, maybe some specialists have to be brought on board, and so on. People, money, time—all of these are part of your resources.

You must have a very clear picture of your goals, and you must be able to communicate those goals to others.

Efficiency and Effectiveness

The resources that you have must be used efficiently and effectively. By efficiently, I mean you want the most output from the least input. In the cage cleaning example used earlier, an employee who cleans three cages in three minutes is more efficient than one who cleans one cage in five minutes. The person who cleans the most cages (output) in the least amount of time (input) is the most efficient one.

As managers, we have to realize that it takes more than people to have good efficiency. Efficiency involves all of the resources we have, and our job is to manage resources. How can a large-size animal facility run efficiently if the cage washer isn't working correctly? What kind of efficiency will there be if there is no cage washer at all? That's bad planning on somebody's part unless you use disposable caging and wash your racks and carts in a special area. All the resources we have must be entered into the efficiency equation.

We also must recognize that it is not only animal care and veterinary technicians who have to be efficient. You, the manager, also must be efficient. Don't expect the people who work for you to be efficient if you're not. Your boss must be efficient too. If your supervisor has you running in circles, you will not be efficient and chances are that he or she will not be either. And if your supervisor's supervisor is inefficient, there is a good chance that everybody below her or him will not be working at optimum efficiency. Clearly, efficiency is an organizational responsibility.

Effectiveness, on the other hand, is a measure of the quality of the output. It describes how well an organization, a part of an organization, or an individual accomplishes a goal. To continue with the cage cleaning example, the manager would expect that regardless of the time required, the cages have to be cleaned properly. If not, the technician cleaning the cages would not be very effective. The manager wants the technician to be both efficient and effective. A good manager—in fact, any good employee—should strive to be both efficient and effective in how he or she uses resources on the way to accomplishing a goal.

Efficiency is getting the most output from the least input. Effectiveness is a measure of the quality of the output.

In laboratory animal facilities, we tend to be very service oriented. To that end, it is somewhat more important for us to be effective rather than efficient. But as I just noted, we try to balance both of these needs.

You do have to be a little careful so as not to fall into the trap of thinking that your primary concern is to manage the efficiency and effectiveness of any one person. As managers, we are managing a facility that has animals, floors, walls, ceilings, budgets, peeling paint, and persnickety people. We have to look at the whole picture of laboratory animal facility efficiency and effectiveness, not just the individual parts. We want the parts to mesh just right, but not just for our peace of mind. They must mesh to help reach the goals of our animal facility and our organization.

What Do Managers Do?

Let's concentrate now on what managers actually do. I don't want to spend any real time talking about managing the number of cages that should be going through the cage washer in an hour or how to write a disaster plan, but I do want to talk about the more general aspects of managing: those things that every manager does, whether in laboratory animal science or skyscraper construction. I suppose the simple answer to the question "What do managers do?" is that managers manage something, but we all know that's a simplistic answer that avoids discussing the details of a manager's job. Earlier in this chapter, I wrote that management is the effective and efficient use of resources in order to accomplish a goal. Therefore, it seems reasonable to reword this a little and say that managers want to use their resources effectively and efficiently to accomplish goals. Now, it is true that managers spend most of their workdays *accomplishing* goals, but so does everyone else who is out there earning a living. The difference is that managers have more flexibility than nonmanagers in *establishing* goals, managers' goals are usually directed at ensuring the quality of the work that is being done, and managers seek ways to improve efficiency and effectiveness. Managers have more resources at their disposal than do nonmanagers, and they typically have a level of authority that allows them to put their goals into practice. This doesn't mean that all of a manager's goals are of the utmost importance or that managers sit around all day thinking about new critical goals. Like everybody else, managers have certain mundane chores that they have to do, such as filling out administrative forms or going to meetings where everybody knows nothing much will be accomplished. One thing is for sure: managers don't come to work every morning hoping that all will be well and they can just sit in their office, praying that the day will end really quickly. Just like the manager of a baseball team, managers typically have part of their day planned in their minds before they get to work. Maybe they'll be at a biosafety meeting making a decision about new equipment, making phone calls to job applicants, researching a new sterilizer, organizing a scientific conference, checking on some work that was done the day before, or meeting with some key people to try to develop a training program. One way or the other, on a daily basis managers use one or more of the various resources at their disposal to reach one or more goals. Once again, these resources are human, financial, capital, information, and time. I'll have much more to say later in this book about resources, but for right now, let's focus on how managers use the resources at their disposal to reach their goals.

Roles of a Manager

A goal is a desired endpoint, and managers most certainly want to accomplish their goals. Developing a new veterinary technician training program or improving the efficiency of cage washing might be two of your goals. In order to reach a goal, managers plan, make decisions, organize, direct, and control each of their resources [13]. These are the so-called *roles of a manager*, and they help us understand why we need managers. There would be anarchy if everybody made their own plans and decisions, and then tried to independently organize, direct, and control the resources they need to get a job done. We need managers, or a team of people acting in unison, to make sure that there is efficiency and effectiveness, not chaos, in getting work done.

Managers plan, organize, direct, control, and make decisions about each of the resources that they use.

Interestingly, the eminent management theorist Henry Mintzberg questioned if managers really perform all of these roles [1]. For example, if a manager goes to a business luncheon, which of the above roles of a manager is he performing? The answer is none of the above. The truth, as I see it, is that managers don't spend every moment of every day performing managerial roles that fit into nice little categories. Managerial roles indicate the outcome of something that previously happened, even if it was moments earlier. For example, if you go to an AALAS meeting to listen to a lecture and then talk with your colleagues, none of those activities neatly fall into planning or directing or any other managerial role. However, if that meeting leads to a plan to build a transgenic mouse facility (the plan is an outcome of the meeting), then one of your managerial roles—planning—has surfaced. Then, if you have to make decisions about the transgenic facility, organize activities, and so forth, even more managerial roles come into play. So although managers aren't always performing classical managerial roles, when they do plan, organize, direct, and so on, they do so to advance their goals for their animal facility. That can happen around a water cooler, in an animal room, or at a desk—it really doesn't matter. There will be times when a manager serves as a referee between bickering employees and when a manager looks at a wall in utter frustration, wondering why she ever got into this line of work. Managers often do what they do as the need arises, but they're always doing something. Taken as a whole, the manager is working with his or her resources to advance the goals of the animal facility, even if the specific work being done at a particular moment in time doesn't fit neatly into one of the roles of a manager.

Let me begin the discussion of how managers use their resources on a daily basis by discussing how each of these roles are performed by managers. Here is a hypothetical case report about the laboratory animal facility at Great Eastern University.

Great Eastern, like many universities, was chronically short on money. The state had cut back on its support of education, tuition was already high, and salary increases for faculty and staff were near zero. The university president had just imposed a 4.5% cut in the budgets of all divisions, including the laboratory animal facility.

Jim Johnson, the manager of the facility, was already working with what he considered to be a skeleton crew and insufficient funding. Now things were even worse because the academic year was about to start and researchers were anxious to begin new projects.

Jim knew what the problem was. He had less money, but he was expected to provide the same amount of service, perhaps even more. He also knew that a major goal would be to find ways to cut labor costs without substantially decreasing efficiency or effectiveness. Now, he had to figure out how to reach that goal. He had to use his resources, but he wasn't quite sure how he would do this.

His first decision was to enlist the help of his senior staff in formatting a plan of action.

1. A Manager Makes Decisions

Granted, Johnson's decision to ask for help was not earth-shattering, but it was logical nevertheless. Jim would use one of his resources (people) to help him. Just because it was a relatively easy decision does not mean that it was an unimportant step.

Managers are always making major and minor decisions. Some typical examples are a decision to hire a person, to have a meeting, to keep or reject a shipment of animals, to purchase equipment, or to discuss a problem with a supervisor. Some of these decisions are routine, while at other times a manager will have to make more complicated decisions that entail certain risks. For example, a decision to have or not have a meeting is usually a routine and low-risk decision. But if you are faced with a choice of lowering costs or losing your job, the cost-lowering decisions that you make carry a higher level of risk, yet you still have to make a decision because "action brings at least a possibility of success, whereas inaction brings none" [14]. Or, as the lottery people say, if you don't play the game, you have absolutely no chance of winning.

Decision making is a crucial managerial role for you, perhaps your most important one. You are often judged by the results of your decisions. Because of this, you don't want to jump too hastily into an important decision unless you happen to be an expert on the subject matter and a quick decision is needed. In most instances, it is better to sit back and make sure you have sufficient information and you have thought about the alternatives available to you, because most decisions need not be made with haste. For example, if you think you need a second cage washer, is it cost-effective to pay people overtime to use your existing cage washer? Can you use an entirely new second shift of people to operate the existing cage washer? Would you be able to find enough people for overtime or a second shift? Should you consider robotics with a cage washer? Equally important, how accurate is the information you used to determine that it might be time for a new cage washer? These important and differing choices require a substantial amount of information and discussion before a final decision is made.

Managers are judged by the results of their decisions.

Decisions have to be based upon your knowledge of how your organization normally works, the requirements of your job, knowledge of laboratory animal science, general knowledge of management, your animal facility and organization's mission, common sense, and many other factors. If you are a first-line supervisor, you must know most of the details of the job responsibilities of the people who report to you. Even if you are a little higher up on the management ladder, you should know the responsibilities of the supervisors who report to you, and so forth up the chain of command that is found in most hierarchal organizations (which, as you know, is typical of animal facilities). There will always be some circumstances where you will have to make a decision by yourself or with only a few other people, and the decision must be made quickly. Dealing with a burst water pipe in an animal room is a good example. When a crisis occurs, there must be an effective and rapid process for gathering and analyzing data, followed by effective execution of your plan. Communications have to be open and honest [15]. But in general, when there is no crisis, it's worth your while to be up front with your concerns and thoroughly debate the pros and cons of a problem with staff members before making a final decision [16,17]. You want to be careful not to allow biases to creep into your decision making. Just because Joe Smith is your friend doesn't mean that his company makes the best cage washer for your needs. Just because Laura Conn made an outstanding impression during her interview doesn't mean that you should brush aside comments from her references that suggest she would not fit in well with your staff. And as a final example, just because the majority of your direct reports suggest that you use sentinel mice from Sharp Point Farms doesn't mean that you should disregard the reasoning of the few people who are telling you that Sharp Point animals would be a bad choice for the vivarium.

One good thing about making decisions in laboratory animal facilities is that they often are based on issues that occur on a somewhat regular basis, where we have previously encountered the question, and we have a framework for a response (e.g., should we buy irradiated or autoclavable food? should we purchase a preventative maintenance contract?). Typically, the decisions you make when you are starting out as a young manager are probably going to be focused on getting a specific job done efficiently and effectively. Those decisions often need only a small amount of information and limited options in order to get a relatively rapid resolution. As you progress in your career, you may find that you have to make more broad-based decisions, such as setting a direction for the entire animal facility. Making that kind of a decision may require additional information, time, consultation, feedback, and so forth. This change in your decision-making process over time is not at all unusual, and as you advance in management, you should embrace the opportunity to incorporate new ways of doing things to reach a sound decision [18].

Very often, the decisions managers have to make are so repetitive that they are incorporated into a set of policies outlining the day-to-day functions of a laboratory animal facility. This set of policies is usually called the *standard operating procedures* (SOPs). SOPs are valuable not only because they standardize the way everybody performs common work-related activities (and standardization of procedures is usually important in biomedical research), but also because, in some instances, SOPs can help limit the number of choices a manager has to make, thereby expediting the decision-making process. Another aspect of SOPs is that they are meant to be living documents. By that I mean they should be regularly reviewed and updated as necessary. Experience suggests that the best people to review and update SOPs are the primary users of them, and often these are the animal care and veterinary technicians. SOPs that are only reviewed and updated by managers may miss important knowledge held by technicians or other daily users. The concept of allowing the employees who use the SOPs to develop and revise the SOPs is a common and recommended practice of Lean management, which is discussed in Appendix 1.

The purpose of making a decision is to establish and achieve organizational goals and objectives.

You cannot routinely avoid making decisions for fear of making a mistake, because if you do, you are not being an effective manager. There are usually some risks, and typically you will have to make some hard decisions—decisions you would just as soon not make. But if you shy away from these decisions, you are avoiding what you get paid to do, particularly if others know that you're shirking your responsibility. You rarely need every bit of possible information in front of you before making a decision, so don't use waiting for more and more information as an excuse for doing nothing. If you want to maintain the respect of your colleagues, earn your salary, and advance your animal facility, you're going to occasionally have to make some difficult decisions, and maybe you'll make some wrong decisions.

Like everybody else, you will learn from your mistakes and move on. You can't dwell on mistakes forever, and there are very few things you will do that you can't reverse. There is nothing wrong with changing horses in midstream if the horse is drowning. In the example used earlier, Jim Johnson wasn't looking for world-shaking input; he just needed some help.

A few years ago, I participated in a seminar on decision making. The speaker was advocating an algorithmic process for making decisions. In other words, you start at Box A and you provide a response to the decision-making question that is in Box A. Depending on your response, you are then directed to go to Box B or Box C, and so on, until you come to the proper decision. If life were that easy, we would all be using little boxes and algorithms, but life and decision making are not that simple. In fact, I think that one of the hardest managerial decisions is to determine which decisions have to be made quickly and which ones can be put on the sideline. You have to think hard about what is an important decision that needs immediate attention and which decisions are lower in importance and you don't have to waste your time and energy with them right now. Even when you sort out the important decisions from the less important ones, you still have to decide who will make the decision. As a manager, you do have to make decisions, but that doesn't mean you have to make every decision. Perhaps another person or group with whom you work is more suited to make a particular decision. You certainly don't want to place everybody's problems on your own shoulders, nor do you want to put your own responsibilities on somebody else's shoulders. Sometimes you just have to sit down, close the door, and work out who should be making the decision.

Decisions are not often made in a logical, linear manner [19]. More often, the process starts, stops, goes backward, forward, maybe sideways, and often requires coalition building, lobbying, and negotiating before a final decision is made. Very often, we unconsciously compare the current situation with past situations when making a decision [20]. The take-home message is that no matter what information we may have, important decisions take some time to be thought through, polished, and finalized. Along the way, there will some bumps, but they are to be expected, not feared.

Sometimes decisions are made that are based on hunches, listening to information that suits the needs of the moment, a desire to prove that we can do something that nobody else has done, and so on. That is, we make the decision and then we do what is necessary to justify having made that decision. This is not a good idea. Unless there are extenuating circumstances, as was noted in the previous paragraph, we should not rush into making decisions that can significantly impact animals or the operation of an animal facility. Very few people do (or should) make major decisions by themselves, and it's the exception when a decision must be made "on the spot." Decisions usually occur over time and involve more people than you might realize, with information that is gathered at the water cooler, during lunch, at formal meetings, and in informal conversations with animal care technicians, researchers, and your own boss. You may be the person who delivers the ultimate decision, but don't think that you, as the facility director or a senior manager, single-handedly make major decisions. More often than not, many people and many pieces of information are involved [20]. The noted management writer Gary Hamel has an interesting view on making major decisions. He believes that "as decisions get bigger, the ranks of those able to challenge the decision maker get smaller … but the danger is greatest when the decision maker's power is, for all purposes, uncontestable" [21]. The lesson to be learned is that we should rarely make a major decision without listening to the opinions of others, and before a final decision is made, we should encourage feedback that encourages those who are at odds with the pending decision to provide reasoned criticism.

I just described typical decision making in most institutions. If you are working in a university, this involved decision-making system is probably something you have often seen. But in Lean management, which is discussed in Appendix 1, decision making is much more formalized and accountability occurs at every level of management, not just at the top of the pyramid.

Some key needs for making a decision [22] are summarized below, complemented by some of my own opinions:

- *Ask yourself if there is truly a problem that requires a decision on your part.* I would supplement this question by asking whether the decision (assuming a decision is needed) should be yours or someone else's. It is not a good idea to accept responsibility for making other people's decisions. There are, for example, many decisions that can and should be made by nonmanagerial employees, such as veterinary or animal care technicians. Other decisions, such as purchasing a piece of property, will not likely be within the scope of your authority. It's important that you understand which decisions are yours to make and which are not.

- *Obtain sufficient information to be able to make an intelligent decision.* Use whatever resources you have, such as spreadsheets, memos, journals, newspapers, discussions, and gossip. You rarely need 100% of all available information to make a good decision. In fact, I am of the opinion, as are many other people, that you can probably make almost all of your decisions with no more than 80% of the information you might think you need, and you can probably make many decisions with perhaps as little as 20%–30% of the total information available. The amount of information needed may depend on the importance of the decision. You have to develop your ability for determining which information is the important information that you need. It's a classic example of separating the wheat from the chaff. And, as I wrote earlier, be aware of personal biases that might improperly influence your decision.

One certainty is that a decision must be clearly understood by everybody. I don't know how many Institutional Animal Care and Use Committee (IACUC) meetings I've attended where some people on the committee had a very different understanding of a problem compared with others at the same meeting. Because of the misunderstanding, the outcome was that any decision made could be wrong or biased, depending on what a person perceived to be factual. One example had to do with multiple survival surgeries using the same animal. The required decision was to either approve a study allowing multiple surgical procedures on the same animal or use more animals, but only allow one surgical procedure per animal. After about 15 minutes of discussion, it became obvious that some people thought all of the surgeries were part of the same research protocol, while others thought the surgeries were on two different but related protocols. The committee chairperson could have done a better job at defining the decision that had to be made.

■ *Every decision requires some basic assumptions.* For example, if you have to decide whether to purchase new surgical lighting, there is an assumption that the existing surgical lights are inadequate for current needs, and there is a means of getting the needed funds to buy the lights. Working with this assumption, you have to develop potential solutions and choose the best solution. Therefore, if the surgery lights are rarely used, it may be wiser to purchase portable lighting than to try to get the funds for expensive overhead lights with prewired video capability. There are almost always choices involved in decision making, and some good advice is to ask yourself, "What *could* I do?" rather than "What *should* I do?" The first question emphasizes that you have to think about alternatives [23].

■ *Consider the risks.* Related to asking "What could I do?" is considering "What might happen if I do this?" As a decision maker, the "what if" scenario forces you to think about what might go right and what might go wrong with a decision. If you believe that there is a greater chance of your decision leading to a negative outcome, then you have to rethink that decision, consider an alternative, or accept the risk.

■ *Make your decisions at the right time.* Tell people about your decision when they are ready to accept it. A decision to close down one of your animal facilities should not come out of thin air two days before Christmas. Your employees should be primed for such a statement.

■ *Keep your decisions flexible.* You should never lock yourself into a decision as you may have to modify it. One mistake that newer managers often make is to stubbornly stay with a decision even though, as time passes, the decision seems to be incorrect. Perhaps they made the decision because it worked under a similar circumstance and they have convinced themselves it will work again, or perhaps they were afraid to admit to a mistake. Either way, sticking with a bad decision, such as sticking with a person you recently hired and who has called in sick every Monday, shows a lack of flexibility that can result in significant problems. Some of you probably know Peter Seeger's song "Waist Deep in the Big Muddy." It's about a military platoon that was crossing a river and wading into deeper and deeper water. The sergeant wanted to turn back, but the captain leading the platoon ordered it to keep moving ahead. The verse is "we were waist deep in the Big Muddy but the big fool said to push on." Eventually the captain drowned and the sergeant ordered the platoon back to shore. If you find yourself in the Big Muddy, flexibility can be your life preserver.

■ *Implement your decision.* Once a decision has been made, do something about it. There is no sense in making a decision and then doing nothing to implement it. The longer it takes you to implement your decision, the more people will become frustrated and demotivated waiting for something to happen [24].

I'm going to add one more comment about decisions: managers should explain the reason behind their decision. People may not agree with you, but at least you have provided important feedback.

2. A Manager Plans

Let's return to the scenario where Jim Johnson had to trim an already sparse budget.

> Johnson was honest with his staff. He organized a meeting and told them about the budget cut and the reasons for it. He also told them that he knew they were doing as much as they could reasonably be expected to do. The question, he said, was, "How could we make our daily operations even more efficient?" A lot of ideas were tossed out, but Randy Adams had the best one. His plan was to place more animals in each room in order to decrease the time needed to maintain many partially filled rooms. There were some drawbacks to this idea, but it was the best they had.

By definition, planning implies that something will be done at a future time. A plan is the method or methods you will use to take you from where you are now to where you want to be at some future time. Once you get to where you want to be, you should look for ways to continually improve the outcome. If successful, Randy Adams's plan would help Jim Johnson get the same amount of work done with less available money. In reality, more than one managerial role was being exercised in the previous paragraph. Johnson made a *decision* to have a meeting and *organized* the meeting. From that meeting came a general *plan*. Nothing special, just something managers do every day.

A plan is the methods you will use to take you from where you are now to where you want to be.

Plans are important because as part of the plan, you will develop the specific strategies that are needed to reach the goal. The plan can give us an idea about the time it will take to reach the goal and the amount of money it will cost. It will also tell us something about the number of people or other resources needed to reach the endpoint of our planning (i.e., our goal).

Depending on your position in your organization, you will be involved in different levels of planning, although there is always some overlap. For example, the vice president for research at Great Eastern University may be involved in developing long-range goals that the university must take in order to compete successfully for limited research dollars. The vice president must have good information about the state of the economy, political realities concerning research funding, and perhaps how other universities (the competition) are reacting to the same situation before he and others can develop a plan on how to accomplish the goals.

Jim Johnson wasn't the vice president for research. He was more concerned about the daily operations of the laboratory animal facility. His planning was directed toward placing more animals in a room to increase labor efficiency. He couldn't do this overnight because he had to talk to the investigators and animal care technicians, and even find out if the heat load from the extra animals would be too much for the ventilation system. Nevertheless, it was a plan that would go forward with both care and determination. For both the vice president and Jim Johnson, their goals reflected a realization that money was hard to get, and plans (and decisions) had to be made to be able to accommodate the fiscal problem that affected the entire university.

Whether you are a vice president for research or an animal facility manager, the basic steps in planning are largely the same. To begin with, we must understand that planning is a continuous process, not something that happens only when a particular need arises. Planning certainly implies moving forward to reach new goals, but it also means that current ideas or past decisions may have to be abandoned in light of new information or new needs [25]. The planning process begins by clearly stating the goals we want to reach and how we will know when the goal is reached. Your goal may be to increase labor efficiency. Or, your goal may be to improve your technician training program (goals don't have to be fancy). The next step is to gather information to determine if the goal is feasible. Assuming the goal has a reasonable potential to be reached, we have to start thinking about the methods we will use to reach that goal. So, if we want to increase labor efficiency, we might consider a strategy of putting more mice in each room. That will lead us to consider (among many other things) if there is the needed electrical supply for ventilated animal racks and if there are any special air handling needs if we put additional animals in the room. Do we have enough animal racks, and if not, is there money to buy racks, or should we just move racks from one room into the other? If we don't have the money, can we get funding from our institution, or are there grant opportunities through the federal government? Do we have to provide matching dollars for that federal grant? If it appears that there are too many roadblocks, we may have to consider alternative methods of reaching our goal (remember, as a manager you have to make decisions), but if our initial plan seems feasible, we will take the next step, which is to develop additional strategies (specific smaller and more detailed plans) needed to reach our short-range goals, and if needed, we will develop yet other strategies to reach our long-range goals. Finally, we will give specific people responsibility for implementing specific parts of the plan, and we will monitor progress (typically weekly or monthly) until the plan is completed, which can take as long as a few months or as little as a few days. During this time, we will make any needed changes to the plan because it is rare that a plan proceeds perfectly with no needed corrections.

General Planning Process

1. **Define your goal and how you will know when it is reached.**
2. **Determine the resources needed to accomplish the goal.**
3. **Develop strategies (specific methods) to reach the goal.**
4. **Assign people to have responsibility for each strategy.**
5. **Monitor the progress toward reaching the goal and make any needed changes in the plan.**

Later in this chapter, there will be a discussion on strategic planning, which is part of the process of setting and reaching long-term goals. You will see that the general planning process just described is essentially the same for strategic planning.

3. A Manager Organizes

The decision was made to proceed with Randy Adams's plan. Jim Johnson decided that each of the supervisors who reported to him would coordinate the combining of animals from 10 holding rooms into 8. They were each given the responsibility of working out the time schedule and implementing the plan. Johnson would help in the coordination of the plan with investigators whose studies would be affected.

It is sometimes difficult to separate organizing from planning or directing. In using the word *organizing*, I am referring to the responsibility to coordinate the activities under your control. For example, Johnson organized the meeting described above. Organizing includes

- Giving direction, such as clarifying everybody's job
- Ensuring that resources are available to accomplish a goal
- Removing roadblocks if they occur
- Ensuring that people are properly trained to accomplish a goal
- Ensuring that proper supervision, if needed, is in place
- The delegation of responsibility when appropriate

When responsibility is delegated, you must also delegate authority. The extent of delegation, or what you delegate, is a decision you must make. Nevertheless, although the immediate responsibility may be delegated to someone else, the decision to delegate was yours, and you will be judged by the results of that other person's actions. There is nothing wrong with holding you responsible. You are the manager. If you make a wrong decision in the way you choose to organize your activities, recognize your mistake and correct it. If you do not, you should and will be held responsible.

Organizing is a critical role because it requires that you pull your resources together in order to make a whole that is greater than its component parts. In our example, Jim Johnson's actions were used to illustrate how planning, decision making, and organizing are integrated into the roles of the manager. Also, as part of his responsibilities, Johnson delegated a certain amount of responsibility to his supervisors. This integration of the roles of a manager, while using resources (mostly human in this example) to reach a goal, is typical of how managers function on a day-to-day basis. Let's continue looking at the managerial roles performed by Jim Johnson.

4. A Manager Directs

When Johnson initiated the meeting with his senior staff, he assumed a leadership role in the meeting. He called the meeting itself, organized it, and then exercised his leadership by discussing the need to increase efficiency in support of the organization's overall need to save money. By discussing the situation with his staff and by getting their input and agreement on the plan of action, they had greater motivation to do the job in an efficient and effective manner. In turn, the supervisors might follow a similar procedure with the employees who will actually do the job of moving the animals. If all goes well, everybody will compliment each other on a job well done. A manager's role in directing can include

- Hiring
- Training
- Delegation of responsibility
- Evaluating the performance of your direct reports
- Leadership

By now, you know that a manager is much more than a foreman who makes sure that a job gets done. Nevertheless, directing people is an important role for animal facility managers. In most instances, when you are directing people, you are making sure that work is done correctly and on time. This should remind you that managers are always looking to accomplish their goals in a more efficient and effective manner. But when directing people, managers are also expected to motivate and lead their employees, not to stand over them with a whip. Many books have been written on managerial leadership and motivation, and for good reason. Your employees will look to you to provide leadership and motivation. In Chapter 7, I'll expand the discussion on leadership, but it is important to note that a good leader incorporates most or all of the roles of a manager, not just directing activities.

5. A Manager Controls

When Johnson gave the go-ahead to combine animals into fewer rooms, he assumed that the move would go smoothly, but he could not be sure of this. He therefore went into the animal holding area to see for himself that there were no problems. This was a simple but effective way of monitoring (controlling) the progress of the plan. When a manager "controls," he or she is comparing the actual outcome to the expected outcome.

A manager has a responsibility to compare the actual outcome of a plan or project to the expected outcome.

You cannot make decisions, plan, organize, and direct, and then sit back and hope that all goes well. In the example I've been using, it was important for Jim Johnson himself to ensure that the plan was proceeding as anticipated. We call this managerial function controlling. That does not mean that you must have your fingers in every nitty-gritty detail of everything that goes on every day. It *does* mean that you should have some mechanism for obtaining periodic feedback about the progress of a project. The feedback can be in the form of written reports, staff meetings, or going out and looking for yourself. You want to compare the progress being made against a standard that was previously set, such as the maximum time to completion of a project. A good manager monitors a project while it is ongoing, and then makes a final check when it is done. Further on in this chapter, I'll discuss controlling with reference to what you can do to track the progress toward reaching your goal.

Try to avoid what has happened to many managers when they fail to get out of their offices and look for themselves at the progress of an ongoing project. A former colleague is a classic example of this problem. He invariably assumed that everybody was doing exactly what they should be doing, but he did not take the time or make the effort to find out for himself how things were progressing. In one instance, a project was delegated to a subordinate and, unfortunately, it was not performed correctly. After all was said and done, my colleague tried to shift the blame for the failure to the subordinate and his employees. Well, perhaps there was an employee failure, but there was also a managerial failure by not having a control mechanism in place. Not long afterward my friend was fired. Managers cannot put the entire blame on their staff if they have failed to monitor a plan's progress. You delegated the responsibility, so you have to be sure the work is getting done properly. The worst excuse a manager can give is "I told them to do it but they didn't." That doesn't hold a lot of water with upper management, nor should it. That's like saying, "I didn't know the gun was loaded." See how far that gets you on judgment day.

This raises another interesting point about controlling your resources. When people are working with you to reach a goal, they have to know what constitutes acceptable performance of a task. Can you effectively exercise managerial control over what people are doing unless they know what is expected of them? The answer is no. You must decide how much leeway to give people in terms of time, money, or methods. Controlling is important, but we must define the limits of acceptable performance before we can control anything.

In order to have proper managerial control over a project, all persons involved must clearly understand what is expected of them and what constitutes acceptable performance.

You may not be an expert at everything, but you should have a clear vision of how all the parts of your operation are integrated. For example, you should be able to follow an animal order from the submission of an investigator's request to the point when the animal is entered into a study. You should also be able to do this from an investigator's (your client's) point of view. If a particular strain of a mouse isn't available, how long did it take for that information to get back to the investigator? What is being done to help? What happens if only part of an order arrives? Do you tell the investigator right away, call the vendor, or wait another day to see if the remainder of the order comes in?

There is an important corollary to your ability to understand how all parts of your operation come together. You also must work to have everybody in your animal facility—in particular the managers who report to you—clearly understand all of the animal facility's goals. That helps everybody, not just you, understand how all the operations come together. You are implementing the organizing role of a manager. As a laboratory animal facility grows, there comes a point where no one person can (or should) exert managerial control over all systems. That is when it becomes important to delegate authority to others, making sure that everyone understands the values, vision, and objectives of the vivarium. They are enabled to fix problems on the spot, rather than sending reports to you and waiting for you to make a decision. Consider what can happen if there is an unexpected increase in the cost of hardwood chip bedding from ABC Distributors. Must the person who purchases bedding pay ABC the new price? Will he or she then have to wait for you to see the monthly budget report, integrate it in your own mind with all other operational control systems, and then make a decision about what to do? Or, does that person routinely receive the animal facility's budget sheets and have the authority to bargain with ABC, change distributors, or take other actions (such as locking in a price at the beginning of the fiscal year) as long as there is no detriment to ongoing studies?

You are the captain of the team. You get the best people you can to do the job, and then, like the manager of a baseball team, you make things mesh. Don't become a tyrant. Have managerial controls over those functions that you believe to be most important to the proper performance of the project or area in which you work, but have everybody understand the facility's goals so they feel a responsibility to the team. Your objective is to have everybody understand their job well enough that the task itself, not the manager, has built-in controls on acceptable performance.

Organizational Mission: The Big Picture

Goals

Let's briefly talk a little more about the word *goal*. A goal, as was said earlier, can be defined as a desired endpoint. As you already have read, we develop plans to reach our goals. Thus, having three cages properly cleaned in three minutes might be your goal. This is a short-range goal and may or may not require much planning. Other goals might be developing space to house guinea pigs, having greater control over divisional spending, developing an employee training program, consolidating food purchasing on a large university campus, improving your utilization of time, learning how to prepare a budget, reading this book, getting money for a security system, improving relations with a union, or developing an automated animal census system.

A goal is a desired endpoint.

It always has seemed to me that establishing goals is a more difficult task when things are going well compared with when things are not going so well. When a problem arises, you usually can deduce, without too much trouble, the options you can use in order to reach a desired endpoint. If, for example, your surgical lights fail, you can decide to repair them, replace them, or perhaps conclude that they are no longer needed. Those are fairly obvious endpoints that can be explored further for feasibility. But if there are no immediate problems and your facility is working efficiently and effectively, what goals do you set? Do you even have to set goals when everything seems fine? You certainly don't spend your entire day thinking about goals, but a good manager (and a good leader) is constantly challenging current assumptions. You should think in a "what if" mode [26]. "What if there is a burst water main and the building water supply fails? How will the animals get water?" Or, "What will I do if the union goes out on strike?" Or, "What will I do if I have to cut my budget?" Another is, "Is there a need for our animal facility to offer surgical training to investigators?" As a final example, consider, "What if there was an easy way to check my mice every day without having to go to every cage and look into them?" Perhaps you will be the person to develop an in-cage video monitoring system. The possibilities for goals are endless. In addition to "what if" thinking, Rowan Gibson [27] offers another way to develop business opportunities. I've adapted for laboratory animal science his four key perspectives for finding goals that can lead to growth opportunities:

1. *Challenge orthodoxies.* In laboratory animal science, certain things that we consider (or considered) to be facts may turn out to be nothing more than opinions or guesses when challenged. For example, with all of the well-deserved concern we have about pain alleviation, does tail clipping to get cellular material for genotyping actually hurt the mouse [28]? Is the dwarf tapeworm, *Rodentolepis nana*, actually a zoonotic infection? Are there data that say it is, or is it just a guess because that tapeworm has been found in humans? Is it a reasonable goal for us to try to answer these questions, or are they better left to others?

2. *Harness trends.* What new products do we see at meetings? What are the hot topics that are discussed in the laboratory animal science literature or at meetings? For instance, many animal facilities are now using individually ventilated cages on ventilated animal racks. Should we do the same? Is this a reasonable goal? Joel Kurtzman [29] correctly notes that leaders (and I'll add managers) must pore over newspapers, magazines, and books and review what's on the web. They have to spend a lot of time searching for ideas. In Chapter 2, I'll elaborate further on how managers consider opportunities and threats that reach their institution and the animal facility through the environment that's external to their institution and, of course, their institution's own environment.

3. *Leverage resources.* Do we have available resources that can be used in new ways? For example, a goal might be to hire a private practice veterinarian if your animal facility foresees a need to employ a veterinarian whose work will be focused on clinical medicine.

4. *Understand needs.* Pay attention to concerns that have been ignored in the past. An IACUC office might set a goal of revising an IACUC protocol form that has frustrated its institution's research community for many years.

All of these goal-related suggestions should be considered in light of a constantly recurring question in laboratory animal science, which is "Will it help my customers?" The question noted above about *Rodentolepis nana* is a good example. Is it a reasonable question for your animal facility to try to answer if you don't have a tapeworm problem or nobody is studying that tapeworm? Perhaps, if we were part of a comparative medicine department, this would be a valid question to ask, but if we are only operating a vivarium, trying to answer that question has no impact for any of our customers and is a waste of our resources.

Your goals are often developed in response to the needs of animals or researchers, who are the two primary customers in most animal facilities. That is appropriate because goals that have no current or potential value to your customers may not be worth pursuing. And, of course, our customers are not all the same. Different species of animals have different needs, and different researchers or educators also have different needs. Providing more or improved fee-for-service procedures is an example of a goal that is typically focused on the needs of most all researchers. If your goal is to build a new centralized animal facility within 10 years, that is a long-range goal focused on the needs of both researchers and animals. To a certain extent, goals are what you want them to be, although common business sense says that there should be a reasonable chance that they can be achieved. Keep in mind that as a manager, you are always developing new goals and devising plans to reach those goals. You are going to use the various resources discussed above while developing your plans, organizing activities, making decisions, directing people, and setting up control systems. All of this is done to reach your goal. More will be said about goals later in this chapter.

Mission Statement

Let me back up a little bit. Goals are what you want them to be, but as I noted, you do have to use a little common sense. You cannot do whatever you want to do or try something you want to try just for the fun of it. Our business goals should be related to what our organization is trying to accomplish in the long run, or what is commonly known as the *organizational mission*. Most organizations, whether they are private research laboratories, universities, or pharmaceutical firms, have what is known as a mission statement. This is a written document defining the overall reason for an organization's existence. For example, a mission statement of a supermarket chain may be "Our mission is to help feed American families by providing high-quality food at reasonable prices."

A mission statement is a written document defining the overall reason for an organization's existence.

Mission statements are actually quite beneficial because they are always there to help us remember the big picture. They help direct us to our goals. They inspire us. They keep us honest. They give employees a rallying point. In general, your animal facility's goals have to be aligned with your organization's mission and vision statements (vision statements are discussed below). Obtaining Association for Assessment and Accreditation of Laboratory Animal Care International (AAALAC) accreditation, hiring an operations manager, or obtaining new caging are typical goals that are most likely aligned with your organization's mission.

I used to think that mission statements were just another piece of academic fluff that had no place in the real world. I was wrong. Sometimes, when I have a problem deciding what information should be included in a lecture or demonstration, I think about the mission of my school or even the smaller mission of the course I'm teaching. It helps keep me focused. Presumably, your organization will take its mission to heart and practice what it preaches. If it does, it helps motivate and retain employees. If it doesn't, a sick corporate culture may emerge. A lofty mission statement is obviously meaningless if the parent organization pays it no heed.

Some organizations don't have a mission statement, yet they manage perfectly well. There is no law that says an animal facility has to have a mission statement or, as we will soon see, a vision statement or, for that matter, written goals and strategic plans. Joe Calloway hit the nail on the head when he said, "Keep the rules that work for you, and break the rest" [30]. The senior staff of my department believes in having a meaningful mission statement because it acts as an adhesive that binds our department together. We consist of veterinary technicians, animal care technicians, supervisors, managers, office staff, and veterinarians. But we have only one mission, which is to provide for the physical and psychological well-being of laboratory animals and to help investigators properly use laboratory animals. We're animal people, and that mission gets repeated as often as possible.

Here is a true story that illustrates how the leadership of an organization kept a focus on its mission, in this case, remaining a not-for-profit institution. I have only changed the name of the institution and its people.

As with many institutions in the late 1990s, Eastern General Hospital, a not-for-profit hospital, was undergoing rapid growth. It had a mission to serve people in its community irrespective of their ability to pay for services. Nevertheless, within a period of three years, its fortunes turned. Although the ship was not sinking, it was beginning to list. Peter Ferrar, the CEO of Western Hospitals, Inc., a profit-making hospital conglomerate, was seen having lunch with Stan Reilly, the CEO of Eastern General Hospital. Soon afterward, rumors began to fly about a possible buyout of Eastern General by Western Hospitals.

When Reilly heard of the rumors, he acted quickly, sending all employees a strongly worded memo reminding them of Eastern General's not-for-profit charitable mission, and assuring them that they would never merge or be purchased by a profit-making organization.

An interesting side effect of Reilly's memo was that concern for a merger or buyout by another not-for-profit organization actually increased. Reilly might have been honest with his employees and true to the mission statement, but he did little to stem people's fears about the loss of their jobs. Eventually, Eastern General Hospital was purchased by and then merged into another not-for-profit organization, with a resultant loss of many jobs. The mission remained intact, the hospital's basic service function continued, but there was no escape from the realities of business.

Mission statements should be brief and to the point. They should state what the organization is in business to do and for whom it does it. Those mission statements that go on for paragraphs or pages are unnecessary and probably reflect a lack of understanding as to what the organizational mission really is or should be.

Mission statements should be brief and to the point, stating what the organization does and for whom it does it.

The other side of the coin is that mission statements can be brief but not tell you very much. Can you imagine a mission statement from a pharmaceutical company that says, "The mission of XYZ, Inc. is to make sufficient profits so that shareholders will continue to invest in the company's future"?

Certainly, there is nothing wrong with XYZ, Inc. wanting to make a profit. It had better do so if it wants to stay in business very long. But is that the company's entire philosophy? Is robbing banks okay? Can they go into whatever business they want? What do they do? Are shareholders its only customers? At the very least, the mission statement of XYZ, Inc. might read, "The mission of XYZ, Inc. is to use its resources to develop, test, and market prescription drugs that help prevent, alleviate, and cure human diseases."

That's a reasonable start. With that statement, if there were an opportunity to purchase a detergent manufacturing plant, XYZ management might say, "It's a good buy and it will be profitable, but how does it relate to our mission?" Here is another example. The mission of Great Eastern University may be stated as

> The mission of Great Eastern University is to accumulate knowledge in a wide variety of arts and sciences, and disseminate this knowledge to its students and the public.

It says nothing of teaching, research, sports, or any of the many other activities of a large university. It doesn't have to. All forms of teaching, whether it involves students in chemistry, medicine, theater, football, or agricultural extension services, are part of the dissemination of knowledge. Likewise, informal discussions and guest lecturers also fall under the heading of the dissemination of knowledge. The accumulation of knowledge can include all forms of research, formal and informal meetings, and even providing "real-life" learning experiences for students, such as student teaching or dramatic presentations.

Nevertheless, the mission statement of Great Eastern might also be written as follows:

> The mission of Great Eastern University is to accumulate and disseminate knowledge in a wide variety of arts and sciences; to instill in its students, faculty, and staff a desire to understand and improve the world they live in; and to encourage and support excellence in scholarship and research for its students and faculty.

In many ways, this mission statement is a substantial improvement over the first, as it gives us more of the university's philosophy. It tells us who its customers are (students, faculty, staff, and even the rest of the world). The university is not just an information warehouse or a glorified library. It is a vibrant institution that wants to use its resources to make a better world for all people. To do that, its resources will be directed toward the encouragement and support of excellence in teaching and research. It will share its knowledge with everyone. Now you have a good "feel" for why this university exists. You know something of its values. Indeed, there are some organizations that have separate mission and value statements.

Here are four hypothetical mission statements. Do you think they are adequate as written, or would you expand on them? Try adding one or two sentences about values to these mission statements.

Hypothetical Mission Statements: Are They Adequate?

1. *The College of Veterinary Medicine of Great Eastern University*: To accumulate and disseminate knowledge about animal health and well-being.
2. *The Department of Veterinary Preventive Medicine, College of Veterinary Medicine at Great Eastern University*: To accumulate and disseminate knowledge about the prevention of diseases of animals.
3. *An organization owning a football team*: To provide for the entertainment of sports enthusiasts.

4. *A laboratory animal facility*: To provide for the physical and psychological well-being of animals used in biomedical research and teaching.

In all of the above examples, the mission statements do not go into great detail. Indeed, the organization that owns a football team never even mentioned football in its mission statement, although it did define its customers (sports enthusiasts). Perhaps it is because the concept of entertaining sports enthusiasts includes putting on football games, selling souvenirs, owning a hockey team, selling sports videos, and other sports entertainment–related items. The animal facility mission statement included animals as a customer, but it didn't say a word about another primary customer—the researchers who use the animals. Writing a mission statement isn't as easy as you might think. Give it a try. What is the overall mission of the organization that you work for? What is the mission of your animal facility? If you don't already know it, write one.

Before moving on, let's reexamine the mission statement of the laboratory animal facility. It stated that it should "provide for the physical and psychological well-being of animals used in biomedical research and teaching." But does it really tell us who our customers are? To a certain extent it does, since animals are definitely the recipients of the services we provide. Nevertheless, as I previously noted, we have to recognize that a laboratory animal facility is a service-oriented enterprise that has at least two customers: animals and investigators. Perhaps a better mission statement for a university laboratory animal facility is the one in use at the Department of Animal Medicine at the University of Massachusetts Medical School.

> The Department of Animal Medicine is dedicated to providing for the physical and psychological well-being of animals used in biomedical research.
> The Department accomplishes its mission by providing high-quality animal care and by aiding and advising the school's research community on the appropriate use of laboratory animals.

Compare this statement to the mission statement of the laboratory animal facility at Yale University:

> Yale University is committed to conducting quality animal research in an ethical and responsible manner to further science and to improve the health of society. [31]

Which of the above two mission statements do you prefer, or do you think that they both are satisfactory? Here are other actual mission statements. Evaluate them for yourself.

Actual Mission Statements of Various Organizations

Cummings School of Veterinary Medicine at Tufts University

> Cummings School of Veterinary Medicine at Tufts University improves and promotes the health and well-being of animals, people and ecosystems we share. [32]

The American Association for Laboratory Animal Science

> AALAS is an association of professionals that advances responsible laboratory animal care and use to benefit people and animals. [33]

The University of Massachusetts Medical School

> The mission of the University of Massachusetts Medical School is to advance the health and well-being of the people of the commonwealth and the world through pioneering advances in education, research and health care delivery. [34]

The Laboratory Animal Management Association (LAMA)

> LAMA is an association dedicated to enhancing the quality of management and care of laboratory animals throughout the world. [35]

The Committee for Laboratory Animal Training and Research

> To aide laboratory animal and comparative medicine trainee career development by providing research mentorship and networking opportunities, facilitating sharing of training resources among training programs, and promoting interaction of trainee-focused committees of national organizations. [36]

American College of Laboratory Animal Medicine (ACLAM)

> The American College of Laboratory Animal Medicine advances the humane care and responsible use of laboratory animals through certification of veterinary specialists, professional development, education and research. [37]

Here's a real mission statement from a company that has nothing to do with laboratory animal science. It's an executive recruiting firm. From its mission statement, what do you know (and what don't you know) about this company?

Our mission is to place top priority on listening carefully, communicating honestly, and working strategically with our candidates and clients. [38]

Once you have a mission statement that defines your overall philosophy, you can develop and focus your goals. As I wrote earlier, these goals must be in line with the mission of both your laboratory animal facility and your organization as a whole, and whenever possible, the goals you set should be of value to either the research community or your animals. However, there will be times when you are told to implement a goal that is so strictly financial (such as increasing *per diem* income by 15%) that you cannot see how it helps either animals or investigators. That is an unfortunate fact of life in corporate and academic America. But whatever the motive behind the goal, it is your responsibility, using the resources you have, to develop strategies to reach short- and long-range goals.

Your goals must be related to the mission of your organization.

Fulfilling Your Organization's Mission

Did you notice how often the word *quality* is stated or implied in mission statements? There's nothing wrong with that if it's true because the most important end product that an animal facility provides is consistent high-quality animal care and healthy animals. A recent article [39] reported the findings from a study on what leads to quality in the workplace. It concluded that monetary incentives, training, and sharing of best practices have little impact on quality. Rather, managers and other leaders have to continually discuss quality and not just have it as a slogan on their mission statements. They have to make the definition of quality very clear to all employees and, when possible, fit the message to what is important to the worker. For example, showing a new employee how to restrain a rabbit is good, but explaining how proper handling can prevent a spinal fracture in the animal adds to the demonstration's importance. This helps the employee take ownership for the quality of her work and the fulfillment of the vivarium's mission.

Sometimes it seems hard to achieve quality when administrators are telling you to do more with less, but that doesn't mean we give up. Rather, we do the best we can with what we have, and when opportunities arise, we try to improve on what we do. Improvement doesn't mean we go out and buy fancier equipment (although that's nice to have), but it does mean that we hire smarter, train people better, and continually sharpen our management skills to help fulfill our mission. Animals and people depend on us for quality care, and we don't want to let them down.

Let's assume that the mission of your animal facility is "to provide for the physical and psychological well-being of animals used in research." That's fine, but how are you going to do that? What are the day-in and day-out general methods you will use to accomplish your mission? Every organization needs to establish a basic operational framework for accomplishing its mission [40]. The Red Cross is a relief agency, and, like other organizations, it needs a daily operational framework to accomplish its mission. Should its work include providing temporary housing for people? Permanent housing? No housing at all? Should it obtain blood from donors once an emergency occurs, or should it bank blood in anticipation of an emergency? If you are a book publisher, your mission may be to publish scholarly books in the life sciences for adult readers, but does that include scientific journals? Will your animal facility be strictly a service department, or will you incorporate research into your activities? Will it be basic research, clinical research, or both? In other words, how will you fulfill your mission of providing for the physical and psychological well-being of animals? This is the operational picture you need to establish to fulfill your mission, not the SOPs that tell you how or how often to change a cage. Every animal facility, and every member of that animal facility, should have a clear understanding of how their facility actually accomplishes its mission statement.

While the mission statement of an organization is typically difficult to change (and it should be hard to change), the broad operational activities that are used to fulfill the mission should be somewhat more flexible to give you the ability to meet changing conditions.

> *The Great Eastern University laboratory animal facility operated successfully for many years by having technicians who performed both animal husbandry and veterinary care. However, as its research endeavors grew in size and experimental complexity, the leadership of the facility began to question if they should continue to have generalist technicians who did both animal husbandry and veterinary services, or if they should have one group of technicians providing animal husbandry and another providing veterinary services. On the one hand, having generalists would allow the technicians to have better familiarity with individual animals and decrease the potential boredom of performing the same functions day after day. On the other hand, having specialists for each function might facilitate a higher quality of animal care.*

The key point in the situation given above is that the mission of the animal facility will not be changed whether there are one or two groups of technicians. The means used to accomplish the mission may change, but not the mission itself. In other words, an animal facility should have flexibility in how to accomplish its mission.

Often, even though an organization doesn't change its mission statement or the overall methods used to accomplish its mission, it may have a certain major long-range goal, such as becoming accredited by AAALAC. This major goal is directly related to the mission statement and usually requires many people to work together, using many available resources. This major goal is called an organizational vision and is the subject of the next section.

Vision Statement

Your mission statement is your overall business philosophy, and it sets a general direction for any goals that are to be set. But in order to chart a course for the next few years, many organizations want to state a major direction or action that the organization (such as your animal facility) will take. This is sometimes called a vision statement, a strategic vision, or strategic intent. A vision statement, even if it appears to focus only on the animal facility (e.g., a vision to obtain AAALAC accreditation), should always have an ultimate goal of adding value to the work of the researchers or educators or enhancing the well-being of the animals. Further, the vision statement should be based on a realistic assessment of your ability to ful-fill the vision [41]. "The ability to form and follow a vision is what leads to creativity, innovation, insights, and brilliant solutions to nagging problems" [42]. On the other hand, pie-in-the-sky optimism sounds good, but if there's nearly no chance of fulfilling the vision, it most likely isn't worth attempting until circumstances change.

Not every animal facility (or even every large research organization) has a vision statement because some managers may honestly feel it isn't needed under their particular circumstances. Furthermore, there are managers who do not have a vision statement because they are content doing tomorrow what they are doing today. They will buy new equipment or even plan for a new animal facility if told to do so, but there's really no leadership to initiate a significant change (such as obtaining AAALAC accreditation) and lead the facility in accomplishing that change. These managers are perfectly happy to figure out the best way to accomplish their mission in a very general way, as was discussed in the previous section, and that's about it. If there's no vision, they don't have to worry about accomplishing it. They may be good managers, but not good leaders. *Leadership implies moving an organization forward* and usually requires a clear vision statement to inspire people and be disseminated throughout the organization. Management, however, often entails improving on what we are doing today. It doesn't always focus on short-term goals. Most animal facilities, especially larger ones, employ people (such as the facility director) who are expected have a longer-term view of the facility's operations and needs. Therefore, if you want to move forward and be known as a good manager, you may need a vision statement. Employees want a rallying point, and the vision statement provides that. It not only informs employees about the direction the animal facility will be taking, but it also informs them about where, how they can fit in and help accomplish those goals, and which goals are the most important to the animal facility. That's a nice way of saying that there has to be clear and realistic priorities incorporated into a vision statement. If everything you do or every need you have is equally important to your animal facility, then nothing is really important and stagnation can result [43,44].

We'll discuss some of the differences between management and leadership in Chapter 7, but there's nothing to say that a good manager cannot also be a good leader, develop a vision, and lead the animal facility toward accomplishing that vision.

A vision statement describes the major direction or action that an organization will take over the next few years. It helps all employees set goals to accomplish the vision.

A vision statement has to impact most of the people working in and with the animal facility in a forceful manner because it's not unusual to have naysayers question its achievability. Therefore, vision statements are typically developed by one or more people at higher levels of management, such as the CEO or the dean of a school. Nevertheless, it should not be written in a vacuum. Even the CEO should get input from senior staff members, and it's not unusual for senior staff members to get some of their ideas from the people who report to them, and so on. The vision statement helps managers and other employees set specific goals that can be used to accomplish the vision. It should be clear, short, and to the point because most of us will not be able to remember or repeat a lengthy vision statement. It should be something that can be pictured in one's mind, like a home run ball sailing over the fence. It should also be repeated, time after time, at meetings, in newsletters, on bulletin boards, and certainly to new employees when they are hired. This constant repetition reminds everybody in the organization that there are specific goals that have to be reached, and it's not just come to work, do the job, and go home. All of us want to be proud of our organization and want to be part of accomplishing its vision, but that's hard to do if the vision statement is casually mentioned one day and we never hear of it again. In the next few pages and in Chapter 7, we'll provide some suggestions on how to actually implement an organization's vision.

Some people talk of long-range visions (i.e., the overall direction for the next 20 years), and others talk of shorter visions, such as the vision for the coming 3 or 5 years. I prefer the shorter vision because it is apt to be more practical for a laboratory animal facility. In fact, I've just about reached the point of believing that planning for more than two years into the future can be problematic, and there is even a suggestion that a short-term approach can encompass as little as three months [30].

As mentioned a little earlier, a vision statement has to be based on reality. For example, the University of Massachusetts Medical School is routinely rated as one of the nation's best medical schools for primary care education. Therefore, there is a reasonable chance that with specific plans, it might be able to become number one in a few years. On the other hand, had the school been rated as number 50, a vision to become number 1 within three years would have been foolhardy.

How might a vision statement's operational guidance work in a real-world setting? Take a look at the 1992 vision of the former Hahnemann University (Hahnemann merged with Drexel University in 2002). The vision was that the school would become one of the nation's 40 best medical universities by the year 2000. To my way of thinking, that would have been a reasonable vision statement. However, the board of trustees went much further and defined certain goals that were to be accomplished in order for the vision to be fulfilled (there's nothing wrong with doing that because goals will have to be set as either part of or separate from the written vision statement). The university's trustees developed goals such as having the colleges of the university act as a unit, defining standards for excellence, revitalizing its facilities, improving public relations, reevaluating programs, and creating an atmosphere of trust for its faculty, staff, and students. The vision statement went on to describe the economic and operating issues necessary to consider when implementing this vision. In other words, the vision statement also included specific goals. The Hahnemann vision was consistent with the university's mission.

In stark contrast to the detail provided by the Hahnemann vision statement is the vision that the president of a health care organization presented to its board of trustees. He said, "My vision for [this organization] is to become one of the best places in the world to receive health care, to get an education, to conduct research, and to work."

What about your research animal facility? Does it have a vision statement? If it does, does it complement your organization's vision? Does your organization have a vision?

The vision statement of the Department of Animal Medicine at the University of Massachusetts Medical School is a little different in that it includes two statements:

■ To integrate ourselves into the academic framework of the medical school through teaching and producing scholarly and practical publications and presentations in laboratory animal science
■ To develop the management skills of the department's administrators to create an enhanced work environment and provide career training for future managers

Most, but not all, vision statements have a time line. For example, the first of the two vision statements from the Department of Animal Medicine might have indicated that it intended to achieve its vision within five years. Now it's about seven years since the Department of Animal Medicine implemented its vision statement, and I can honestly say that it largely has been fulfilled. The challenge now is to continue improving.

The vision statement is the starting point for developing the specific goals that must be reached to fulfill the vision. Within the context of asking where are we now, where do we want to be at some point in the future, and how we will get there, the vision statement focuses on the last two, that is, where we want to be and the means of getting there. This requires goals and specific strategies to reach those goals. I will elaborate a little more on this in just a moment, but for now, I want to emphasize that while establishing these goals, don't lose sight of your customers (animals and investigators). It's very easy to get caught up in organizational bureaucracy to the point where fulfilling your goals becomes more important than your mission.

Let me give you an example of what is likely to happen if you thoughtlessly put your needs ahead of your customers' needs.

> One of the goals of Steve Shannon, the operations manager for the Great Eastern University rodent facility, was to get all cages changed in the early morning to allow the animal care technicians to have sufficient time in the afternoon to tend to other needs. Without giving much thought to consequences, Shannon instructed the technicians to start cage changing at 7:00 a.m., which is when the room lights came on and the workday started. This was not a problem for most technicians, but it was for the technician who cared for the animals used by Dr. Sam Gordon. Gordon's research staff always worked with his rats between 7:00 and 9:00 a.m. Previously, the animal care technician would change cages in the room beginning at 9:00 a.m. so as not to be in the way of the research staff. Shannon was told about the conflict and took it upon himself to tell Gordon's technician of his new policy. He politely asked the research technician to either start the experiments a little later or carefully work in the room while the animal care technician was busy changing rat cages. In other words, Shannon put his needs ahead of the needs of his customer.

> There's probably no need to tell you what happened next. Gordon complained to the director of the animal facility and the director in turn called Shannon on the carpet and reminded him that a reasonable need of an investigator cannot be undermined by his desire to get cages changed a little earlier in the day. Shannon was told that in the future, he was to consider the potential consequences of his actions before implementing a new plan.

The large majority of our laboratory animal facilities are owned and operated by its parent organization. In other words, we have minimal competition. More often than not, within our organization we are a monopoly. When we start to believe that we are indispensable or more important than others, as Steve Shannon appears to have done in the above example, we are asking for trouble. Nobody is indispensable. If fact, we are all replaceable. Be careful. Remember your customers.

> **Never allow the fulfillment of your goals to become more important than your mission and customers.**

Whether the vision statement comes from the president of a university or the manager of a laboratory animal facility, it is that person who must lead the way. "Whether managers express their vision of the future in numbers or metaphors, they must use it to mobilize the talents and energies of people in the company, as well as suppliers and vendors" [45].

As a person who has spent most of his professional career in academe, there is no doubt in my mind that vision statements are at least as important as mission statements. If you dissect the mission statement of almost any university, you will find that they are similar. There's usually a reference to teaching, research, and service. It is the vision statement, though, that differentiates one university from the other and one animal facility from the other. It is not the only step in reaching goals, but I believe it to be one of the most important. Let me give you an example of how a vision statement can be used in a real-world situation:

> The laboratory animal facility at Great Eastern University operated like many others. That is, it has some basic goals (such as hiring an additional animal care supervisor and getting additional rabbit cages to meet immediate needs), but there was no unifying vision for the facility. The new director of the facility recognized that there was a significant difference between putting out the day-to-day fires that every animal facility has to deal with and establishing an overall direction for the facility. (The overall direction is what we have been calling the vision of the animal facility.)

> The new director met twice with other employees of the animal facility and developed a vision that they believed they could reach with some hard work, that was important to the animal care mission of their facility, and that required a few years to accomplish. Specifically, the vision statement was "to achieve full AAALAC accreditation within 5 years."

Why would Great Eastern University choose this vision? Well, to begin with, the research at Great Eastern was growing rapidly, and it was to the school's benefit to have AAALAC accreditation. Additionally, the committee developing the vision wanted its coworkers to share in the recognition enjoyed by institutions having AAALAC accreditation. Finally, the new director wanted to pull everybody together and create a spirit of cooperation by working toward fulfilling a vision that affected all aspects of the animal facility's operations; AAALAC accreditation would meet that need.

Strategic Planning and Long-Range Goals

How might the animal facility at Great Eastern University fulfill its vision to become AAALAC accredited within five years? The animal facility managers will have to develop specific goals and plans to reach those goals. You might recall that planning is one of the roles of all managers, but it is usually a good practice for a manager to obtain input from coworkers up and down the chain of command concerning most all aspects of goals being considered and the means of reaching those goals. That includes frontline employees (e.g., most technicians). They should be included in many (perhaps all) of your strategic planning discussions because they are likely to know more about the daily work processes than anybody else, and it also is a strong morale builder when frontline employees know that their opinions and expertise are sought out and valued.

Once the short- and long-range goals are in place (and setting goals is a part of the planning process) you begin to develop specific methods (strategies) to reach those goals. For example, one goal might be to replace all cracked and discolored polycarbonate cages with polysulfone cages, and one of the strategies might be to have people on the clean side of the cage washer remove and discard all of those cracked and discolored cages as they exit the cage washer. The overall process of developing goals and specific strategies for activities that require a relatively long time to complete has been called either long-range planning or strategic planning. Some people use those terms interchangeably; it probably doesn't make that much of a difference. The important thing to remember is that you have to develop one or more strategic plans to reach your goals. It just doesn't happen magically. If you do reach your goals without any planning, consider yourself lucky, not smart.

Establishing long-range goals and developing strategic plans to accomplish those goals is often thought of as a somewhat exotic upper management function that doesn't involve middle- or first-level managers. That's not necessarily true. As often occurs, upper management has final decision-making responsibilities (although even that's changing in some organizations), but middle- and first-level managers are often asked for their opinions, and the agreed-upon final plan may be a summation of many smaller plans. Middle- and first-level managers certainly have responsibilities that require both long- and short-term planning. Even as middle- and first-level managers, we have to look forward and decide what strategies we have to initiate now to reach a future goal. I will reemphasize this point a little later.

An organization must have one or more long-range goals in order to remain competitive. In industry or academia, you can't live for today only. You must look to the future. You must be prepared for what might happen or what will probably happen. You try to prepare for and take advantage of the future. If you don't devote time to setting and accomplishing goals, your competition will eat you up or you may find that your service is no longer needed.

All of us should set short-range and future (long-range) goals, whether it is in our personal or business lives. In business, these goals should be of value to the users of the animal facility or to the animals themselves, consistent with the organization's mission and vision, measurable, and attainable in a reasonable time span (perhaps two to five years for a laboratory animal facility). Sometimes, what appears to be astute strategic planning is nothing but pure luck, or a desirable result that came through trial and error. In many successful companies, this may happen far more often than we realize [46].

Goals should be

- **Important to your organization**
- **Of value to your customers**
- **Consistent with your organization's mission and vision**
- **Attainable in a finite but reasonable time span**
- **Easily understood by everybody**
- **Measurable**

Once you have determined what your long-range goals are, then, and only then, you should devise strategies (specific plans) to reach those goals. As I said earlier, goals and their strategies, when taken together, are your *strategic plan*. At this point, I have to make a clarification: just because we use the term *long-range planning* does not mean that we can procrastinate and do nothing for a couple of years and then, at the last minute, go into high gear to fulfill our goals. In business, some parts of a long-range plan are fulfilled as soon as possible, some parts are fulfilled a little later, and maybe some are fulfilled toward the end of the time period. However, the total plan should be completed within the time that you specified. In the example I used earlier, obtaining AAALAC accreditation is certainly not something that can be accomplished at the last minute. One goal was to discard some cages, and that could probably be accomplished very quickly. Another goal within the long-range plan might include resurfacing all the floors with poured epoxy, and that might take somewhat longer. Finally, a third goal might include fixing the building's humidification system, and that might take still longer. Each of those individual goals requires planning and the use of resources. When all of the goals and all of the strategies needed to reach the goals are taken together, we can call it our long-range plan or our strategic plan.

One of the keys to developing successful strategies is to keep them simple [47]. That's good advice for most anything, be it a vision statement, goals, strategies, SOPs, or communications. Most of you know how often long, complicated emails get read in their entirety versus becoming a casualty of the delete key. As a manager in an animal facility, one goal may be to have poured epoxy floors in the cage wash area (or perhaps the entire facility). This goal should have been discussed previously in terms of its practicality relative to alternatives, such as doing nothing, mending the current floors, or using a different flooring material. Hopefully, there was a really good discussion on why the new floor is needed because there's an old maxim in research that says that if you don't ask the right question, it doesn't matter what the answer is. Focus on how to get the resources (usually money) to get the job done, who will be the lead person from the animal facility, how you will coordinate the needs of the flooring contractor with the need for clean cages, and so on. You don't have to worry about the details of tearing up the old floor and laying the new floor; that's somebody else's problem. You will also have to lay out your priorities. For example, is it more important to have the floor renovated as soon as possible, or is it more important to coordinate the process to ease the research-based concerns of investigators? All through this process you are performing the key roles of a manager: planning, organizing, directing, making decisions, and controlling. Controlling? Yes, somebody has to make sure things are being done as planned, a crucial role of a manager. In Chapter 2, I will elaborate on some of the factors that can impact your organization or laboratory animal facility and influence your choice of strategies. These factors are usually known as environmental factors, and they include influences from inside and outside of your organization.

Don't be confused if you see phrases that do not precisely fit the definitions I've given you. Also, don't worry if you read an article that draws fine distinctions between strategies, strategic goals, tactics, objectives, and strategic objectives. As long as you understand the overall concept of what is happening, you will do just fine.

Long-range goals are specific endpoints your organization wishes to reach in the future. They are often part of a vision statement.

Strategic plans are the specific actions your organization will take to accomplish the long-range goals.

The entire process is often called *strategic planning*.

Mission
↓
Vision
↓ ↓
Short-range goals Long-range goals
↓   ↓
Short-range   Long-range plans
(operational) strategies (strategic plans)

Figure 1.2 Relationship between an organization's mission, vision, goals, and strategies.

Long-range goals may include obtaining AAALAC accreditation, building a state-of-the-art laboratory animal facility, establishing an animal disease diagnostic laboratory, creating a Division of Comparative Medicine, or replacing scores of primate cages. Clearly, these are not simple budget items that can be accomplished overnight or even in a year. They are long range in nature.

We can visualize the planning process by diagramming the relationship between our mission and goals as in Figure 1.2. Remember: Not all goals must be fulfilled a few years in the future. We have many short-range goals (such as buying a small number of cages) that we may be able to accomplish in a few days. Plans to accomplish short-range goals are often called *operational plans*, and they can be components of a long-range plan or just stand by themselves as short-range plans.

Some years ago, there was a series of television commercials with Dan O'Brien and Dave Johnson. They were promoting athletic shoes. Both were "shoo-ins" for the 1992 Summer Olympics decathlon. The question was, which one of them would be the Olympic champion?

Unfortunately, Dan O'Brien didn't even make the U.S. Olympic team. It was a shock to everybody. But in February 1993, about six months after the games had ended, I heard O'Brien being interviewed on television. He said that what had happened, happened; he couldn't change that, but he was not giving up. He had a long-range goal to become the 1996 Olympic decathlon champion. His strategy involved a specific training timetable, entering certain track meets along the way, and lecturing to youngsters about not throwing in the towel because of one setback. A positive attitude, he said, was what it took to succeed in life. The footnote to the story is that he became the 1996 Summer Olympics decathlon gold medal winner.

O'Brien's planning was classic. He had a long-range goal and long-range strategies to reach that goal. In other words, he knew exactly where he was and where he wanted to be in 1996, and he had a plan on how to get there. More important, he began to follow his plan. It should be obvious that if you do not actually follow your plan, all the planning you have done becomes meaningless.

Strategic planning isn't always as easy as O'Brien's plan makes it seem, but the essentials are always there. In laboratory animal facilities, as I indicated earlier, strategic planning should probably not be for more than five years into the future. I base this on the fact that most research grants are rarely for more than five years. However, given the erratic nature of academic grant funding, faculty turnover, rapid advances in many fields of research, and changing research objectives, as I also said, planning ahead for no more than two or three years seems reasonable. In industry, the rule of thumb is for three years. Of course, product development, as occurs in pharmaceutical companies, usually extends for more than five years, but more often than not, the laboratory animal facility manager will have adequate notice if modifications or other significant changes are needed in the facility's strategic plan. In laboratory animal science, your best-made plans may have to be quickly revised due to changes in funding, research priorities, and many other factors. Strategic plans are plans, not biblical commands, so be flexible, observe the economic and political environment of your institution to try to predict what changes may be coming, and don't be surprised or exasperated if and when they come.

Before proceeding, let me note that I'm making certain assumptions. I assume that your long-range goals are part of your mission and vision, that they are truly necessary, and that there is a reasonable probability of achieving them. I'm also assuming that they are approved by your superiors. This is not always the case. Look at this actual example, in which I have only changed the name of the organization.

> *The vice president for research for Great Eastern University went to a meeting in the mid-1980s. At that meeting, he became convinced that transgenic animals were the wave of the future, and his faculty could not afford to be left behind. He told the director of the laboratory animal facility to conduct a survey to determine the perceived need among the faculty to build a transgenic animal facility.*

> *The facility director went to the deans of those colleges that had potential use of the facility and told them that the vice president looked favorably on building a transgenic animal facility if the need was there. He then asked the deans if they believed such a facility would be helpful. All the deans said yes, and the facility director reported this to the vice president. With the subsequent concurrence of the deans, the vice president had this construction project placed near the top of the list of the university's capital improvement projects.*

> *There was only one problem. In the rush to jump into the twenty-first century, nobody bothered asking the faculty if their long-range research plans involved, or might ever involve, transgenic animals. The deans were simply asked if such a facility would be helpful. They said yes, but that was a gut reaction*

and it really wasn't unexpected since it wasn't coming out of their budget. They did not ask the faculty either.

As it turned out, the faculty had minimal needs at that time for such a facility. That did not mean that there never would be a need for such a facility, but at the time, it did not belong near the top of the capital improvement list for the current strategic planning session. Even today, with the explosive use of genetically modified animals, not every institution uses them. Obviously, long-range planning requires accurate information about long-range needs.

But what if the faculty *had* wanted and needed a transgenic animal facility? The university would then have to think about where to build (capital resources), how to get the money to build it (fiscal resources), the needs of researchers and laboratory animal care personnel (human resources), what goes into a state-of-the-art transgenic facility (information resources), and whether it could be completed in five years (time resources). That is, the university knew where it was now and where it wanted to be, and it was thinking about the resources it needed to accomplish this long-range goal.

Let's assume that the single most important factor in building a new transgenic animal facility is obtaining the money to build. That is not an unusual occurrence, and therefore raising the needed money will become a goal for the planning team. If we can solve that problem, the rest may fall into place. We will establish a planning team that cuts across job categories. There will undoubtedly be representation from upper management, research, laboratory animal facility management, and many other parts of the university. This team may begin by brainstorming. That is, they may sit around the table and suggest a variety of ways to raise money, but before doing that, we will clearly define the goal to ensure everybody is on the same page and does their homework. What homework? Once the goal (or the problem) is defined, the team has to gather potentially relevant information and share that information with each other to make brainstorming a worthwhile process. Without that information, it will be difficult to determine which ideas have the most potential [48].

These are the initial stages of developing strategies. Examples of ways to raise money include floating a bond, having a fund-raising drive, using the interest from an endowment, using the principal on an endowment, increasing student tuition, increasing animal care charges, writing a government grant, or a combination of these. I think brainstorming, using people from within or external to the institution, is a fine way to encourage thinking, but it does need some direction from the manager or leader; otherwise, it can get so out of hand that it wastes time.

Once the team agrees on the best way to raise the money (i.e., once strategies are agreed upon), the next step must be even more specific. We will assume that the methods chosen will be to apply for a government grant and also to use some of the organization's own fiscal resources (this type of an approach is most appropriate for not-for-profit organizations). Now we have to decide which grant to apply for, who will be the principal investigator, what deadlines have to be met, whether we can develop a preliminary floor plan in time, and so on. We might even have to develop secondary plans to raise money if the grant application is unsuccessful. In other words, there is a good amount of detail involved.

You cannot set long-range goals without including the strategies necessary to implement them. Without having the needed strategies to fulfill a goal, there is no real plan. It's the old story of the devil being in the details, and for all practical purposes, the details are incorporated into the strategies. The strategies can be phased in over the entire life of the plan, but they have to be there.

Like most of us, I know of more than one organization where the concept of setting long-range goals meant very little. In one not-for-profit organization, when asked to present a five-year strategic plan (remember, we are using the term *strategic planning* to mean incorporating long-range goals and the associated strategies), each department head simply stated the direction that they thought their research would take over the next five years. That's not a strategic plan or even a goal. At best, it forms part of a plan. It does not say what their goals are; it simply states what they think they will be doing. Is that really a goal? It does not state how they will reach their goal (whatever it may be); it simply assumes that a way will be found. It's unfortunate that in many organizations, this is more of the norm than the exception.

The Need to Balance Long-Range and Short-Range Goals

Strategic plans are the blueprints for accomplishing long-range goals. But as I wrote earlier, not all goals are long range. In fact, a good part of your work, perhaps most of it, will involve short-range goals and their associated strategies. I called those operational goals and strategies. In management, you are always balancing long-range and short-range goals. All of us have heard stories about executives who considered only short-range profits or goals, and left the company just before it became obvious to everybody that the business was in significant trouble. I have no idea how often this actually happens, but when it does, it is a tragedy. You need short-range operational goals and their associated strategies because without them you may not have an organization in the long run. It often takes a lot of managerial experience to learn how to balance both needs.

Reevaluating Goals and Plans

Long-range goal setting and developing strategic plans to achieve those goals are not static processes. Both must be periodically reevaluated to see if they are pertinent to your organization's mission and vision. In fact, it has been argued that

CEOs should be more concerned about building strategic options [49]. That is, those in more senior management and leadership positions should be developing different strategies, based on organizational needs and environmental realities. That way, there is a contingency plan if the primary plan has to be changed. But for now, we'll focus on the needs of middle managers and first-level managers. You may have heard of "strategic planning retreats," which are usually meetings held in a slightly isolated place (to reduce interruptions) and often used by organizations to reevaluate their mission, vision, goals, and strategies. If the mission is changing, which can happen, the vision, goals, and strategies must change as well. I'll be honest with you and somewhat sheepishly add that I have never been a big supporter of strategic planning retreats. The ones I have attended seem orchestrated to reach certain predetermined goals, and the final decisions on which goals to implement and how to implement them are rarely, if ever, made at these retreats. I would say that in my experience, retreats have been an exercise in futility.

If an organization's mission changes, then its goals and strategies also must change.

It makes sense that you have to change your goals and plans if the mission changes, but let me bring up two caveats. First, missions and visions need a certain amount of stability. You simply cannot change your mission every few years and run a successful pharmaceutical company or laboratory animal facility. Likewise, you should not willy-nilly change your vision because in the first three months of the fiscal year you have fewer animals than anticipated. Have a little patience.

Second, don't try to accomplish all of your major goals at the same time. There is a good chance you will not do a good job on any of them. Stick with one or two major goals at a time. If animal handling needs to be improved, stick with that goal; it's an important one. Do not concurrently push people to do change cages faster, become AALAS certified, and have everybody get to work before 8:00 a.m. Once you are satisfied with animal handling progress, move on to the next major goal.

Keep in mind that *it is the vision statement, not the mission or goals, that guides strategic planning.* Although I just said that it is nice to have stability in the vision, it is far more important to be flexible if necessary. If new opportunities appear, you may have to do your homework and see if they will be of any importance to you. In fact, you may have to change your vision. The good news is that even if your institution's vision changes, there is a strong possibility that the vision of the laboratory animal facility will remain the same.

Finally, and as I mentioned earlier, do not believe anybody who tells you that strategic planning is for upper-level management only. That is nonsense and reflects a total lack of understanding of managerial processes. There are differences in the extent of planning and some of the details when we compare upper managerial versus lower managerial strategic planning, but the concepts are the same. Every part of an organization, and most certainly the laboratory animal facility, must have long-range goals and plans that complement those of the parent organization. Without that forethought and direction, we cannot be a vibrant and contributing part of our parent organization.

Developing and Measuring Productivity Goals and Strategies

So far in this chapter, we've had a lot of discussion about goals, goal setting, resource usage, and being both efficient and effective. But all the visions, goals, and strategies in the world don't mean a thing if they're left on paper or go no further than talk around the water cooler. An effective organization has to accomplish its mission, vision, and goals. Therefore, we now have to ask, how do we know that we're actually accomplishing our mission and vision? How do we know that we are really making progress toward our goals? Is there some way to measure our progress? The answer is yes, but a little work is needed to understand the process because measures of effectiveness are usually harder to develop than are efficiency measures, and in animal facilities, we often want to put effectiveness just a little ahead of efficiency.

Appendix 1 presents a fairly detailed description of the procedures used to develop productivity goals and strategies and how to measure progress toward those goals. At this time, it behooves us to at least outline the process in order to complete the discussion on the fundamental concepts of management. Simply put, we use the word *productivity* to indicate the sum of the efficiency *and* effectiveness an organization has. Therefore, an animal facility that is both very efficient and effective has high productivity. Because the goals that we set should lead to better productivity for our animal facility, we want to measure our progress toward those goals. To do this, the goals that we set should have certain characteristics whenever possible. These characteristics are

- The goals to be measured should be aligned with our mission statement (and vision, if appropriate) and should bring value to our customers.
- The goals to be measured should be specific, not vague.
- There should be a limited number of financial and nonfinancial goals if we are trying to achieve a vision that requires many individual goals to be achieved.
- Each goal should be associated with clearly defined strategies.
- We should use numerical measurements to track our progress toward our goals.
- The number of items to be measured should be limited to those that are central to enhancing productivity.

If we can meet these basic criteria, we will be well on our way to being able to measure the progress toward our goals. As shown in the second item above, all the goals should have a specific numerical endpoint (e.g., decreasing overtime by 10%, increasing the number of cages washed per hour by 15%, or increasing customer satisfaction by 20%). Then, we develop measurement systems. If we want to decrease overtime by 10%, we can measure progress through payroll or other work records. If our goal is to increase customer satisfaction by 20%, we can measure our progress by having surveys sent to our customers (e.g., investigators) and asking them to numerically rate their satisfaction with various items. We also have to meet regularly to review our progress (where we are now) and make any necessary adjustments to get to where we want to be.

Now might be a good time to read Appendix 1. See if you can implement the goal measurement process into the functioning of your animal facility.

Final Thoughts and a Summary of This Chapter

Let's see if we can tie everything together. Management is the efficient and effective use of resources deployed to accomplish goals. Managers are those people who have the authority to use those resources in a manner that can significantly affect their organization's functions. By organization, we mean an animal facility, an academic department, a department in a pharmaceutical company, or even the company as a whole. The resources that managers use to help them reach their goals include people, money, capital equipment, information and time. In order to use these resources effectively, managers have to make plans, make decisions, organize activities, direct people, and check to make sure that everything is moving ahead as expected.

Most companies have a mission statement that informs its employees and the public of what the company does and for whom it does it. Almost anything a manager does should have relevance to the company's mission. Each department within a company (or school) may have its own mission statement, but it has to be related to the mission of its parent company. Every organization also has to determine the general methods it will use to fulfill its mission on a day-to-day basis. Furthermore, many companies have a vision statement that informs their employees and the public of the main goals the company wants to accomplish over the next few years. Here again, managers must be sure that their department's vision is aligned with the company's vision. Managers establish their department's or work area's goals to help fulfill the mission and vision statements. They also develop goals to help accomplish the daily operational challenges that always accompany the management of a large or small organization.

When trying to develop a broad vision (or even individual goals), managers ask three key questions: Where are we now? Where do we want to be in the future? How are we going to get there? The answers to those questions typically require the development of specific goals using the input of many people (human resources) and the potential use of all our other resources. Some goals that are developed are short range and can be fulfilled fairly quickly. At other times, there are long-range goals that take a few years to accomplish fully. All goals, whether long or short range, require specific strategies (plans) to fulfill them. A strategic plan is a combination of the goals and the strategies needed to fulfill the goals.

All goals should be routinely monitored and their progress measured to help ensure that they are actually progressing as intended. To measure progress, we first have to decide for ourselves what measurements are truly critical. The measurements we eventually decide to use should be limited in number, include financial and nonfinancial parameters, be very specific, include clear strategies, and be numerical so that they are easier to understand.

Managers set goals because they try to plan for the future even if there are no obvious problems to resolve. They are thinking about new opportunities that might allow them to provide a better service, capture a certain share of a market, or be more efficient and effective in general. In laboratory animal facilities, we rarely worry about market shares, but we do concern ourselves with providing better service, better animal care, or new services, even if we believe we are doing a good job.

In practice, managers spend only a small percentage of their time developing new goals. Most managers do routine work. Some of this work is directed toward accomplishing previously defined goals, and some of it is the daily work necessary to accomplish their mission (such as general animal care). I will emphasize again that you should not set your goals only in response to problems. You should always be on the lookout to find ways of having work accomplished in a more effective and efficient manner, even if things are going fairly smoothly now. Look for those opportunities and act on them.

Even if your daily operations are proceeding smoothly, they should be routinely reevaluated to determine if they can be made more efficient and effective.

When you think that a change may be needed, the change should be made only after you have analyzed the situation and you feel that the change would be beneficial. Never be afraid to try out a new idea if you have given it due deliberation. There are very few decisions that you will make that are so earth shattering that you cannot reverse them.

If you do identify specific problems in your facility, you should ask yourself if the problem is just a symptom of a larger problem. Try to look at the big picture. What if you have an investigator who wants to start a large primate study as soon as possible, but you don't have any cages and any more money in your budget for caging? How will you get the money? Is that really the problem you have to solve? Is there a bigger problem, such as an organization that does not want to do primate research or a staff who has never worked with primates? Whatever the problem or opportunity, look at the big picture before you set your goals.

Look at the big picture. Is your problem just a symptom of a larger problem?

Some new managers have not had the opportunity to evaluate the operations and managerial concepts used in other animal facilities. This is particularly true of managers who have only worked in one animal facility and are not aware of more productive ways of running a facility. It's particularly difficult to think of new ways to do things if you've never seen them done in any other way. As noted by Kenneth Kinnamon, "we usually handle the new item with reasoning, which consists in finding justifications for keeping on believing and doing things the way we always have" [50].

So how do managers reevaluate and use their resources if they have limited experience? There is no simple answer for this. Your job is to get information, ask for feedback, and set and fulfill goals. You must use all the resources at your command to learn what is right for your facility and for you. Information resources such as this book, professional management associations, visiting other facilities, and going to meetings may help you understand how to use your resources, but you still must use them. It is not enough to read a book or go to a management seminar, and then forget about what you read or heard. Remember, you have your job because someone believes in your ability. Don't disappoint them. I can assure you that you are not alone if you are a new manager with cold feet. Many others have been in your shoes, and most have survived.

References

1. Mintzberg, H. 1975. The manager's job: Folklore and fact. *Harvard Bus. Rev.* 53(4): 49.
2. Beck, R., and Harter, J. 2014. Why great managers are so rare. http://www.gallup. com/businessjournal/167975/why-great-managers-rare.aspx (accessed March 3, 2014).
3. Hill, L.A. 2003. *Becoming a Manager: How New Managers Master the Challenges of Leadership.* 2nd ed. Cambridge, MA: Harvard Business School Press.
4. Thomson, R., Arney, E., and Thomson, A. 2015. *Managing People: A Practical Guide for Front-Line Managers.* 4th ed. Boca Raton, FL: Routledge.

5. Zenger, J.H., Folkman, J.R., Sherwin, R.H., and Steel, B.A. 2012. *How to Be Exceptional: Drive Leadership Success by Magnifying Your Strengths.* New York: McGraw-Hill.
6. Colvin, G. 2009. *Talent Is Overrated: What Really Separates World-Class Performers from Everybody Else.* New York: Penguin Publishing.
7. Marchant, D.C., Polman, R.C.J., Clough, P.J., Jackson, J.G., Levy, A.R., and Micholis, A.R. 2009. Mental toughness: Managerial and age differences. *J. Manage. Psychol.* 24: 428.
8. Sauer, S.J. 2011. Taking the reins: The effects of new leader status and leadership style on team performance. *J. Appl. Psychol.* 96: 574.
9. Simons, R. 2005. Designing high-performance jobs. *Harvard Bus. Rev.* 83(7): 55.
10. Miles, S.A., and Watkins, M.D. 2007. The leadership team: Complementary strengths or conflicting agendas? *Harvard Bus. Rev.* 85(4): 90.
11. Bryan, L.L., and Joyce, C.I. 2007. *Mobilizing Minds: Creating Wealth from Talent in the 21st-Century Organization.* New York: McGraw-Hill.
12. Leavitt, H.J. 2005. *Top Down: Why Hierarchies Are Here to Stay and How to Manage Them More Effectively.* Cambridge: Harvard Business School Press.
13. Hitt, M., Middlemist, R., and Mathis, R. 1983. *Management: Concepts and Effective Practice.* St. Paul, MN: West Publishing Co.
14. Rosenzweig, P. 2014. *Left Brain, Right Stuff: How Leaders Make Winning Decisions.* London: Profile Books.
15. Tichy, N.M., and Bennis, W.G. 2007. *Judgment: How Winning Leaders Make Great Calls.* New York: Portfolio.
16. Goldstein, N.J., Martin, S.J., and Cialdini, R.B. 2009. *Yes! 50 Scientifically Proven Ways to Be Persuasive.* New York: Free Press.
17. Wiseman, L., and McKeown G. 2010. *Multipliers: How the Best Leaders Make Everyone Smarter.* New York: HarperCollins.
18. Brousseau, K.R., Driver, M.J., Hourihan, G., and Larsson, R. 2006. The seasoned executive's decision-making style. *Harvard Bus. Rev.* 84(2): 111.
19. Roberto, M.A. 2009. *The Art of Critical Decision Making.* Chantilly, VA: The Teaching Company.
20. Roberto, M.A. 2009. *Know What You Don't Know: How Great Leaders Prevent Problems before They Happen.* Upper Saddle River, NJ: Pearson Education.
21. Hamel, G. 2011. Management is the least efficient activity in your organization. *Harvard Bus. Rev.* 89(12): 50.
22. Natale, S., Libertella, A., and Rothschild, B. 1995. Decision-making process: The key to quality decisions. *Am. J. Manage. Dev.* 1(4): 5.
23. Beshears, J., and Gino, F. 2015. Leaders as decision architects. *Harvard Bus. Rev.* 93(5): 52.
24. Wells, J.R. 2012. *Strategic IQ: Creating Smarter Corporations.* Hoboken, NJ: John Wiley & Sons.
25. Drucker, P. 2008. *The Five Most Important Questions You Will Ever Ask about Your Organization.* San Francisco: Josey-Bass.
26. Birla, M. 2013. *Unleashing Creativity and Innovation: Nine Lessons from Nature for Enterprise Growth and Career Success.* Hoboken, NJ: John Wiley & Sons.
27. Gibson, R. 2015. *The Four Lenses of Innovation: A Power Tool for Creative Thinking.* Hoboken, NJ: John Wiley & Sons.
28. Silverman, J., and Hendricks, G. 2014. Sensory neuron development in the mouse coccygeal vertebrae and its relationship to tail biopsies for genotyping. *PLoS ONE* 9(2): e88158.

29. Kurtzman, J. 2010. *Common Purpose: How Great Leaders Get Organizations to Achieve the Extraordinary*. Hoboken, NJ: John Wiley & Sons.

30. Calloway, J. 2013. *Be the Best at What Matters Most: The Only Strategy You Will Ever Need*. Hoboken, NJ: John Wiley & Sons.

31. Yale University. www.yale.edu (accessed February 22, 2016).

32. Tufts University. www.tufts.edu (accessed February 22, 2016).

33. American Association for Laboratory Animal Science. www.aalas.org (accessed February 22, 2016).

34. University of Massachusetts Medical School. http://www.umassmed.edu (accessed February 22, 2016).

35. Laboratory Animal Management Association. http://www.lama-online.org (accessed February 23, 2016).

36. Committee for Laboratory Animal Training and Research. www.clatr.org (accessed February 22, 2016).

37. American College of Laboratory Animal Medicine. http://www.aclam.org (accessed February 22, 2016).

38. Insight Recruiting. www.insightrecruiting.com (accessed November 16, 2015).

39. Srinivasan, A., and Kurey, B. 2014. Creating a culture of quality. *Harvard Bus. Rev.* 92(4): 23.

40. Rangan, V.K. 2004. Lofty missions, down-to-earth plans. *Harvard Bus. Rev.* 82(3): 112.

41. Koch, C.G. 2007. *The Science of Success: How Market-Based Management Built the World's Largest Private Company*. Hoboken, NJ: John Wiley & Sons.

42. Blair, G.R. 2009. *Everything Counts! 52 Remarkable Ways to Inspire Excellence and Drive Results*. Hoboken, NJ: John Wiley & Sons.

43. Lencioni, P. 2005. *Overcoming the Five Dysfunctions of a Team: A Field Guide for Leaders, Managers, and Facilitators*. Hoboken, NJ: Jossey-Bass.

44. Sullivan, C. 2013. *The Clarity Principle: How Great Leaders Make the Most Important Decision in Business (and What Happens When They Don't)*. Hoboken, NJ: John Wiley & Sons.

45. Champy, J. 1995. *Reengineering Management*. New York: HarperBusiness.

46. Collins, J., and Porras, J. 1994. *Built to Last*. New York: HarperBusiness.

47. Gregory, D., and Flanagan, K. 2014. *Selfish, Scared & Stupid. Stop Fighting Human Nature and Increase Your Performance, Engagement and Influence*. Hoboken, NJ: John Wiley & Sons.

48. Burkus, D. 2013. *The Myths of Creativity: The Truth about How Innovative Companies and People Generate Great Ideas*. Hoboken, NJ: John Wiley & Sons.

49. Raynor, M.E. 2007. *The Strategy Paradox: Why Committing to Success Leads to Failure (and What to Do about It)*. New York: Doubleday.

50. Kinnamon, K. 1993. Lessons mostly from Dale. *LAMA Rev.* 5: 8.

Chapter 2

The Organizational Environment

Insularity, rigidification, and failing to adjust to change in the environment are forces that destroy organizations.

Joel Kurtzman

Organizational Culture

Think back to the first day of any job you ever had, or even to your first day in a new school. Along with anticipation, there undoubtedly was a degree of fear. You wanted to know how things were done, how much you were expected to do, what the policy about coffee breaks was, where you could find out about preparing a budget for your division, and so on. Can you imagine how you would have felt had you showed up at your first senior staff meeting dressed in jeans and a turtleneck shirt, only to find everybody else in suits or dresses? If you had known about your organization's culture, you would have been spared that embarrassment.

It's obvious that the organization's culture is important to any manager. By *culture*, I am referring to those written and unwritten policies, procedures, norms, and values that tell you how your organization does things and how your boss and other members of the organization may react to any given situation. Culture covers phrases such as "It's a friendly place," "It's very regimented," or "Make sure you get it in writing." It also includes how people are expected to act at meetings, whether decisions are handed down from top-level managers or if input is solicited from lower-level managers, and how the inevitable mistakes are handled. Culture even includes items such as the pictures people hang on the walls, the friendliness of those you talk to, and even the body language that is used. At its core, an institution's culture is its brand name. It's how people, particularly those who work for you, view their work and the work environment. As you read through this chapter, it is important to recognize that your animal facility has a culture of its own. There is no right or wrong culture because the right culture for one organization and its goals (e.g., a pharmaceutical company) may be the wrong culture for another (e.g., a university). For the people who work with you, the animal facility culture is as important as, and probably more important than, the overall organizational culture.

Organizational culture is the written and unwritten policies, norms, and values that inform you about how things are done where you work.

An organization's culture can impact employee morale and retention and influence overall productivity [1,2]. Culture is very strongly influenced by its senior managers because they set the tone and give the verbal and nonverbal cues to everyone else [3,4]. By referring to employees as coworkers rather than subordinates or a similar term, a senior manager is sending a clear signal about the organization's culture and how people are expected to treat each other. If the boss has an "open-door" policy but people know better than to go in without an appointment, that speaks tomes about the organizational culture. If the vice president for research cares about the quality of laboratory animal care and emphasizes its importance to the company, then you can be reasonably sure that the general culture of the organization will reflect that concern. If the vice president doesn't care, the opposite happens and the animal facility personnel may have an uphill climb. Some years ago, I was watching television and listening to David Gergen, a former presidential advisor, describe his impressions of president Gerald Ford and his administration. Gergen was quite eloquent and positive in characterizing how Ford's basic method of operation was to be truthful and open. He described Ford as being "congenitally honest" and commented that Ford's own leadership style, which was one of honesty, caring, and a deep desire to do what was right for the nation, filtered downward throughout his administration. What a wonderful way to emphasize that an organization's culture begins with its leadership and then spreads to everybody else!

An organization's leadership has a powerful impact on its culture.

As expected, culture can have an impact on the efficiency and effectiveness of an organization. The truth of the matter is that the people you hire, taken as a whole, probably aren't much better or worse than people hired by any other animal facility. However, over time (and we all know this) some animal facilities function better than others. A good part of these functional differences emanate from the corporate culture—how people interact, how they are trained, and how they are treated. Most people want to succeed; therefore, as managers, our job is to evaluate and improve on the culture in the *animal facility*, while doing what we can to improve the overall *organizational* culture. Although every part of an organization may have its own special subculture, as a general statement, the animal facility culture is often an extension of the larger organizational culture.

But, what type of culture is typically found in successful organizations? Pennington [5] has identified certain cultural traits shared by successful organizations, and as you will see, they can readily be applied to laboratory animal facilities. I added the last trait.

1. *The culture of successful organizations includes telling the truth and valuing candor and honesty.* This is done within and outside of the organization. In Chapter 3, I'll talk more about communications and trust, but at this junction I simply want to emphasize that honesty with our own staff and researchers is only part of the equation for establishing a positive corporate culture. The other part is being introspective and not sugarcoating any weaknesses. A positive organizational culture faces up to reality. If we have multiple complaints from investigators about understanding our monthly invoices, we can't simply assume that all researchers are mathematically challenged. The honest answer may be that our bills are too complex for most people to understand.

2. *The culture of successful organizations includes pursuing the best path rather than the easiest path.* Here's an obvious example: it's easier to intentionally skip a 3:00 a.m. analgesic injection than it is to give it. Skipping the treatment is the easiest path, but not the right path. A positive culture emphasizes doing the right thing and doing it correctly. Let everybody know that your animal facility isn't one to take inappropriate shortcuts.

3. *The culture of successful organizations includes leveraging the power of partnerships with employees.* People can be our single greatest asset or our single greatest detriment. Part of our job is to work with our employees, not against them. Can we envision their future and provide the training and other opportunities for them to reach that future? Do our own coworkers trust us? Do we treat our coworkers as adults? Do we fight with unions or work with unions? Is there any shared decision making with employees (as there should be) or is it top-down management?

4. *The culture of successful organizations includes continually focusing on its core values.* We are very fortunate that animal facilities across the nation almost invariably have a stated core value of providing high-quality animal care.

Nevertheless, the implementation of that statement can vary. Think about what you really do, not what you wish you were doing. For example, when you develop your budget, do you ask what can be included that might make your animals more comfortable? When you have a general staff meeting, do you reiterate your mission and vision so that everybody clearly understands them? Are short-range goals understood by all those affected? Does everybody receive adequate feedback on your progression toward your facility's goals? Basically, do you really do what your mission, vision, and goals say that you are going to do?

5. *The culture of successful organizations includes having the courage to be accountable.* At the level of individual accountability, we have to understand that people will make mistakes, but the culture must allow for people to own up to their errors without the constant threat of reprisal. Repeated poor performance is another story. At the organizational level, managers must ensure that tasks are completed properly, not just completed any old way. This is the concept of management control that was discussed in Chapter 1. Managers simply can't sit back and accept below-average performance as being "good enough." There are all kinds of excuses that we can develop for accepting poor performance, but few will ever withstand legitimate scrutiny. An animal facility that expects high-level performance, supports high-level performance, and constantly emphasizes high-level performance by rejecting low-level performance is establishing a positive organizational culture. We all like to play on a winning team.

6. *A successful organization promotes a culture of civility (respect) toward others* [6]. A successful animal facility should not tolerate and certainly should not foster a culture of incivility. Incivility (e.g., rudeness) to investigators or research technicians can decimate our animal facility's reputation. Within the animal facility, incivility can rear its head when a supervisor publically chastises a coworker or when an employee makes malicious remarks about another over the lunch table. And incivility can become infectious and lead to incivility by other employees, which in turn can lead to morale and performance suffering [7]. Even one uncivil person who is knowingly allowed to continue being disrespectful can result in lowered production and lesser quality work by others. It is always a responsibility of managers to set the tone for civil discourse. For example, during a recent performance review, I complimented and thanked a manager for courteously and professionally disagreeing with me. I meant what I said, as providing an employee (me in this case) with constructive and civil feedback is a cultural characteristic of many successful organizations and managers. An additional discussion about civility can be found in Chapter 3 under the heading of "Difficult Conversations."

You can get some insight into an organization's culture by simply looking around, by reading formal organizational literature, by reading memos or similar documents that hang on bulletin boards, and by speaking with employees.

Nowadays, you may be able to use social media such as LinkedIn to locate and communicate with people who currently or recently have worked for the organization you are considering. As a practical matter, when you first consider a new managerial position, the more you know about an organization's culture, the better off you are. If you don't think you will fit in, don't take the job. If you do accept the position, by knowing the culture, you can avoid making statements or establishing policies that are contrary to it, thereby saving you potential problems in the future. Knowing the organization's culture helps you to fit into it in a relatively easy manner. Equally important, it gives you some of the unwritten corporate guidelines on how to use your resources.

Not only does a new employee have to take some initiative to understand the corporate culture, but as a manager, you have a responsibility to ensure that a newly hired person—in particular a newly hired manager—can gain any needed exposure to your company and its people in order to succeed. This may mean that you will have to set up meetings for the newly hired person with key people, in particular those people who control the resources that the new person requires [8]. A good part of your job is to help people succeed, so get to it. Did you let the appropriate people know ahead of time about the new person's pending arrival? Did you distribute basic information about the new person? Does your animal facility have a fact sheet that tells a new person how to dress, what the department stands for, what its mission and vision are, and so forth? Understanding the corporate culture and how to get things done is critical to the success of any new hire, and as existing managers, we can make a significant difference in that person's success or failure.

Changing the Organizational Culture

Let's assume you recently moved into a position as manager or supervisor of a group of people. It doesn't matter if they are animal care technicians, veterinary technicians, veterinarians, or others. After settling in for a few weeks and getting a feel for the existing culture of your group, you found that people were undermining each other and trying to finish their work as quickly as possible without much attention to the quality of their work. How do you begin to change that culture? Is it even possible to change an organization's or a group's culture?

Yes, cultures can be changed, but it's not easy. When I was employed by Hahnemann University, it merged with the Medical College of Pennsylvania (MCP), and it was quickly obvious that two different cultures were coming together. Neither culture was bad, but they were clearly different. The leadership of the newly merged schools, which was largely derived from the former MCP, had considered many financial, space, and teaching matters, but there was no consideration of the cultures of the two medical institutions. Presumably, we were all intelligent adults and could handle that change. Not so. For the next nine years that I worked at the merged schools, there were tangible tensions between the former Hahnemann and MCP faculties and administrators. From my keyhole observation of the merger, many of the problems emanated from a lack of trust caused by unfulfilled promises, which in turn led to employee discontent. Granted, this was not a true cultural "change" (it was more of a poor adaptation to merged cultures), but it made me acutely aware of how ingrained cultures can be and how important trust is in all aspects of management.

What do you have to do to effect a culture change? First, make sure a change is needed. You may see a clear need to change or modify the existing culture, but would any such change lead to more or less job satisfaction among employees? How would a change impact efficiency and effectiveness in either the short or long run? Would a change affect how other managers in the animal facility interact with the people they manage? Would a change make a difference to the investigators or even to the well-being of the animals? Don't make a change unless there is a good reason to do so. Any culture change should be made only after you have done your homework and you truly believe that a change is in the best interest of the people you manage, your department, and the customers you serve. Also, before you go too far in initiating a change, make sure you have the support of your boss. If you cannot obtain that support both verbally and through her actions, you're going to find yourself behind the eight ball before your start [9].

Organizational culture, as you know, encompasses the values that the organization has. If you believe a change to these values is required, the change should be done gradually; never try to totally change a culture overnight. When you begin to implement a change, it is important that all members know what the new culture will be (i.e., the new values and operating methods) and why the change is needed. A "new culture" need not mean that everything has to change. Sometimes, only a few key changes are needed. You can communicate these values via a values statement [9] that stands by itself or is a supplement to a mission and vision statement. A values statement was mentioned in Chapter 1 as part of the discussion on mission and vision statements. It is simply a written statement that says what the culture will be. For example, a value statement might be "We believe that all of us contribute equally to the well-being of the animals we care for, and we will work together as a group to make life better for those animals." That simple sentence emphasizes a common endpoint (making life better for animals) and the need for intragroup cooperation to reach that goal. You can consider obtaining feedback about the best wording to describe a new culture from your own supervisor, from those you manage, or from any other persons you believe would be helpful. Once the value statement is developed, make sure it is prominently posted in many places and make sure you frequently talk about the new culture and why its implementation is important. I suggest that you talk informally about the new culture to coworkers during lunch or small group meetings. I also suggest that you remember to interact with the thought leaders; those are the people in the animal facility that other people tend to follow no matter what their job title might be. They can be technicians, supervisors, veterinarians, or office personnel. If additional training is needed as part of the culture change, see that it occurs. Here's a final thought:

consider having a measurement system to help you evaluate if your coworkers have implemented the new culture. Appendix 1, which discusses productivity and suggests methods to assess productivity by using metrics (numerical measurements), can easily be modified to help determine if a cultural change is being implemented.

Initiating a change in culture requires leadership from you. In Chapter 7, where we discuss leadership in greater detail, there is a section entitled "Framing a Situation." The section emphasizes that we have to approach different leadership needs by using different frames of reference (i.e., the leadership methods we will use to help change an existing culture will differ, depending on the particular situation). You might want to take a quick look at that section and see if you should be framing the desired culture change as (1) structural (e.g., establishing a culture where it is important for everybody to have the needed skills to perform their work at a high level of competency), (2) human resource oriented (such as establishing a culture that aligns the management of the vivarium with the needs of its employees), (3) political (such as establishing a culture that advocates respect for each person's work contributions), or (4) symbolic (e.g., establishing a culture that values the welfare of all laboratory animals). Of course, you can combine two or more of these frames of reference if it is appropriate to do so.

As a senior manager, you have to walk the talk; you cannot expect others to accept a culture change if you don't take the lead in accepting that change [10]. As successes occur, you should publically thank the group for each of the little accomplishments that are achieved on the way to establishing the culture change. You will quickly learn that having people *want* to follow you as you initiate a culture change can be a significant leadership challenge.

External Environment of Your Company

It's easy to fall into the trap of thinking that the most important relationships are those that are among people or divisions of the same company. This is particularly true when you are managing an animal facility where you may not even have a window to look out at the rest of the world. Considering activities outside of your building—or even outside of your organization—may be the last thing on your mind, yet you should be aware of many external considerations that can affect your day-to-day operations. Let's take a look at some of these outside forces, the so-called forces of the *external* environment. The *internal* environment (organizational culture, budget information, personnel policies, and many related items) is discussed elsewhere.

The external environment consists of those factors outside your organization that can affect your operations, yet you have little or no control over them.

To be successful, you must function as an integral part of your organization, not in an isolated world that consists only of rats, rabbits, mice, and monkeys in the basement of a research building. Some of these interactions are obvious. For instance, you might meet with vendors to obtain the best price for a needed item. Other interactions may not be quite as obvious. Changes in tax laws could cause a private corporation to decrease its charitable support of some biomedical research organizations. Your organization may be one of those that lost some of this support. If some of this lost money had been earmarked for your animal facility, you can readily appreciate how forces outside of your control can directly affect you. It therefore becomes important for you to try to be aware of the many factors outside of your area of immediate responsibility that can affect the way you manage your resources. You should constantly attempt to incorporate your knowledge about your organization and its environment into your daily operations and your short- and long-range goals. You should periodically go to meetings, other animal facilities, vendors' warehouses, and so forth to advance your own understanding of the external environment and establish personal relationships whenever possible. Even though you may not be able to control these outside factors, simply knowing that they *may* influence your operation gives you an edge.

At this point, I'm going to address some of the key factors in your organization's external environment and try to show you how they might affect your goals and plans. You can never know everything about the external environment (or the internal environment for that matter), but if you want to be a better manager or move up the corporate ladder, the more you know, the better off you will be.

Social Environment

I think of the external social environment that we work in as composed, for the most part, of the general public, people we meet at professional gatherings or socially, and the vendors we deal with. Not only are vendors excellent sources of information about their own products, but they also are fine suppliers of information about other institutions and competitive products. Good working relationships are needed on both sides. If you are thinking of purchasing a new style of primate cage, perhaps you would like to know who else is using it. Ask the cage vendor. This may be a simple example, but you have just defined two goals (purchasing cages and finding out who is using them) and used two resources (human and information) in the external environment to help you make a managerial decision.

In other parts of this book (e.g., Chapter 7), I spend a little time talking about the importance of social networks, but let me just say a few words here as well. The personal contacts that a manager makes within and outside the animal facility can have a major impact on your short- and long-range planning. From these contacts, you learn about new products, new trends, new management techniques, new threats, new opportunities, and so forth. It helps your career by allowing you to call on people when needed, to get placed on committees, and so forth. You can always take the high road and argue that good management, like good science, speaks for itself; however, that's simply not true in many instances. You will probably find that you have to do a good deal of socializing to make the jump from being an acceptable manager to becoming a good manager. But personal perks aside, your job is all about being a good manager.

One aspect of the social environment that substantially affects laboratory animal facilities is the continuing controversy over the use of animals in research, education, and product testing. This battle has been raging since biblical times, but the issue of animal experimentation gained momentum in Victorian England, and subsequently has become an emotional issue in many parts of the world. In recent years, many animal facilities, particularly in the United States and England, have been vandalized and have had animals stolen, all in the name of their humane treatment and liberation. The animal rights controversy actually affords us a wonderful learning opportunity because there are intelligent people on both sides of the debate who look at the same facts, yet reach different conclusions. In the eyes of some people, the use of animals for research purposes is morally unacceptable, whereas others see animal use as a moral necessity. Later in this book, when I talk about communications, I will emphasize the need to step back and understand that people view the world through different lenses, based on their life's experiences, and good managers understand that and consider that when talking with people.

How do you, as an animal facility manager, contend with this external environmental factor? What are you supposed to do, since you're only one person? The answer has nothing to do with changing the world, because no matter how hard you try, you will not be very successful. But you do have personal contacts, journals, meetings, the Internet, social media such as Facebook, newspapers, and other ways of finding out about the "pulse" of that world. You should look at your own facility and evaluate its activities in terms of societal standards and research needs.

From a practical perspective, certain areas must be addressed. To begin with, you must be aware of all pertinent federal and state laws, policies, and regulations that are applicable to the care and use of laboratory animals. The Animal Welfare Act, its regulations, and the Public Health Service policy on animal care and use should be prime considerations. Both you and nonmanagerial employees must be aware of the recommendations of the *Guide for the Care and Use of Laboratory Animals* [11], the *Guide for the Care and Use of Agricultural Animals in Agricultural Research and Teaching* [12], and any other pertinent documents. More important, there must be continuing efforts in educating all animal facility employees, all individuals coming in contact with animals, and all institutional administrators outside of the animal facility that animals are to be treated as humanely as possible. I cannot emphasize this strongly enough, for there are still some individuals who look upon animals as if they were test tubes.

Taken a step further, if there is a reasonable chance that changes might be proposed to the Animal Welfare Act or its regulations, you might suggest holding off buying replacement cages because new size requirements may be in effect in the near future. If this is the case, you have been influenced by the external social environment. You have not changed it, but you have used information about it to influence a purchasing decision.

How does the Institutional Animal Care and Use Committee (IACUC) fit into this discussion of the external environment? If your committee is similar to many, only the attending veterinarian and the committee chairperson have more than a cursory knowledge of the external environment. Still, the committee must be kept aware of factors that can affect the operations of your animal facility. This can include new and upcoming legislation, community sentiments, notices from regulatory agencies, and if appropriate, the sentiments of your organization. In addition, the committee must have a member of the local community as part of its required composition. By having this member, you have incorporated a portion of the external environment into your organization.

Politico-Economic Environment

When speaking of groups who oppose or favor the use of animals in research, education, and testing, it becomes apparent that there is an overlap between the social and political external environments to which laboratory animal facility managers must relate. Similarly, the political and economic environments are frequently so closely intermingled that it is convenient to discuss them together.

Some states have passed laws prohibiting the sale of dogs from shelters or pounds to laboratory animal facilities. However, animals from pounds and shelters cost considerably less than dogs purchased from private vendors. Thus, the cost of doing certain types of research or teaching or product testing can be increased when pound dogs are not used. Nevertheless, had you been monitoring the political-economic environment in 2014, you would have known that the National Institutes of Health (NIH), a major source of federal research grants, began to implement a program that in 2015 prohibited researchers from using NIH funds to procure or support the use of dogs from dealers who obtained their animals from shelters, pounds, individuals, and other "non-purpose-bred" sources. Currently, dogs used in NIH-supported research may only be obtained from licensed "purpose-bred" dealers.

What did that mean to you if dogs were used in research at your institution? What should you have done as a manager? Truthfully, there was probably very little that you personally could have done about this type of legislation once it was in place. Nevertheless, as a manager, you could have and should have considered a number of things. You should have made plans for getting dogs from licensed purpose-bred dealers. To that end, you likely would have used information and human resources and planned on using fiscal resources. You may have needed different types of quarantine facilities (or even considered eliminating quarantine); thus, you would have considered the use of your capital resources. And all of this had to be done quickly, because research grants don't wait. Therefore, you also had to consider your limiting resource—time. Finally, you would have discussed the needed change with your vivarium staff and your organization's investigators. By taking all these steps, you would have used all your resources, and while using them, you planned, organized, and managed efficiently and effectively.

Political priorities can affect entire research programs. For example, beginning in approximately 2005, federal funding for most biomedical research grants was greatly diminished, in part because of financial priorities necessitated by ongoing wars in Iraq and Afghanistan. Many of the larger federal grants that funded laboratory animal facility construction in universities and other not-for-profit organizations had dried up. However, during that same period, federal funding increased somewhat for research focused on bioterrorism, which had become a national priority. The astute animal facility manager would not have been planning grand expansions of the animal facility unless it was abundantly clear that the institution would be able to develop the needed construction funds in a manner that was different from typical past practices.

Politics can even influence specific diseases. Between 2011 and 2014, the estimated NIH budget for funding spinal cord injury research remained nearly unchanged, whereas funding for Alzheimer's disease increased by about 25%, along with increases in funding for a limited number of other diseases, such as childhood leukemia and infectious diseases. This was a time of budget reductions for the NIH, and grant and contract spending for most diseases and conditions either decreased, stayed the same, or barely increased [13]. If your institution was putting an emphasis on Alzheimer's and related dementias, you probably did not have to construct an entire new wing for your building, but you did have to ensure that any unique needed space and equipment (such as might be needed for behavioral testing) was available. Now, you may ask me to return to reality because you don't believe that a laboratory animal facility manager should have to investigate funding trends between Alzheimer's and spinal cord injuries. In reality, you may not have to if you are a first-level manager, but you should stay on top of those changes if you are directing the entire facility. This is easy to do because the federal funding for different diseases and conditions is readily available online, and the funding figures include proposed budgets for near-future years [13]. However, as a first-level manager, you may simply be told that a new program for the study of Alzheimer's or infectious diseases, such as West Nile Virus, plague, or some other disease related to bioterrorism, will begin soon. That's when you have to act. Ask investigators about West Nile, read about West Nile, and go to seminars about West Nile. Do the same with Alzheimer's disease. Investigators want their study to start on time and go smoothly; they don't worry about the things you worry about. I believe good managers are not simply reactive. They're proactive. They plan for the future by finding out as much as they can about what the future may bring.

Good managers are proactive. They plan for the future by finding out as much as they can about what the future may bring.

Physical Environment

When one university campus has a large number of animal facilities, the managers of each may be able to decrease the cost of purchasing supplies by ordering large quantities at a discount and then distributing them to their respective facilities. From this simple example, we see that the physical environment of an organization refers to its location relative to all the other resources it has and the type of work it does. Laboratory animal facility managers and directors who are located in the same geographical area may encourage their animal care staffs to meet on a regular basis to discuss management or other areas of mutual interest. This is another example of how the external physical environment impacts the use of resources.

The physical environment can also affect the operations of an isolated animal facility. The manager may have to order supplies far in advance because there may be no local suppliers. Also, in the isolated facility, the manager may have to initiate some innovative recruiting programs, as there may not be many skilled (or unskilled) people who can be hired. Animal facility managers spend an inordinate amount of time on issues involving the physical environment where they work, such as managing problems with room temperature and humidity, water supply, and floors that need to be replaced. Managers in animal facilities must be knowledgeable about the many topics that impact their daily work.

As with most aspects of management, the experienced manager usually knows how to adapt to the physical environment. If you are a new manager, you should think about the impact of the external physical environment on your resource utilization and determine how to deal with it both efficiently and effectively.

Information and Technical Environment

Finally, you must consider the information and technical environments that surround your laboratory animal facility, as well as your organization. Your organization's abilities to acquire and use needed equipment, have library resources, and be able to send members to continuing education meetings are but some examples of the information and technical environments that impinge upon your work as a manager.

Even if you cannot change these factors in the external environment, you must work hard to manage them. If, for example, you can afford to send only one person to a meeting, you can ask that person to report to your entire group on what was covered. By doing this, you will have many people benefit from a meeting that only one person could attend. You will be using your resources as best you can, and therefore, you will be managing as best you can. Pat yourself on the back.

Relating Your Goals to Your Organization's Goals

Throughout this discussion I have attempted to show you how the external environment impacts upon your organization as a whole as well as on your laboratory animal facility. I want to remind you once again not to lose sight of the fact that the responses you make to situations should be consistent with your organization's mission and goals.

The responses you make as a manager should be consistent with your organization's mission and goals.

Let me give you an example of what I mean:

Kathy Morrison was the supervisor of surgical nursing for Great Eastern University's laboratory animal facility. A nurse left for another position and Kathy wanted to hire someone to replace her. One of her goals was to hire individuals who had been trained in an accredited academic program of laboratory animal science and who also had experience in surgical nursing. Such individuals would have to be paid a salary that was commensurate with their training and experience.

Unfortunately for Kathy, John Ross, the vice president for research, reminded her that the university was going through a difficult economic period and she had to be frugal and delay hiring a new nurse. Kathy was a realist and knew that she would have to wait a bit longer, assuming animal care was not compromised. This decision did not mean that she did nothing about the nursing position. First, she reevaluated the current situation to see if she could get by without filling it. She then questioned the doctors who used the facility as to what their short-range surgical research plans were.

She found out that because of the widespread budget cuts, her surgery facility users would also be cutting back on their research, and Kathy would not need to replace the nurse who left until research activities resumed on a larger scale.

The important point here goes beyond the fact that Morrison's actions were consistent with short-term corporate needs. After speaking with John Ross, she developed a goal (reevaluating whether she needed a nurse) and a plan (speaking to the doctors). There was not too much organizing, controlling, or directing to do with the resources she used (essentially human and time), but not everything managers do is earthshaking. There was a simple problem, a simple goal, and a simple plan to reach the goal. It was her responsible management that evaluated the animal facility's goals relative to the university's goals.

What about long-range corporate goals and plans to reach those goals (the process we call strategic planning)? How does the external environment impact the strategic planning of your entire organization, as well as the laboratory animal facility? As a convenience, let's concentrate on the animal facility, although these comments are pertinent to the entire organization.

In Chapter 1, I spent a good deal of time discussing strategic planning. I indicated that your animal facility's mission and goals must be in accord with those of your organization. From your mission, you derive a vision, which is a broad summary of where you want your goals to place you at some specified future time. The vision then demands that you establish specific goals and strategies to accomplish it. This entire process, I said, is often called strategic planning. In my earlier discussion, I intentionally avoided including environmental forces that can influence the development of specific strategies. My intent was to help you understand the overall process, but as you have seen in this chapter, the external environment can influence your long-range goals in a multitude of ways. As an example, let's look at Great Eastern University's vision for its laboratory animal facility: it wanted to become a model for animal facility efficiency and effectiveness by the year 2020. This was to be accomplished by defining standards of excellence, revitalizing facilities, improving public relations, reevaluating programs, and so on. That's fine, but what are the environmental issues that will influence the choice of strategies used to accomplish these goals? Let's look at just one such issue.

Assume that I am the animal facility manager at Great Eastern University. In the past year, Great Eastern reevaluated its programs and concluded that, in the future, a large portion of its research effort will focus on neurobiology. A substantial part of that research will require the use of rats, cats, and primates. At this time, Great Eastern is lacking space for cats and primates. Therefore, as part of my plan to revitalize, should I request funds to expand or renovate the animal facility? Animal holding and procedural space is needed. But are there environmental factors that must be considered? Yes, there are.

As the individual responsible for long-range animal facility planning, I always try to determine which external factors, if any, will affect my operations. Here is my interpretation of why facility expansion, although desirable, will not be a realistic goal in the next few years.

In past years, Great Eastern University, like many other hospital-associated universities, depended heavily on income from Great Eastern University Hospital. The hospital produced the majority of the operating income for the entire university. As far back as 1995, the Clinton administration chose to emphasize primary health care over the typical tertiary and quaternary health care provided by university hospitals, and hospitals began getting less money and had to reevaluate their financial projections. The decrease in government reimbursements for treatments and diagnostic studies has extended into the twenty-first century, and Great Eastern, like many other hospital-affiliated universities, found itself in a financial pinch. Future financial prospects did not look good.

The timing would not be right for animal facility expansion. The politico-economic environment is telling us that. To provide animal care and use service for neurobiological research, we must be fiscally conservative, and perhaps set our goals and strategies toward remodeling and reallocating the space we have.

I'm sure that you can think of your own examples of how the external or internal environments can shape the way you approach your strategic planning. I "survived" a university going bankrupt, faculty leaving in droves, salary increases that never came, and understandably, a staff that was very depressed. My long-range plans for renovating the laboratory animal facility and expanding services had to be completely dropped. My short-range planning focused on how to consolidate our six animal facilities down to four, then down to two, and how to retain as many of my coworkers as was reasonably possible.

With the understanding you now have about the importance of the internal and external environment, let me redraw the flowchart on strategic planning to factor in these influences. We must always consider the opportunities and threats engendered by the environments surrounding our laboratory animal facilities. Wherever possible, we should take advantage of the opportunities and turn away from the threats. By doing so, we can respond to the environment in a manner that is optimal for our needs. Figure 2.1 illustrates this relationship. In order to establish and maintain long-range goals (and often short-range goals), the astute manager is constantly sniffing the air for opportunities that can aid the planning needed to reach her goal or evaluating threats that might create a problem and require a change in her plans. This ongoing process of evaluating environmental threats and opportunities helps a manager develop the strategies needed to reach the goal or even to decide to abandon a goal entirely. Management is not a static process. You can't keep going down a path that probably leads to a dead end. In Chapter 1, I gave you some good advice: keep your decisions flexible, or you may find yourself drowning in the Big Muddy.

Figure 2.1 Influences on strategic planning.

References

1. Emmerich, R. 2009. *Thank God It's Monday: How to Create a Workplace You and Your Customers Love.* Lebanon, IN: Pearson Education.
2. Greenwald, J. 2014. How to get the wrong people off the bus. S+B Blogs. January 31. http://www.strategy-business.com/blog/How-to-Get-the-Wrong-People-Off-the-Bus?gko=5516d (accessed February 26, 2014).
3. Kurtzman, J. 2010. *Common Purpose: How Great Leaders Get Organizations to Achieve the Extraordinary.* Hoboken, NJ: John Wiley & Sons.
4. Krapfl, J.E., and Kruja, B. 2015. Leadership and culture. *J. Organ. Behav. Manage.* 35: 28.
5. Pennington, R. 2006. *Results Rule! Build a Culture That Blows the Competition Away.* Hoboken, NJ: John Wiley & Sons.
6. Spreitzer, G., and Porath, C. 2012. Creating sustainable performance. *Harvard Bus. Rev.* 90(1–2): 93.
7. Sidle, S.D. 2009. Workplace incivility: How should employees and managers respond? *Acad. Manage. Perspect.* 23: 88.
8. Johnson, L.K. 2005. Get your new managers moving. *Harvard Manage. Update* 10(6): 1.
9. Heathfield, S.M. How to change your culture: Organizational culture change. http://humanresources.about.com/od/organizationalculture/a/culture_change.htm (accessed March 10, 2014).
10. Schein, E.H. 2009. *The Corporate Culture Survival Guide.* 2nd ed. Hoboken, NJ: John Wiley & Sons.
11. National Research Council. 2011. *Guide for the Care and Use of Laboratory Animals.* 8th ed. Washington, DC: National Academies Press.
12. Federation of Animal Science Societies. 2010. *Guide for the Care and Use of Agricultural Animals in Agricultural Research and Teaching.* 3rd ed. Champaign, IL: Federation of Animal Science Societies.
13. U.S. Department of Health and Human Services. Estimates of funding for various research, condition, and disease categories (RCDC). http://report.nih.gov/categorical_spending.aspx#legend4 (accessed July 22, 2015).

Chapter 3

The Management of Human Resources

If employees are so important, why do so many companies entrust them to bad managers?

Rodd Wagner and Jim Harter

Consider, for a moment, the title of this chapter. Instead, of "The Management of Human Resources" I could have called it "Personnel Management" since it pretty much means the same thing. There is a difference, though. The former term gives more credence to the importance of people as a resource. The latter is just a cold administrative concept. Of course, calling people a resource might be taken to suggest that people are the possessions of a company. That's not true. People cannot be treated like machines or buildings; they have personalities, desires, and needs, and they leave the workplace to go home at the end of the day. Machines do not. Just how important are people? It's obvious that we cannot run an animal facility without people, and right now about 60% of my operating budget goes to staff salaries and benefits. Consequently, give or take a little, human resources represent about $5 million of my annual expense budget. If that amount of money was capital equipment, I would be very careful about its use, so if we actually were to think of people as machines, people are probably the most important pieces of capital equipment we have in our animal facilities. This analogy between people and equipment would have intrigued René Descartes, the philosopher who centuries ago characterized animals as machines of nature. Today's enlightened manager recognizes that human resources and the animals we care for are anything but machines. The people with whom we interact, whether inside or outside our organization, are critical resources who want and should receive the same respect and consideration that you and I want. As managers, we must recognize that these are our colleagues, not pawns to be sacrificed as needed. As managers, we must recognize that employees (including ourselves) who have a higher level of job satisfaction will produce a higher level of customer satisfaction [1], and keeping our customers satisfied is one of our major responsibilities.

Chapter 2 noted that human resources included the employees who work with us, other managers, vendors, friends, acquaintances, and many other people. I also stated that a resource is rarely, if ever, used by itself. Rather, resources are used together, and you should never put so much emphasis on any one resource that the others are obscured. This point is particularly important to remember relative to human resources, as it is easy to convince yourself that managing people is your first and only responsibility.

It is not unusual for an animal facility to be directed by an individual with a professional degree, such as a veterinarian or a PhD. Working alongside this person may be other college graduates, people who have no training beyond high school, and, at times, even less formal education than that. Because of this educational (and sometimes social) stratification, cliques may form, particularly in larger facilities. These cliques can easily hinder productivity. What can we do, then, to help ensure that everyone knows what our short- and long-range goals are and that we have to work together to reach those goals? Even more important, what can we do to make sure everyone we interact with feels like a significant contributor? Remember what I wrote earlier: A pyramid will not long stand with a top of gold but a base of sand.

Communications

In my opinion, the two considerations that are the most crucial for the efficient and effective use of human resources are the need for clear communications and the need for mutual trust between you and all the people with whom you interact. To me, these are the keys to successful management of human resources. They are concepts that are pervasive throughout human resource management. We communicate to share information, make decisions, influence people, organize our work, motivate coworkers, and maintain work relationships [2]. Communicating isn't a special skill you use when directing, planning, or performing any of your other managerial roles. It's a continuous process. If a manager cannot or will not communicate, he or she is going to have a hard time succeeding. Think about this for a moment: How can you expect employees to speak truthfully to a manager if they have no idea how the manager will respond [3]? A manager has to communicate frequently, honestly, clearly, and in simple language. Long, drawn-out speeches or emails will lose your audience and your message. Although managers and leaders may try to influence people with their communications, they do not try to manipulate them by providing false information or half-truths. Managers should be able to provide the reasoning behind the mission and vision statements, along with the importance of a person's job to the overall operations of the animal facility [4]. They have to be able to communicate the reason behind a decision, especially if it was a difficult decision. Equally important, and perhaps even more important, managers have to communicate on a regular basis with each of their direct reports to ensure that goals are being met and the resources to reach those goals are available.

Unfortunately, there are people who are very talented in many ways, but communicating is not one of their talents. Perhaps they're shy; perhaps there is a language deficit or a speech defect. There are also a small number of managers who feel that they don't have to communicate; you're supposed to be able to read their minds. All of these people may have their areas of expertise, they may even remain managers, but without effective communication skills, we have to question if they'll ever be highly successful managers. Now let's take a look at some aspects of communicating with people.

Clear communication and mutual trust are the keys to the successful management of human resources.

Verbal Communication

Communicating clearly is not always easy. In fact, I'm willing to say that it's out-and-out hard to do it right if you don't think about it. You must consider that the perceptions of the person you are talking to can differ from your own. People may interpret words in different ways, based on their life's experiences. In management circles, you may hear about different people having different *filters*. That means they interpret words differently, depending on their life's experiences. Another term you may hear is that people have different *lenses*. That's also the same as saying that people interpret a person's words based on their life's experiences. No matter how it is phrased, the bottom line is that the words you say may mean one thing to you and another thing to others. Therefore, try to adjust your communication style to fit the person with whom you are talking because if you (the communicator) and the recipient (your intended audience) do not have the same idea about what is being communicated, then the communication is of little or no value.

It doesn't matter if the person with whom you are communicating is your superior, your subordinate, or your mother; your job is to communicate successfully. To quote Henry Spira, the late animal rights activist, "People's perceptions are the political reality." That is, what *you* think you mean is not as important as what other people think you mean. There's no sense in pounding people with the facts as you see them because if that's not what they believe, they're just going to dig their heels in deeper, reject your beliefs, and amplify their own. As an example, think about what might happen if there were a conversation between somebody who fervently believes in using animals in research and someone who fervently opposes their use. What you perceive to be a reasonable argument for using animals might well be perceived as a reason not to use animals by your counterpart. If you want to get along with this person, begin by finding areas of commonality, such as the affection for animals that you both have.

Some barriers to clear communications (whether verbal, written, or "body language") include a person's age, personality, past work experience, the way in which he or she learns, his or her cultural background, the corporate culture of your organization, academic training, emotional state, fatigue, accents, and hearing impairment. You may not be able to compensate for each barrier, but you can often focus on the major ones to facilitate communication and understanding.

> **Adjust your communication style to fit the person to whom you are talking.**

Let's look at some examples. What is actually meant when a veterinarian says that a dog that was severely injured when hit by a car has to be "put to sleep." The veterinarian probably means that euthanasia is necessary. The owner may think that the dog is going to be anesthetized for surgery. A misinterpretation such as that can be disastrous. Should a pet owner be expected to know what "put to sleep" means to a veterinarian? If the pet owner is from Mexico or Japan, will that make clear communication even more difficult?

If you are interviewing people who have never worked in a laboratory animal facility, do you really think they understand what you are talking about when you mention cage washers, monolithic floors, or transgenic animals? If it's important that people know what a cage washer is and does, show them the washer and how it works. By doing that, you have created an image in the person's mind so he or she can be listening to you, rather than trying to figure out what a cage washer is. It's nothing more than a picture being worth a thousand words.

As I mentioned above, a person's age and cultural background can affect his or her communication style. A senior executive in her sixties is likely to have a somewhat different vocabulary and manner of talking than a woman in her twenties. A woman in her twenties may not think twice about her audience reading messages on their cell phones during her presentation. A man leading an organization where all of the other men wear ties and jackets to work may feel more at ease disseminating information by group emails, newsletters, or other somewhat formal means than by talking directly with employees. There is a time and place for different communication styles, but in general, talking directly to employees in a friendly, informal manner often wins the day. The bottom line is that you have to be understood to get your message across, and similarly, you must be able to understand the person who is communicating with you.

You probably experienced the need for clear communications when you were a student taking a test. Were you ever absolutely frustrated because you knew the subject matter but you couldn't figure out what a question really meant? This is an example of poor communication. As long as I'm on the subject of tests, let me bring up the concept of learning styles. It's no secret that different people learn in different ways. For example, some people are visual learners. That is, they either have to see something being done or picture it being done in their mind. I'm one of those people. I'm bad at statistics because it's too conceptual for me; I can't visualize the process in my mind. On the other hand, I understand anatomy quite well because I can see it, even if it's a mental picture. I can even do well in physiology because I can conceptualize in my mind where and how all the cellular processes are occurring and how they affect the entire organism.

So, I'm a visual learner. Some visual learners do better if they're told a little story about the issue you're discussing with them. In fact, throughout this book I use little stories to help put a concept into a real-world perspective. Maybe you're a learner who has to read something over and over again. Or perhaps you are a hands-on person. You actually have to try something for yourself before it really sinks in. What does all this have to do with management of an animal facility? I'm hoping the answer is obvious. An excellent manager is excellent at knowing the people around him or her. This isn't a party trick or magic; you have to be observant and look at and listen to people. Once you start to get an idea how a particular person learns, then you have an idea about the best way to communicate with that person.

Sometimes people hear what they expect to hear, or what they want to hear. Let's take a simple phrase that a manager might say, such as "Clean the mouse cages now and then clean the rat cages." If there is an employee who likes to do most of his work in the afternoon, the manager may find the rat cages being done in the afternoon, long after the mouse cages were cleaned. Another employee, one who wants to get his work done as soon as possible, may interpret your statement as meaning you wanted the rat cages cleaned immediately after the mouse cages were cleaned. What did you mean? If you wanted the rat cages cleaned immediately after the mouse cages, you should have said that. If you need a project done before noon, say that. If you want your child to clean up the basement, don't say, "We have to get this basement cleaned up," because if your kid is like my kid, it's never going to happen. There's nobody in my house named "We." Be specific and say that he or she has to clean up the basement before Saturday. (Then sit back and pray that it will actually happen.) Of course, if it doesn't matter to you when the work gets done as long as it gets done within a reasonable time frame, then don't get upset with a person who didn't understand what you actually meant. A little flexibility on your part will be appreciated by most people.

No matter what your goal is, if you cannot clearly communicate your wishes to your staff and superiors, you will have trouble fulfilling that goal. Think about how frustrated you become when you are not sure of what is expected of you.

People must have a clear understanding of what is expected of them before they can properly work toward fulfilling a goal.

Communication not only means that you should speak your sentence clearly; you must also consider how individuals and groups will respond to that sentence, which is what I meant a little earlier when I wrote that perceptions are the political reality, and based on their life's experiences, different people may have different filters that come into play. Choose your words carefully, don't make snide remarks, remember to whom you are speaking, and try to be sure that you will be understood. I learned that lesson the hard way when I gave a lecture that contained biblical quotes about animals. I received three notes from students who were upset that I quoted the Bible out of context. They were right; I did quote the Bible out of context, but I never communicated that was my intent. I just sort of assumed that it was quite obvious and everybody would have known that. My wife does the same thing. She'll start a conversation about something I have little chance of knowing anything about, then, when I ask her what she's talking about, she realizes what has happened and always says (with a smile), "How come you don't know what I meant? I know what I meant." If there is any doubt in your mind that a person did not understood you, it is often helpful to politely ask that person to restate what you said. It's just asking for feedback. If the circumstances are not conducive to feedback (as when you are teaching in a lecture hall), consider clarifying your intent to the audience before wading into sensitive waters. You shouldn't be hesitant to sit down with your staff while they're taking a short break, talk casually with them, and try to get a feel if they truly understand whatever it is that you feel is important, such as the mission and vision of your animal facility.

Much of the above discussion is telling you that you have to be a flexible communicator. You are communicating with a real human being who is different than you are, not a little Mini-Me out of an Austin Powers movie. Don't even think of trying to make employees look, work, and act like you do because it will not work. You have to accommodate for the person you are communicating with, and that person has to make some accommodations for you.

Let me make some additional comments about communicating with people. First, you will generally find that if a request to a person is made as a question or a polite statement, it will be accomplished quicker and with fewer problems than if an order is given. Simply consider the difference between "Clean the cage now,"

"Please clean the cage now," and "I would appreciate it if you would clean the cage now." Along similar lines, I learned a long time ago that it helps quite a bit to ask a person to help you, rather than demanding something. As an example, try calling your financial department and ask someone to help you understand how a strange charge arose on your monthly budget report, rather than telling them that you need an explanation of a charge. In the former instance, you are asking for help, while in the latter instance, you are demanding it. A little bit of sugar can go a long way. Here's another example: when you would like to speak privately with a person who directly reports to you, consider the difference between "John, I'd like to talk with you. Please come to my office" and "John, I'd like to talk with you about buying a new cage change station. Please come to my office." The former statement might scare the life out of a person, while the latter one is far more welcoming [5].

Next, be careful about casual remarks, or things said in jest. People, and your staff in particular, listen to your words and observe your actions more carefully than you might imagine and often take you quite literally. A little humor in your conversation is acceptable and often valuable, but you will be amazed to find that some people take what you say seriously, even if you believe they never would or could. I know of a supervisor who had quite a problem undoing the damage he did when he jokingly told a researcher that a small rat is the same as a big mouse! You can most certainly joke or fool around, but there is a time and place for everything. Know your audience.

Words said in jest are often taken more seriously than you would imagine. Be careful about what you say.

I'll give you another piece of advice about knowing your audience, and it's an important one. Do everything you can to remember people's names. If there's one thing that just about everybody likes to hear, it's that you remembered who they are. Salespeople, hairdressers, and clergy people all know how important it is to address a person by using their name. It builds and strengthens bonds between people. It doesn't matter if the correct etiquette is to use a person's first name or if it is more appropriate to address a person as Mister, Misses, Senator, or Reverend; remembering a person's name will help your career as a manager.

Remembering peoples' names will help your career as a manager.

Interestingly, I have the opposite problem. I remember peoples' names but, no matter how hard I try, I can't remember faces unless I am very familiar with the person or they have some unique feature that I can remember (such as a man who always sports a full beard). If a person comes to talk to me and doesn't mention their name, I'm in trouble. But as soon as I hear his or her name, I can often recall almost every conversation I had with him or her.

Here's a final hint: try to use *we* in your communications as much as possible, especially when communicating to a group of people. The word *we* brings your audience into the discussion. Just consider for a moment how you would react to a vivarium manager saying, "We may have different job titles, different salaries, and work different shifts, but as a team we're doing a great job taking care of the animals that depend on us." That's a pretty positive statement. But the same statement might have been "I think all of you are doing a great job taking care of the animals that depend on you." Both are positive statements, but the first is more personal and heartfelt than the second, has a stronger motivational potential, and can subtly raise your esteem in the eyes of your coworkers.

Communicating with People from Different Cultures

Over the years, there has been an influx of people from differing nationalities into laboratory animal science, and of course, there are many researchers and research staff who are from different cultures. When communicating with people who aren't fully acclimated to the American way of doing things, we not only have to be tolerant of their cultural differences, but as managers, it also helps to try learning about those differences ahead of time. For example, if we take a prospective French-background employee to lunch at a fine restaurant and she dawdles over her food, it may have nothing to do with the quality of the food. We Americans like a lot of food and we gobble it down quickly. Quality takes a back seat. The French are more content with smaller portions, fine presentation, and fine taste. They take their time.

A little earlier in this chapter, I made the argument that if you have something to say, it should be clear and to the point. That's what is expected in the United States. But that expectation doesn't hold for all people we meet, especially those coming from other cultures. In many countries, such as China, Japan, South Korea, India, and Saudi Arabia, things that are said are not necessarily specific and to the point. You have to read between the spoken (or written) lines to get the full context of what is meant. Not only do you have to listen quite carefully, but also you may have to ask clarifying questions (which can be quite important) and observe the person's body language to make sure your message has been understood or, of course, to see if you are fully understanding what the other person is saying [6].

A similar concern arises when you have to provide negative feedback to a person coming from certain countries where the culture is to avoid conflicts (once again, Japan, China, India, and Saudi Arabia are examples). Until these individuals are acclimated to the American way of communicating, you may want to be much more diplomatic than you usually are, using softer words of criticism preceded by specific praises for tasks that were well done. For example, if you were to say to a person born and raised in the United States, "You're doing a good job overall, but I would like you to improve your animal handling during the coming year," you might change this statement somewhat for a person who was born and raised in China and is new to the United States. You might say, "I'm pleased with your work and it would be appreciated if you would work even harder learning how to handle mice and rats." Conversely, when speaking with a person having a Russian background, where direct and clear comments are expected, you may want to say something like "Overall your work is good, but you have to improve your handling of rats and mice" [6]. All of this discussion goes back to the basic concept that good managers know that each of the people with whom they interact are different, and you have to adjust your management style for different people.

Written Communication

Many, and perhaps most, managerial communications are still verbal. Although some people equate communicating with speaking, a large amount of communication is written (such as memos or this book), electronic (such as email or fax), or even through body language. Those of us who have worked in clinical veterinary medicine and laboratory animal facilities know how crucial written communications are. No research supervisor or laboratory animal veterinarian would ever consider not having written records. There are, of course, legal reasons to keep written records. For example, many animal facilities work under the Good Laboratory Practices Act, which requires records to be kept in a specified manner. Likewise, the Animal Welfare Act regulations and the Public Health Service Policy on Humane Care and Use of Laboratory Animals all have various sections on record keeping. Almost every animal facility has standard operating procedures (SOPs) that spell out in some detail how the facility will operate. These and other written communications are part of the information resources you are expected to manage.

Written communications require the same care as do oral communications. You must consider the person or group of people who will be receiving your messages. If you are leaving weekend instructions for a technician on how to care for an animal, don't forget to consider that person's background. Does he know that you mean "twice a day" when you write "b.i.d."? How about "asap"? If you want to be called if problems arise, be sure it is clear to others what constitutes a problem. If an animal should die, there is obviously nothing that can be done to help it. Should you be called? Is a dead animal a problem?

Be as careful with your written communications as you are (or should be) with verbal communications. If you are upset about something, calm down before sending an email that you will regret forever (this is what some people call a nastygram). Another good bit of advice is to double-check the "To" heading on your email to make sure that you are not sending it to the wrong person or, worse, to an entire mailing list. One of my former colleagues unintentionally sent out a lewd email to a popular mailing list instead of to one person on that list, and he had to make copious apologies. But if that wasn't enough, a few months later he made the same mistake. There is no excuse for that kind of tastelessness, but I hope the underlying message is clear: for any email, make sure the recipient is correctly identified in the "To" heading.

Leaving written instructions is frequently necessary (most animal facility managers routinely do this). You might consider leaving specific instructions to call you if an animal does not eat all of its food within two hours, if there is reddish fluid seeping from a surgical wound, or any other criteria you believe are important. And please, try to be brief when leaving written instructions. People don't like to be confused, and long memos are confusing. If an email or memo must be detailed, break it down into numbered paragraphs. It's simply easier for people to read that way.

Equally important, stay away from jargon that only you and a few associates understand. Not only can *you* forget what it means, but a new person will have no idea at all as to what you mean. One of my colleagues always wrote "chloro" on his medical records when he wanted to abbreviate the antibiotic chloramphenicol. This worked well until a new employee dosed a dog with Clorox. That's not likely to happen again because chloramphenicol is infrequently used nowadays, but there are always new drugs or procedures where inappropriate shorthand communications can lead to long-term problems.

To go a step further, whenever possible, leave written, not verbal, instructions for the animal care staff. Use SOPs, formal memos, or whatever you believe is appropriate. Similarly, ask all investigators to give you instructions, in writing, if they want to modify an animal care procedure (email makes this fairly easy). If you don't get it in writing, don't be surprised if what you are doing is not what an investigator really wanted done. It will come back to haunt you.

> *I vividly remember the time an investigator stopped me in a hallway and asked to have the animal facility technicians begin treating his animals every other day, not every day as we had been doing. This was before there were IACUCs [Institutional Animal Care and Use Committees]. I dutifully told the supervisors what had to be done, and they in turn told the technicians. Two months later, the investigator finally looked at the experimental records and became infuriated. He ran into my office and called me and the animal facility staff every foul name known to humankind. The problem was that he wanted his animals treated three times a week, not every other day. He swore that he said treatments were to be on Monday, Wednesday, and Friday.*

Since that time, I have insisted that any changes to research protocols be made in writing and signed by the principal investigator or a person previously designated as having authority to make changes. If there is a problem with an investigator who does not want to take the time to put a change in writing, then don't be afraid to write a memo that begins, "As per our discussion of July 19, 2016, the following changes are being made in study number 12345…"

Having expounded on the importance of written communications, I would not want you to spend the better part of your day writing out everything imaginable, just to protect yourself. There's a time and place for everything, and common sense has to come into play. One administrator I knew (actually, I think bureaucrat is more accurate) would send me the most preposterous memos you can imagine. I once received a two-page memo from her, asking for the number and species of nonhuman primates in our facility. It's beyond me how she managed to take two pages to ask that question, but she did. Most of it was a detailed rationale explaining why she needed the information. She had her secretary prepare a chart to enter the information, which was placed on a computer disk that was sent along with the memo. All this happened one day after she had toured our primate holding areas. It took me longer to read the memo than to respond to it. I left a message on her voice mail: "Cheryl, we have 12 rhesus monkeys, 6 pig-tailed macaques, and 8 squirrel monkeys. Call me if you need any more information."

Key Points for Written Communication

- Know and understand the background of the people who will be reading your communication. Write so they can understand it.
- Do not use jargon. Be clear.
- Be brief. Say what you have to say, and no more.
- Always write out changes in animal care or related procedures. Verbal messages often get lost. IACUC approval may be needed.
- Ask investigators to put requested changes in animal care or other procedures in writing, or follow up their verbal request with a written confirmation.
- Not everything must be in writing. Use common sense.

Keep Communications as Open as Is Practical

Whether you communicate verbally, by handwriting, electronically, by body language, or by any other means you can think of, one guiding principle is that communications should not be restricted. Most modern organizations want and encourage communications at all corporate levels. Not every university president or corporate chief executive officer has an open-door policy whereby you can come in whenever you please and say whatever you please. Nevertheless, there is usually a mechanism by which you or any other employee can make your thoughts known to that person in a relatively easy manner. With open communications, you will rarely have to go over anybody's head to reach the president or chief executive officer with either your suggestions or complaints. You, and everyone else in the organization, should be confident that their thoughts will, at the very least, be given consideration at the appropriate levels of the corporate structure.

One simple but effective way of fostering open communications is to hold a weekly meeting with the immediate staff that works with you. Keep them abreast of what's happening. Let them know about problems as you see them, and ask for their comments. Get your supervisory staff involved in goal setting. They are not there solely to carry out your plans and directives. Encourage them to disagree with you. If you have a small and congenial staff, discuss work matters during or just after lunch (which you should occasionally have with your staff).

Similarly, supervisors who work for you should routinely meet with their staff. The nonmanagerial staff should know which studies will be starting and what problems have been encountered with ongoing studies. These problems don't have to be "things that are being done wrong." If, for example, the research itself is not being productive, let your staff know this. Ask for their opinions. Pull them into the research. They should not be uninformed bystanders.

Along these lines, before a study begins, you might consider arranging a meeting between your staff and the investigator, so your staff knows the objective of the study and what type of special care (if any) is needed. Again, the main purpose is to involve your staff in the study and therein increase morale and enhance the overall research effort. There is also a secondary purpose, and that is to break down some of the usual aloofness that occurs between groups, particularly when there are differences in educational levels. As one investigator said to me (referring to researchers), "We're all kings in our own little kingdoms and we don't like to have anybody tell us what to do." Face-to-face meetings help dissipate this attitude. It brings us into their kingdom in a positive manner. Similarly, some organizations have learned that one truly likeable person who knows everybody and gets along with everybody can be an invaluable asset by helping to bridge the gap between different groups within an organization, such as animal care and researchers, or even between animal care and veterinary services. They are the troubadours who go from village to village, making people happy with news and song. They can facilitate getting people together, making people feel good about each other, and in general, they can help keep communications open when everybody else is hitting a stone wall. Sometimes we have to keep this type of a person on board, even if they are not a superstar, because of their importance in greasing the wheels of our operations [7]. But do we always have to bend over backward to have meetings when everybody already knows what has to be done? Some research or testing techniques are redundant. They are the same ones that have been used for many years. The only change may be that a different chemical or drug is being used. Is it necessary to meet *again* with the same investigator? There's no pat answer to this. You might first find out the significance of the new chemical or drug, and then you can determine if it's worth another meeting or if, perhaps, you can do the explaining yourself.

Most modern organizations want and encourage communication at all levels. This type of open communication is to be supported.

Here's a reality check. Some organizations are still run like a military base. Subordinates speak to supervisors only when asked, and supervisors speak to subordinates only when necessary. People from one division rarely speak to people in another. And heaven help you if you go over your supervisor's head—that's the same as high treason. This attitude is all part of a sick corporate culture—not the kind to which most organizations should strive. The implied message is that you are a number, not a thinking and capable human being. It tears at the heart of communication and trust. In my opinion, it is a morally bankrupt tactic for anybody managing a laboratory animal facility. Just as communications within a division should be as free as it is practical to be, communications between divisions of the same company should also be reasonably open.

By this point, I hope you can appreciate that open communications enhance morale by giving people a sense of worth and security. Few things are as detrimental to organizational morale as rumors and the feeling that management doesn't care about employees. If you want to stop rumors, tell people the truth. Answer their questions. We all have an inherent fear of the unknown, and most of us fear change, even if the change is to our benefit. So, if there is going to be a change, for better or worse, let people know as soon as it is reasonable to do so and make a strong case for the need of the change. If your animal population is reduced and has been low for the past year, don't you think your employees know this? Don't you think they are worried about their job security? Why not talk about it? Let them know what you know and tell them how any planned operational changes will help in the long run. Doing this will rarely cause a problem, and more often than not, it will elevate your status as a manager. Even better than letting people know about the change and talking about it, why not give the people who will be affected by an event a means of providing their input on how the situation might be handled. They may have a valuable perspective that nobody else considered, or they may have an interpretation of the change that is reasonable, but not what upper management intended to convey.

We probably all know supervisors who withhold a certain amount of information to maintain some degree of "power" over their subordinates. They tell you only as much as they think you need to know. They think that there is no need to keep you informed about anything that does not directly affect you, or in some instances, they feel that keeping this information to themselves makes them appear valuable to the organization [8]. One way or this other, this approach is what I call national security thinking. It's a "need-to-know" philosophy that has little place in most biomedical organizations. We all know that not everything can be made public to all employees, but most information can be. Everybody should know the organizational mission and its current goals, certain financial information about the animal facility and the entire organization, studies that will be beginning or ending, and so on. Yet, not everybody has to know, or should know, the details about a new antihypertensive agent being studied.

Try to keep communications as open and informal as possible with both your staff and the investigators with whom you work. It gives people a sense of worth and security and enhances your status as a manager.

Even a little newsletter with animal care guidelines, facility management problems, or anything else you deem to be important is worth its weight in gold.

Sounds of Silence

Sometimes people are afraid to speak up because they don't want to look foolish, be the sole voice of dissent, or offend a person who is higher up on the corporate hierarchy. This is certainly understandable—but nevertheless unfortunate—because a lack of communication about possible problems or solutions forces these issues to remain hidden and unaddressed. Much of the solution to this problem has to do with the corporate culture, as well as your managerial style. Let me give you an example: If I belittle every suggestion made to me that goes against the way I like to do things, people will quickly learn to keep quiet. If they continue to speak up, will I block their promotion? Can I lower their salary increase? Even if I want people to speak their minds, how can this happen if I continually (and perhaps unknowingly) create a blockage to open communication? Managers and leaders have to be careful about their reaction to communications that are not what they might have wanted to hear, as inappropriate responses can lead to organizational silence.

Is there another way of communicating if the culture of your organization or if an individual manager tends to inhibit communication? Yes. Consider communicating as a group. Here's an extreme but true example, with changed names and details.

> *Rob Gray was the manager of animal care at Great Eastern University. He was a congenial man, well liked by his boss because of his personality and "can do" attitude. In fact, his boss frequently and publicly praised Gray for the fine job he was doing. Gray's power within the university continued to increase.*
>
> *The only problem was that Rob Gray had been receiving a kickback from American Cleaning Supplies, the primary cleaning supply company for the animal facility. Any supervisor who tried to use another vendor was clearly and firmly told by Gray to use only American Cleaning Supplies. After a little while, it became known to some of the supervisors that Gray was being bribed; however, at first nobody had the nerve to confront him or his boss. After some time, one of the supervisors did tell Gray that they should not be locked into one vendor for no good reason. He was told to mind his business and do his work. The supervisors felt used, betrayed, and powerless. At first, they were even afraid to talk to each other about the problem. Still, as the bribery continued, they did begin to talk more to each other and then, as a group, they wrote a letter to Gray's boss documenting their concerns. They scheduled a meeting with the boss, walked in as a group, handed him the letter, told him that a copy had been sent to the vice president for research, and demanded that Gray be fired. By the end of the week Gray was out of a job.*
>
> *The moral: When silence is enforced, the power of a group may be able to overcome a problem that no individual person can resolve.*

Listening and Feedback

Good communication includes good listening. It is very important that you take the time to listen carefully to your employees. Do what you can to restrain yourself from interrupting a person before he or she has finished speaking. Don't tune people out because you disagree with what they are saying or even because of a speech impediment. Few things are as frustrating to people as the feeling that you aren't listening to them at all or you're more interested in listening to your own opinion than theirs. It may be necessary for you to sit down and take notes when an employee or other person is talking to you. This also indicates interest on your part. Don't be afraid to occasionally interject a summary statement such as "You said that the maintenance department hasn't responded to any of your requests?" It shows that you have been listening and that you're interested. Ask questions if you have to, but the important thing is to get feedback from your employees and give them feedback as needed.

Good communication includes good listening.

If this sounds like textbook rhetoric again, it is. Not that I've read all of them, but I cannot imagine any communications instructional guide that suggests attentive listening isn't an important managerial function. But attentive listening is only part of the equation. What do you do when you're finished listening? Do you pat yourself on the back for being a good listener and then walk away? Of course not. You have to do something. I was involved once in a situation where somebody routinely did nothing. This person was the executive director of a health research organization, and he diligently took notes whenever I spoke with him. At first, I was very impressed, but when I realized it meant absolutely nothing, deep disdain set in. I felt used. Don't let this person be you. Trust is important. If you go through the motions of listening and then take no action or make no decision, you will lose whatever trust you may have built up. It's what has been called the futility factor, and this can lead to a total block of future feedback: "the biggest reason for withholding ideas and concerns wasn't fear [of reprisal] but, rather, the belief that managers wouldn't do anything about them anyway" [9].

When we interact with investigators, listening is paramount, as it can help clarify what is expected of us and help build trust with the investigator. Let's be honest; if it weren't for investigators, we wouldn't have jobs. Our function in laboratory animal science extends beyond taking care of animals. We also have to take care of investigators. We have to elicit from them what they want, not what we are willing to give them. The reality may be that we cannot always provide what they want, but if we don't ask and listen, how can we plan for the future? How can we set goals to correct existing problems if we don't know what they are? Imagine the bad feelings you would generate if you did a service satisfaction survey of investigators using the laboratory animal facility and then did nothing to correct their concerns. Laboratory animal science is a business. Just as many other businesses are driven by what the consumer wants or needs, consumer desires are also part of our business.

Laboratory animal science is a business. As other businesses do, we must listen to our customers and try to provide them with the services they want.

There's another benefit to listening: you might learn something. You may become empathetic enough to see the other side of the coin. How many of us in laboratory animal science have listened carefully to what animal rights advocates have to say? Are they just crazies that we have to live with? Not at all. Listen carefully, and you may find you have more beliefs in common with them than you thought. At the very least, begin to understand their feelings.

People must feel that they are part of an organization, not just 9-to-5 employees. Give them the chance to speak up. Sometimes this requires that you stop speaking and start listening to what they are saying. This may not be easy, since it implies giving up a little of your own authority, particularly if a person has a valid criticism or suggestion. But it is precisely this ability to accept suggestions from those you work with that can increase your stature as a manager and pull people into the organization. It increases people's trust in you.

Getting feedback may be inhibited by a number of not-so-obvious problems. For example, do you remain behind your desk while talking to a coworker who is trying to give you feedback? If you do, you're sending a message that you're the big boss and she better not forget that. Get out from behind that desk and sit alongside your coworker, even going so far as having both of you using the same-size chair. Do you put your arms behind your head when listening to a person? If you do, you're sending a message that you're "bigger" than the other person. Do you proactively ask people for feedback while in their work area? If you don't, consider doing that, as it makes you appear as a team player, not just their manager [9]. All of these little actions, perhaps seemingly insignificant by themselves, add up to a more relaxed atmosphere and increase the probability of obtaining feedback.

If you encourage feedback and open communications, it must be more than a textbook exercise. If you want to continue to get feedback, you have to give feedback. Make your feedback short, concise, and clear. Most people want to hear feedback in a timely manner, not only during an annual or semiannual performance review. *Timely* doesn't require that all feedback has to occur within minutes of something happening. Sometimes it's better to sleep on a problem for a night and rethink the incident the next morning, but you certainly don't want to put feedback on the back burner. To repeat what I said earlier, you must follow through in some way. This doesn't mean that you must accept everybody's problems as your own or take actions that you believe to be detrimental. It does mean that some form of response must be forthcoming. This response can range from "I see your point but I think if you try ..." to "This is something you should handle as you see fit." It might also be "This is a serious problem and I'll get back to you within the week" or "I appreciate your suggestion. Why don't you try it out and let me know how it works?" Communication is a two-way street. If you expect people to communicate with you, you have to communicate with them.

If you expect people to communicate with you, you have to communicate with them.

Selective Listening

Over the years, I've noticed an interesting problem with some managers, myself included. It's called selective listening or, to be more specific, listening to what is said but not rooting out the cause of a problem. An analogy is the doctor who puts a pressure bandage over a bleeding arteriole, rather than tying it off. There is a chance that you might solve the problem, but there is also a good chance that it will recur as soon as the pressure is released. Here's an example.

> *John W. is in charge of the cage wash area in your facility. You try to be a good manager with open communications, so you routinely meet, individually, with each of your supervisors. On this day, you ask John if there is anything that would help his operations. He responds that he knows everybody else is busy, but if they could get the cages to him by 3:00 p.m., he would not have to pay overtime and have some employees become disgruntled because they are often asked to work late.*
>
> *Soon after, at a managerial meeting, John's problem is discussed and a mutually agreeable solution is found. You pat yourself on the back for being a good manager and continue on your winning ways.*

It's always a good feeling to be able to solve a problem. But did you go far enough? Did you dig deep to find out why the cages were not in the wash area by 3:00 p.m.? Would John have actually told you if it meant revealing that another supervisor took two-hour lunches? And why did it take so long to find out that there was a problem? Why didn't John come to you without prompting? Is he afraid that you are always so positive and cheerful that he's unsure if it's "safe" to talk to you about these matters? Whatever the reasons, look below the surface if you find a problem to try to determine if there is an equally or more significant underlying problem. Truly open communications include not being afraid to speak the whole truth.

Here's another example. It involves an employee who is not carrying his own weight. You all know what happens—your better employees initially pick up the slack, but after some time, they start complaining. First, it's behind your back complaints. Then, it's "in-your-face" complaints. What are you going to do?

A friend asked me to meet with him as soon as I could and discuss the issue of pain recognition in laboratory animals. He seemed a little uncomfortable, but I disregarded that at first. Over lunch, he made it clear that his goal was to make DVDs of animals in pain, side by side, on the same screen with pictures of animals that were not in pain.

I gave him my initial thoughts on his idea (including the fact that others had already done this), but then I asked him why he wanted to do that project, since he had limited expertise in pain recognition. He told me that he just thought it would be helpful to laboratory animal medicine and it was something he was interested in. Then I asked him what was actually going on, since his body language and attitude just weren't his usual easygoing self and he had hardly touched his food. At first, he was inhibited, but then he began to talk. He actually seemed relieved to be able to tell me that his boss told him that the other veterinarians in the animal facility were annoyed at him because they felt he was spending too much time on insignificant paperwork. Therefore, he was going to develop a project to prove his worth. I asked him if he thought the veterinarians were right. Again he hesitated, but eventually related to me that his life was being torn apart by marital problems.

There's no need to go into the rest of the story. The important point is that there was a problem behind the problem. Many personnel problems at work have a nonworkplace root cause that eventually manifests itself as a workplace problem. In the above example, we know that marital problems were the root cause because I told you that, but if you were the boss who had to listen to the complaints of the other veterinarians, what would you have seen as being the real gripe? If you said that the veterinarians were annoyed that a colleague was doing too much paperwork, you've missed the underlying reality. The veterinarians' actual complaint was that they had to pick up what should have been another person's work. They probably wouldn't have given a second thought if it was an animal care technician who was doing the paperwork because it wouldn't have affected them in any real way. And they probably would have been much more understanding and accommodating if they knew that their fellow veterinarian was under tremendous emotional duress.

Getting to the root cause of a problem isn't easy because we're not trained as clinical psychologists, nor should we try to be one. However, we can train ourselves to be better listeners and ask pertinent business-related (not personal) questions. Sometimes, we have to ask for the assistance of other professionals for help, and most organizations can do this through their offices of human resources, equal employment opportunity, employee assistance, and so forth. There's no shame in asking for assistance; it's actually to your credit that you do so.

Difficult Conversations

Most of us want the trust of other people as long as it doesn't cost us anything. One cost is giving up a little bit of authority when we listen to a criticism about us. Another cost is our fear (or at least dislike) of criticism. Sometimes, we become so entrenched in our position that we think we're immune to criticism, or we're so used to being told how good our work is that we can't believe a person actually has the audacity to criticize us. The best advice I can give is that you shouldn't sit on your laurels because you may be in for a very negative surprise someday. I've been in laboratory animal medicine for about 40 years, and I've been slammed around enough that you might think my skin is so thick that I can't feel any criticism. That's not true. I fully understand the need for constructive criticism. I actually go out of my way to encourage that kind of feedback, but when it happens, I still don't like to hear it. I get annoyed with myself, not with the message bearer. Nevertheless, I've learned not only to accept criticism, but also to evaluate it and take action as needed.

Let's generalize and subdivide difficult conversations into two categories. The first is when you are the recipient of criticism that crosses the boundary of a civil discussion. The second is when you are the manager and have to talk to a person about his or her performance or a related problem. We'll briefly consider uncivil criticism first.

Feedback from employees or your own boss that is presented courteously can be considered a form of constructive criticism. Sometimes, constructive criticism doesn't occur and the feedback is better characterized as destructive criticism. Destructive criticism (i.e., incivility) is more than just hurting a person's feelings. There is a long list of management-related problems that can affect the criticized person, such as decreased productivity, decreased work quality, impaired commitment to one's work, and quitting [10,11]. Unfortunately, there's a core of people who have trouble being civil with their communications and actions toward others, and they can be offensive, inaccurate, irrelevant, or one sided with their feedback. My own way of dealing with this has always been to hold my annoyance in check, speak calmly, listen a lot, and do what it takes to turn lemons into lemonade. Others have researched this problem and have provided additional information on how to turn toxic feedback into useful information [12]. The key suggestion is to maintain your temper and not to start an argument with the person talking to you.

In October 2015, Hillary Clinton underwent questioning by a congressional committee investigating the death of four U.S. embassy employees in Benghazi, Libya. She remained calm, collected, and articulate throughout her nearly 11 hours of testimony. Some of the questions directed at her were less than civil, but the aftermath of the hearing was that Clinton received much praise for her deportment and knowledgeable responses.

Like Clinton, we have to work hard and, if necessary, train ourselves to stay composed, maintain a neutral tone of voice, try to determine the person's perceptions that led to any negative feedback, and then use what we have learned to calmly deal with the problem without turning the conversation into a "me versus you" dialog. None of this came naturally to me; it took lots of practice by using mock situations and, unfortunately, some real ones. You have to concentrate on the accurate information and postpone any action on inaccurate information. To be more colloquial, it's a matter of practicing to separate the wheat from the chaff, with the wheat being information we can potentially act on and the chaff being emotionally driven conclusions and opinions. Accurate and relevant information can help us deal with the immediate issue. This does not come easy to most people because not too many of us are fully adept at accepting criticism, even though we may try to be. This problem can be reflected in our responsibility to provide feedback, particularly if that feedback may have a negative tinge to it. Nobody really likes to have difficult conversations with colleagues or employees, and many people shy away from negative comments during performance discussions. However, we have to face up to the need for giving and receiving feedback; if we don't, any existing problems will continue.

When dealing with a person giving you abrasive feedback, stay calm, use a neutral tone of voice, listen carefully, and try to determine what led to the person's perceptions. Focus on the facts, not on guesses or irrelevant opinions.

The more common form of feedback that you will encounter is when you are placed in the position of evaluating and commenting on another person's work or actions, particularly when that feedback may incorporate some issues potentially requiring improvement. Therefore, before we move ahead with this discussion about difficult conversations, we have to agree on one thing, which is that it takes at least two people who can talk respectfully to each other to have a conversation that has a reasonable chance of accomplishing anything. This is of obvious importance, and I'll mention it again in a moment. If tempers are hot or a desire for a conversation is lacking, you have very little chance of resolving a problem by wishing, waiting, or complaining to your friends. Madeline Albright, the former secretary of state, said it best when she commented on the conflict that was ongoing at the time between Israel and the Palestinian group Hamas. She said, "Until there's some political will between the parties, there is very little that can happen" [13].

> **To have a useful difficult conversation between two or more people, the parties must act and speak with civility toward each other.**

One of the most important suggestions (that I will often repeat) is that you, as a manager, must remain calm and courteous at all times. As I mentioned earlier, when talking about our own ability to receive negative feedback, remaining calm is not particularly difficult to do, but it does take some practice. People want to know what to expect of you, and being consistent in how you approach a problem (i.e., staying calm) is a positive managerial trait. Yet, in spite of your best efforts, the person with whom you are talking may be so wound up, so irritated, and so anxious to state his case that he does not hear a word you are saying and, in many cases, does not want to hear a word you are saying. There may be accusations, yelling, finger-pointing, and other forms of maladaptive behaviors directed at you or others, and if that happens, for practical purposes, there is little chance of having a mature dialog. In those extreme circumstances, it is better for you to tell the other person that you can see how upset he is and you would like to continue the conversation later on, once he regains his composure.

There's a little book called *Difficult Conversations* [14] that you should consider reading because it provides helpful "how to" suggestions to use when engaging in difficult feedback. The basic ideas of the book are that during potentially stressful conversations, we have to calmly seek out the facts, be able to say how these facts affect our own feelings and the feelings of the person we are talking to, and try to understand the impact of the problem on others or on the work to be done. However, we can't just blurt out something such as "Your handling of animals makes me sick." Yes, in that phrase you certainly put your feelings on the line, but you have also set the stage for a "me against you" scenario. In other words, if I push you, you're going to push me back, or if I accuse you of a bad trait, your first reaction is going to be to push back by denying that accusation. Take a look at the following conversation. It's a good example of a typical, but not an ideal, feedback session.

Peter: Good morning, Pat. Thanks for coming by. How is everything going?

Pat: Fine, thanks. You said you wanted to talk to me.

Peter: Yes. I wanted to talk to you about how you handle animals while cage changing. I saw your technique this morning, and it was rough on the hamsters. I'd like you to tell me what you can do to avoid that in the future.

Pat: I wasn't rough, and it was the way I was taught to do it so I could get everything done by the end of the day.

Peter: Okay, but in the future, try to be easier on the animals. You're stressing them out and may hurt them.

In this scenario, Peter has accused Pat of rough handling and, as expected, Pat denies the accusation. In other words, although Peter was calm and somewhat polite, he pushed Pat and Pat pushed him back. Peter wasn't particularly clear about what constituted rough handling. The meeting quickly moved from conversation to confrontation. Peter then tells Pat what to do, avoiding any discussion that Pat might have been poorly trained, which is a potentially serious problem. Rather than building a bridge, Peter has built a wall between him and Pat. Look how different it is when the tone of the conversation is changed:

Peter: Good morning, Pat. Thanks for coming by. How is everything going?

Pat: Fine, thanks. You said you wanted to talk to me.

Peter: Yes. I wanted to talk to you about handling animals while cage changing. I saw your technique this morning, and in most cases, it was fine. But it seemed to be a little rough on the hamsters because they all squealed when you lifted them by the scruff to change the cages. I wanted to get your opinion about this.

Pat: It was the way I was taught to do it so I could get everything done by the end of the day.

Peter: I see. I was wondering if you felt the animals were getting stressed out or might get hurt when the work has to be done so quickly. Hamsters usually only squeal when they're very upset, and cage changing shouldn't cause that much of a problem. Do you think we should do things differently? Should we reevaluate our training procedure?

In this conversation, which is a good way of initially approaching most difficult conversations, Peter was not at all accusatory, even though Pat gave the exact same responses as in the first scenario. In the second conversation, Peter was explicit with his concern (i.e., he presented an apparent fact) that the handling seemed rough because the hamsters squealed when lifted), but rather than actually accusing Pat of roughness, he asked for Pat's perspective. Peter was also quite clear in his communication. Pat didn't have to try to read between the lines to try to figure out what Peter meant. Peter also asked Pat for his opinion about the animals' stress level or potential for injury, again not telling Pat that the animals are getting stressed out, but asking Pat for his feedback. Then Peter acknowledged Pat's comment about his training by once again asking for feedback on how things might be done differently. He also shifted the discussion from putting the blame on Pat to suggesting that the training procedure may need to be refined. Finally, and perhaps most importantly, Peter told Pat the expectation was that animals had to be handled gently (i.e., they should not be squealing when handled for cage changing).

By approaching this issue as an unbiased observer, not as a biased accuser, Peter is helping to put Pat at ease and get discussion and feedback. Approaching feedback within a difficult conversation from an unbiased point of view, expressing your feelings in a clear manner, and listening to and understanding the other person's feelings takes time and practice, but it's well worth the effort because so many of us shy away from these situations. Even before you enter into the conversation, take a step backward and ask yourself what your goal is. If it's only to correct a person's behavior, you have a tough road ahead because your mind is already made up and changing a person's behavior is difficult. It's wiser to hold off on assigning any blame, try to get the facts, think about what the other person's perceptions might be, and think about questions and statements that are going to elicit clarifications, not arguments. Toward the end of this chapter, you will learn that the same nonaccusatory, civil approach to conflict can also be used when there is a group problem.

Provide feedback as if you were an unbiased observer, not a biased accuser.

Now let's put the discussion about listening and feedback into another context. As we learned earlier, when we have a difficult conversation, it's important to try to focus the discussion on the core issue, not on superficial issues. If you were the boss who had to talk with the veterinarian who was doing the excessive paperwork, you should be discussing the fact that the other veterinarians are becoming unhappy because they feel they are being penalized for somebody else's transgression. The extra workload on the other veterinarians, not the paperwork, becomes the core issue of the discussion. Focusing on the extra work is appropriate the first time a discussion is required.

How do you approach this discussion without becoming confrontational? First, make sure you are using the basic communication techniques that are appropriate to the situation. This is something I discussed earlier in this chapter. For example, if you know a person is a "bottom liner," then get to the problem as soon as possible. If the person is always nervous and needs to calm down a little, spend a couple of minutes on small talk. If the person is an animal care technician, don't try to impress or intimidate him with advanced veterinary concepts. Then, get to the reason for the meeting. Just as Peter was nonaccusatory with Pat in the second example of their conversations, we have to try to be nonaccusatory and view the problem from the perspective of an outside unbiased observer, even if it has affected you directly. Make sure you have factual information, clearly state the problem and why it is important to you or others, and follow through with questions and suggestions rather than with accusations. That's what transpired between Peter and Pat in the second of their two hypothetical conversations. The goal, of course, is to resolve the issue in a collegial manner so there are no winners or losers.

What happens if that first discussion, no matter how civilly it was conducted by both parties, doesn't resolve the problem? Over the years, I have used—with reasonably good success—the approach suggested by Patterson and colleagues [15].

The basic process is (as mentioned earlier) to first do your homework (by listening, by observing, by talking to people) and try your best to understand the core problem, which may differ from the stated problem. In one of the examples I've been using, the core problem (from the viewpoint of the veterinarians) was the extra workload on the other veterinarians, not the time their colleague was spending on paperwork. I suggest that part of your preconversation homework includes knowing exactly the key points you want to make and even how you want to verbally express your thoughts. Unfortunately, sometimes I get tongue-tied, and even though I know the general context of what I want to say, I mess it up. To help avoid that problem, I will sometimes practice out loud what I want to say before I say it. Once you know what you want to say and how you want to say it, you can move ahead with a three-tiered process [15].

Before entering into a difficult conversation, have clear goals about the key points to be discussed. Write them down if you have to.

---

1. *The first time a problem occurs, talk about what happened.* In our example, you might say, "There's a potential problem I'd like to discuss and get your input. Your colleagues have said that you're doing an extraordinary amount of paperwork lately, and it's really taxing on them because they feel they have to cover your clinical work. I was wondering if you had the same perception?"

   Generally, I'll make a few seconds of small talk before getting down to business. Usually, I don't like dancing around the issue and talking about everything that is happening in the building before getting to the main issue. More often than not, if the person you are talking to doesn't already know why he is meeting with you, he's going to be suspicious and possibly nervous about the meeting, so you may as well just get to the point without being accusatory. The exception is when the person I'm talking to needs some "calm-down" time or prefers a little small talk. Notice that, as in the example of the conversation between Pat and Peter, Peter was not confrontational, but he was very clear about the reason for the meeting. It was his perception that the animal handling was somewhat rough.

   Although some people suggest that after this initial discussion it is advisable to document, in writing, the problem that occurred and the plan for preventing its occurrence in the future, in my opinion this has to be approached on a case-by-case basis. If your focus is on documentation rather than coaching for success, trust is diminished and communications suffer [16]. In certain circumstances, documentation can be humiliating and punitive, whereas in other circumstances, it may be appropriate. Either way, the conversation should not end until both persons concur on the actions that will be taken to prevent future concerns. Invariably, if I don't make formal documentation of the details of the discussion, at the very least, I will make a brief personal written notation that the discussion occurred on a certain date and, of course, the main points of the conversation.

Accurate written documentation of what was said and agreed to is often, but not always, an important follow-up to a difficult conversation.

---

2. *The next time the same problem arises, discuss the pattern that is occurring.* For example, "John, we had a discussion last week about the amount of paperwork you've been doing and the effect it had on your colleagues. You said you had been overwhelmed for a while, but things were back on track. But from what I'm being told, very little has changed. What's happening?"

   In this conversation, you have reminded John about the promise he made to you. The burden is on his shoulders, but again, you're not out-and-out accusing him of slacking or not keeping his word. You have done two important things. You are making a clear statement that a pattern of behavior is developing and that all is not well. But, you are still asking John for his input.

3. *The third time the same problem happens, discuss your relationship with the person.* For example, "John, we've had two discussions now about the other veterinarians having to cover your areas, and twice you promised to resolve the issue, but to date it appears that there hasn't been any resolution. The clinical logs show that your colleagues are covering for you. I'm wondering if I can really count on you to make the needed changes."

   Your tone is becoming firmer and your conversation is now focused on trust. Can you trust John in the future? Give John a little pep talk and remind him how important it is for you to be able to trust him to complete a job that he is capable of doing. If this doesn't work, John may still be a talented person, but those talents may have a better fit elsewhere and you can remain on good terms with him by helping him find another position.

   When I first read Patterson's description of the three steps I just described, it sounded reasonable to me. The first time I used it, it actually worked well because I didn't have a continuing problem with the same person. The next time I used it, it involved a repeated problem, and although I kind of fumbled a little, it went fairly well. By the third time I used it, I was getting reasonably good at remembering not to go backward from step 2 to step 1 (i.e., not to discuss the immediate problem when I should be focused on the pattern of what was happening). I'm fortunate in that I haven't had any really major personnel problems to deal with as of late, but I feel confident in recommending the method. For me, the approach I follow when I have a difficult conversation is to try to combine the methods I just outlined above with the clear and short statements advocated in the *One Minute Manager*, which is that you should be able to clearly and effectively say what you have to say (whether it is praise, a reprimand, or stating a goal) in a very short amount of time [17]. The combined approach allows a manager to clearly and quickly express his concerns and feelings, and also focus on the actual problem, not spin-offs of the problem.

Another Difficult Conversation: Performance Reviews

Annual or semiannual performance reviews are inevitable in many organizations, and the mere thought of doing a difficult performance review is enough to send some managers running to the restroom. Heen and Stone [18] wrote that only 36% of managers complete performance appraisals thoroughly and on time, and one in four employees said their biggest dread at work was their performance evaluation. Noted management writer Samuel Culbert published a blistering article on why we should get rid of the traditional performance review and replace it with a better one [19]. But why? What is the problem with most performance reviews? Let me quote Culbert: "The boss wants to discuss where performance needs to be improved, while the subordinate is focused on … issues such as compensation, job progression and career advancement." The review, says Culbert, is "intimidation aimed at preserving the boss's authority and power advantage." "It creates tensions that carry over to … everyday relationships." I can add that traditional performance reviews assess what happened in the past, and often spend little time discussing what should be happening in the future, and when the employee's day-to-day work serves the boss's needs, unsatisfactory performance in other areas may be overlooked. I can go on and on, but the bottom line is that many performance reviews are confrontational even if the intent is to provide helpful feedback. The good news is that performance reviews do not have to be confrontational, and everything I wrote earlier about how to conduct a difficult conversation also applies to performance reviews. I'll expand upon this a little.

■ Performance reviews need not and should not be limited to semiannual or annual meetings. Although a more formal annual (or semiannual) performance review may be required by your institution, a good management practice is to have frequent scheduled meetings (even weekly, if you think it to be necessary) where performance goals are tracked, questions are asked about the availability of resources to meet goals, expectations are revised or reemphasized, and any other necessary topics are discussed. There should be very little that comes to light during a performance review that is new or unique to you or the employee you are reviewing.

■ Don't assume you are going to change a perceived behavioral problem simply by telling an employee that he or she has to change. It will not work because behaviors are very difficult to change, and your coworker may think that it is you who has the behavior problem. The big problem is that he may be afraid to talk honestly with you if the meeting has a confrontational tone. Remember two things: First, go into the meeting with an open mind, and don't assume that your interpretation of a problem is the only interpretation worth discussing. I prefer to ask questions (such as, how would you gauge your overall performance since our last review?) rather than blurt out a preconceived opinion. Second, if there is a perceived behavioral problem, then make it clear how the behavior is affecting you and others, as well as the employee.

■ Think twice about writing out your performance evaluation and handing it to your employee to read before you meet with her. If it's anything less than what that employee expected, by the time you have your formal meeting, that employee might be angry, depressed, vindictive, noncommunicative, and so forth. However, my opinion on this subject is not universally accepted. Some human resources departments and managers encourage giving employees their written review prior to the formal meeting, even if salaries are not part of the discussion. They then discuss the quality of the review at the formal meeting. The rationale is that it gives employees time to gather their thoughts before the formal session, and therefore makes the formal meeting more useful.

■ I do think that an employee self-evaluation form can be a valuable starting place for a performance review. It provides employees with an opportunity to showcase what they consider to be their contributions since the last review period and what they like or dislike about their work. If there were problems, they can use the self-evaluation to help explain them. I also have been impressed at how honest most people are about discussing areas where they think they can improve. Self-evaluation forms are often used in universities as part of the process of evaluating research faculty, and they can readily be adapted to nonfaculty.

■ I do recommend planning before holding a review. Consider the key points you believe will be helpful to your employee's career and how you will say what you want to say. As I mentioned earlier in this chapter, you might even want to practice out loud to hear how you will sound to your coworker. Very often, I will repeat our mission and vision and emphasize how the person's work contributes to the same.

■ Because performance reviews are based on the goals that were mutually agreed upon with your employee, it's usually a good idea to start performance reviews by summarizing some of the accomplishments you believe the person made since the last performance review. Not only do people like to hear positive comments, but it also makes it somewhat easier to provide constructive feedback, if needed, later in the conversation. If there are any areas where performance improvement is needed, don't go overboard. You really can't read off a list of problems, some important and some trivial, and expect a person to remember all of them and act on all of them during the year. Focus on no more than two or three items and develop a plan for improvement with the employee [20]. As I wrote a little earlier, you might want to ask for the person being reviewed to comment on what she believes to be her own strengths and weaknesses before you get into a deeper conversation.

■ If there are recurrent problems, what do you do? Sometimes people just don't work well together and your style of management may not be right for that person. But, if you can get enough honest feedback, you can try to change your style somewhat to match your employee's needs. Sometimes, you will have to use the three-tiered process discussed earlier (i.e., if a problem was previously discussed, you should now discuss the pattern that is occurring or that your trust in the person is faltering).

■ When performance reviews discuss a needed behavior change, one of the biggest problems I've experienced occurs when comments from a manager assume the employee knows what is expected. This is a communications breakdown. If, for example, you expect a veterinary technician to examine more animals each day than is currently being done, can you clarify what *more* means? What is the standard? Is it twice as many animals? Is there anything in that person's job description that clarifies an expectation? If not, it will be up to you to work with the technician to set a measureable standard (if that's even possible), but you certainly have to be clear as to what is expected and why that expectation is reasonable. "It is the effective combination of feedback, goals, and rewards that is most likely to influence employees to change their behavior in ways that result in improved performance" [21].

■ One of the objectives of most performance reviews is to set goals for the upcoming year. This is not necessarily a "difficult" conversation, but it certainly is an important one, and one in which the employee should be given the opportunity to contribute. Goals must be aligned with three documents: the organization's mission and vision and the employee's job description. As you will see in Appendix 1 (and as also discussed in Chapter 1), goals should be

- Important to the employee's job
- Obtainable (i.e., possible to fulfill with reasonable effort)
- Specific (i.e., unambiguous)
- Measureable (i.e., progress toward reaching the goal can be easily evaluated)
- Whenever possible, quantifiable (i.e., numerical, such as saving a certain amount of money, getting a certain number of cages per day through the cage washer, or attending a certain number of continuing education sessions)

■ Lastly, and there is some repetition here, don't assume your perspective is the right perspective. I don't know anybody who is infallible. It's always better to stand back a little, be impartial, and ask a lot of questions to try to understand your employee's point of view. It's certainly better than having a conversation that is largely full of accusations and counteraccusations. Make sure, before the performance review ends, that you ask your employee what you can do to help him, in terms of the way you relate to him, training he might want, and so forth. You have a responsibility to provide good supervision. To that end, I agree with Culbert [19], who wrote, "I'm sick and tired of hearing about subordinates who fail and get fired, while bosses, whose job it was to ensure subordinate effectiveness, get promoted and receive raises in pay." If we expect our employees to do a good job, then they have a right to expect us to do a good job.

Once the performance review is completed, you should promptly write up the key points from your discussion. This should include your perceptions of the employee's performance, the perceptions coming from the employee you were speaking to, areas of strength and weakness, goals and plans for the upcoming year, and all similar items. If there is agreement that a person is somewhat of a square peg whose job requires working in a round hole, you can try to determine if a different position or physical location in your animal facility would be better for that person.

The performance review should be signed by you and the employee with the understanding that the employee's signature only indicates that the review was conducted; it does not mean that the employee is in full (or even partial) agreement with the outcome of the review. Any lingering disagreements from the employee should be added to the report by the employee.

360-Degree Feedback

Before leaving the subject of performance reviews, it's worth noting that I have never worked at an institution where 360-degree feedback was used for performance reviews or any other purpose. I mention this to clarify why there is only a short discussion on this somewhat controversial form of feedback. From speaking with colleagues, I know that some vivariums do use 360-degree feedback for performance reviews. However, many of those who have written about 360-degree feedback are quick to point out that this form of feedback is *not* the same as a performance review. Rather, it is a review designed to improve a person's professional skills, such as a manager's overall performance as a manager. If a person has a specific performance problem (such as tardiness or bullying), then a more standard performance discussion with the employee should occur.

Why is the 360-degree method controversial? This is because there are mixed literature reports about whether it is any more effective in improving a person's managerial skills than are the standard performance reviews I described earlier.

A 360-degree evaluation focuses on a person's managerial skills and not how the person does his or her daily job.

In 360-degree feedback, a person (such as a supervisor or manager) receives direct but anonymous performance feedback on her skills as a manager from his subordinates, his own manager, and peers who are familiar with the person's work. This process usually occurs yearly or twice a year. The feedback received is not in the form of face-to-face discussions. It typically is written comments or ranked numbers (e.g., 10 is best and 1 is worst) on a preprinted form (computerized or hard copy) that has a series of questions and possible responses. However, the feedback also can be obtained by a neutral third party who interviews other people and records their responses. Obviously, confidentiality must be assured. Often, the responses are on a scale of 1–10, poor to good, or a similar scale, but the bottom line is that the questions are designed to evaluate the key management skills that the person needs in the workplace. Some of these skills may include her ability to communicate, her ability to listen, her trustworthiness, her overall leadership skills, her friendliness, whether she is willing to delegate tasks, whether she sets goals in collaboration with her direct reports, whether she is a good planner, whether she supports the company's culture, whether she is receptive to new ideas, and whether she is a team player.

After the surveys or interviews are completed, they are summarized (and often averaged) and discussed with the person being evaluated. This discussion can be with the neutral person who conducted interviews or collected the data, or a supervisor of the person being evaluated. At some organizations, the person being evaluated can see the results obtained for coworkers having the same job responsibilities (the identification of those coworkers is kept anonymous). Once again, the goal is to help the person improve her management skills, not to be used as an instrument for deciding salary increases or promotions.

One specific area where I believe that 360-degree feedback can prove helpful is when a new manager has been subjected to interpersonal mistreatment (i.e., incivility). Interpersonal mistreatment can include bullying by other managers, bullying by direct reports (which can happen), exclusion of the manager from social events, and not providing the new manager with needed information. Some new or younger managers may not have yet developed the needed political skills to avoid this type of mistreatment, and if the reasons for this abuse can be revealed through 360-degree feedback, so much the better. Once the reasons are known, the new manager can use developmental methods to help change her interpersonal interactions. Methods can include receiving mentoring, formal coaching, role-playing, and other forms of managerial training designed to increase managerial effectiveness and the personal well-being of the manager [22].

The Most Difficult Conversation of All: Apologizing

At one time or another, all of us have been privately or openly humiliated, or wrongly accused by our parents, a spouse, a friend, a teacher, a boss, or others. It hurts, and very few of us can simply smile and shrug off such an incident. Now, place yourself in the position of being a manager who has unintentionally humiliated or wrongly accused another employee and you recognize that what you said was wrong both in fact and in the manner it was said. Should you apologize? It may seem like the ethical and proper action to take (which it is), but that is not the perception of all managers. A few years ago, I was at a management meeting in which we were discussing the topic of apologizing. One of the supervisors present said, without hesitation, that a good supervisor should never apologize because it shows weakness and would lead to a lack of respect from the people she supervised. The individual leading the discussion was Dr. Aaron Lazare, a noted psychiatrist who wrote and lectured about the importance of apologies in medical practice. Lazare acknowledged the supervisor's viewpoint but offered a different view—that an apology shows strength, maturity, and self-confidence, not weakness. Weakness is a fear of apologizing, of not being able to admit that a mistake was made.

A person who has been humiliated or otherwise wronged does not readily forget the incident. They think about how they allowed it to happen and what they could have done differently. They may think about it for months or years. Some may consider revenge, while others may become distracted from their work [23]. This is not how we should treat a coworker or any other person, and it highlights the need to try to right the wrong, even if just a little bit, by a sincere apology. But for many people, apologies are not easily made, or perhaps it's more accurate to say that many of us do not know how to make a sincere apology even if that is our goal. Lazare and Levy [23] describe several aspects of apologies that resonate positively with those on the receiving end. Sincerity in what you say is the singular most important part of an apology, and an honest expression of remorse is almost as important. For example, you might say, "I'm really sorry for what I said. It was foolish and I should have used my brain before I used my mouth." An apology should ask for forgiveness (e.g., "I hope you will forgive me because what I said and what I know are not the same"). And you should acknowledge the offense ("When I said you were a dope, I should have realized that I was really the dope"). It has been suggested that once you properly apologize, there is no need to delve into any additional details about why you are apologizing, as you only risk saying something that will dilute your apology [24]. Needless to say, once the apology has been made, it is incumbent upon you to not repeat the mistake.

Organizations also may have to apologize; more often than not, this occurs when there is a product or service failure. I will not go into further detail about this (it's beyond the scope of this book), but the same principles hold. That is, the organization's spokesperson (who is usually a senior executive) has to be sincere, remorseful, clearly acknowledge the problem, and indicate a desire to change so the problem does not recur.

Apologies can be difficult conversations at first, but they can also help mend broken fences and be emotionally cathartic for the person making the apology.

Here are some key points about difficult conversations.

Approaching a Difficult Conversation

- Focus on the core (root) of the problem.
- Approach the problem as an unbiased observer.
- Do not be accusatory; rather, use questions and suggestions.
- Clearly express your concerns and how they affect you.
- If a single meeting does not resolve the issue, next discuss the pattern of activity that is occurring.
- If that still does not resolve the issue, discuss your relationship with the individual.
- Do not assume that your point of view is the only valid one. Your subordinate may have a different perspective than you have, but it is nevertheless reasonable.
- Make a sincere apology if you have made a mistake.

Difficult People

In laboratory animal facilities, as with life in general, there are going to be people with whom we interact but consider difficult to deal with. They may be so difficult that there is little opportunity to have a conversation with them. They often are outwardly hostile (such as an enraged investigator whose experiment and animals were wasted because the incorrect diet was fed by your staff). Others may not be outwardly hostile but may be passively aggressive (a second in command who wants your job and tries to undermine you behind your back), throw-you-under-the-bus people (the same second-in-command person who undermines you in front of others), workplace bullies, and so on. A lack of civility is the common thread with many of these people. Many readers would like to know, "If they do [enter your favorite gripe here], how should I respond?" I'm going to disappoint some of you by not addressing that question directly because "how to" is not the key intent of this book, and there are so many different variations on the theme that a fairly large book can be written on the subject (and many have been). One thing is for sure; when incivility raises its ugly head, you have to act decisively to stop the problem from spreading. One research group found that 25% of people who have been treated uncivilly will intentionally cut back on their work effort [11].

A recent book that I found easy to read and quite useful was Gill Hasson's *How to Deal with Difficult People* [25]. I will give you some generalities that come from Hasson's book, supplemented with my own training and experience in directing vivariums. I've already mentioned some of these items elsewhere in this chapter, but it doesn't hurt to repeat them.

■ Keep calm. Staying calm is a learned skill, and it's often not easy to do in a tense situation. Nevertheless, staying calm will help you make a more rational decision [26]. At the very least, even if you are not calm on the inside, try to project a calm appearance on the outside. Hiding your true emotions, as opposed to dealing with them, should be the exception, never the norm, but about the worst thing you can do when in a difficult situation with a difficult person is to lose your composure and act out your annoyance or frustration. It usually has the same impact as pouring gasoline onto a fire.

■ If a civil discussion is not possible, for whatever reason, say that you would like to discuss the situation when everybody is a little calmer and walk away, if necessary. Do not expose yourself to possible violence.

■ If a civil discussion is possible, discuss the action, not the person, in order to help calm the waters. For example, you might say, "I heard that the wrong diet was fed to some of Dr. White's mice. Did that actually happen?" That's much better than saying, "You fed the wrong diet to Dr. White's mice. Right?" The earlier discussion about difficult conversations provides additional depth on this topic.

■ Listen attentively and make sure you have the facts right. Repeat back to a person what he or she is concerned about "just to make sure I have it right."

■ For passive–aggressive problems, explain the consequences of the person's actions. If, for example, there is a supervisor who typically agrees to do something and then intentionally does something else, let him know what can or will happen to him if that behavior continues (and be sure to follow through if it does continue; otherwise, it will become worse).

■ Allow passive–aggressive persons to voice their opinions, but once a possible agreement is reached, document that agreement.

■ For many problem people, such as those who are incurable bullies or continually negative people, either you or they may have to leave. That's not the best solution for our egos, but it may be the best for our mental and physical health.

Consider Having an SOP for Difficult Conversations

So far in this section, I've provided a series of suggestions on how to handle difficult conversations and related kinds of conflict that involve animal facility employees. Sometimes, a little conflict is good in that it brings legitimate but opposing opinions to the forefront, where they can be openly discussed and resolved. Sometimes, conflict is bad if a festering problem is not resolved or if tempers are about to explode. Interestingly, many animal facilities have SOPs for dealing with almost anything under the sun, such as cage changing, entrance into the animal facility, anesthesia machine maintenance, and euthanasia methods. My own animal facility also has SOPs for many of our business operations, a vacation policy, and even a policy on travel to professional meetings. But the one SOP we don't have is a policy on the general steps we will take when a difficult conversation is required. Since some IACUCs have a general policy on how to investigate allegations of protocol noncompliance or animal abuse, it seems reasonable for an animal facility to consider having a general policy on how to approach difficult conversations that arise during the daily operations of the facility. As most of you know, many human resources departments have a standard approach to conflict that usually begins with a verbal warning to a person if prior informal discussions did not resolve a conflict. This then advances to a formal written warning if the verbal warning does not resolve a problem, and then a last-chance final warning, which is the last warning before a person's employment is terminated. Unfortunately, this entire process is adversarial in nature and doesn't help the manager or the employee with the actual interactions that they must have with each other to even try to resolve the problem in a nonconfrontational manner. So let me make a suggestion: consider having an SOP or similar departmental policy for difficult (or crucial) conversations that acts as a template to be used by any of the employees of an animal facility. It might read something like this:

Standard Operating Procedure: Conflicts in the Workplace

The following general procedures for conducting civil conversations when a conflict arises reflect our department's belief that every person should be able to voice his opinions or concerns in a respectful, nonconfrontational manner and receive feedback in a respectful, nonconfrontational manner.

1. *There will always be some conflicts at work. If you do not want to address the conflict in a timely and direct manner, it is better to drop the issue so that it does not become a chronic frustration and make you more and more resentful. Conflicts are best resolved personally, not via phone or email.*
2. *Do not engage in a potentially difficult conversation until you are calm and collected. Remain calm and respectful throughout your meeting.*

3. Base your discussion on facts, not what you heard in a hallway or locker room.
4. Do not criticize the person you are talking to; criticize, if necessary, the person's action.
5. Be prepared to explain how the action has affected you (has it impacted your job, or is it a more personal hurt?). Do not speak for others just because you assume they will agree with you.
6. Whenever possible, bring forth one or more suggestions to resolve the problem. Have you tried anything so far to resolve the issue? Try to find a win–win solution, but be prepared to compromise if that is the best outcome available.
7. Make sure that all involved persons have a clear understanding what future steps (if any) will be taken to resolve the problem and how and when they will be done.
8. If additional meetings will be needed, they should be scheduled at this time.

Communicating with Your Own Supervisor

It's important for you to keep your supervisor apprised of significant events or trends that could affect his or her work. That's good politics on your part and good information for your boss. In animal facilities, many of the senior managers and all of the veterinarians have a science-based background. Therefore, you can make an initial assumption that your boss is an analytical thinker who expects you to provide him with data that you have gathered and analyzed, or at least attempted to analyze. If there is a problem, he will expect you to provide him with a succinct summary of the issue and possible solutions you tried or would like to try. Your presentation should be logical and succinct because *you* may know the entire story and it may be very clear to *you*, but if your boss hasn't heard it before, he or she needs to hear it clearly and concisely. Putting sugar on a problem or twisting words around usually results in more harm than good. While these suggestions may seem like common sense, some managers try to handle everything by themselves, without letting their boss know what is happening. Taking the initiative is fine (and often expected), but you still have to let your boss know about things that can affect her. In practice, you may find that your boss may want to know everything, very little, or something in between. You will just have to learn your supervisor's management style and work with her. I'll discuss this some more in just a moment. No matter what management "style" your supervisor may have, you will find that one thing is the same with all supervisors: they don't like surprises (being blindsided) any more than you do. Don't wait until a situation is out of hand before talking to your supervisor. In fact, if a problem is starting to build and you haven't been able to handle it yourself, it's almost always a good idea to converse with your supervisor before the problem gets out of hand. Otherwise, giving your boss aggravation is going to cause aggravation for you, and your value to your boss is going to diminish. My suggestion is, if possible, to try to develop one or more solutions to your problem and discuss them with your boss. That demonstrates your initiative and insight into the situation.

If blindsiding your boss isn't enough of a mistake, here are some more foolish moves you might not want to make in front of your boss. Try coming unprepared for a meeting and stumble over his questions. Not enough? Here's a related one. When your boss asks you for facts you don't have, make up a story and hope she buys it. On a more serious note, you have to appreciate that you are working with intelligent people, most often scientists trained to ask significant questions about significant problems. So, if you tell your supervisor there is a desperate need to come up with $10,000 for rat cages, you'd better be ready to justify the need and the cost. If you tell your boss that rabbits and guinea pigs should not be housed in the same room, you should know why.

The suggestions I just made about providing accurate information, not blindsiding your boss, and being prepared to defend your statements are generic and should be considered as baseline actions every manager should take with every boss. There are additional communication and other interactions with your supervisor that can also be considered generic. Here are just a few of them [27].

1. *Get involved.* This was discussed above. It's a matter of taking the initiative when you see a problem, rather than waiting for your boss to handle it, waiting to be assigned to it, or waiting for it to escalate before you speak up. You can also take the initiative for the larger organization where you work, not just the animal facility. As I'll note below, there are some bosses who might want to use you as a pawn and expect you to take action only when you are ordered to do so, but most supervisors are more realistic and will appreciate your initiative.

2. *Come up with ideas.* Part of your job as a manager (as it is for every employee) is to move the animal facility forward by providing suggestions, not waiting until your boss gets an idea and asks for your input. This goes hand in hand with the concept of getting involved. The worse that can happen is that your idea is rejected, but at least you have demonstrated an effort to contribute.

3. *Collaborate with others.* It's not unusual for animal facilities, particularly larger ones, to be divided into sections such as animal care and veterinary services. These are not mutually exclusive kingdoms, and your supervisor has a right to expect that the interactions between two or more groups within the animal facility will facilitate, not hinder, the facility's operations. Results reported from one study indicated that only 9% of managers stated that they can rely on their managerial colleagues all of the time, and only 50% reported

that they can rely on other managers most of the time. This can easily lead to dysfunctional behaviors that can undermine ongoing work [28]. Wherever you find yourself, be a facilitator for the entire animal facility or your entire organization, not a recluse within your own immediate area of concern.

4. *Stay current.* In Chapter 2, there was a discussion about the various environments that impact an animal facility (e.g., economic and information). Your boss has every right to expect you to know what's happening in the field of laboratory animal science. For example, I expect the associate director for animal care to know more about animal care issues than I do. When I have an animal care question, he should either know the answer or be able to get an answer quicker than I can. I invariably tell new veterinarians that part of their job is to be current with the laboratory animal veterinary literature. In fact, that is written into their job descriptions. The need to stay current is reinforced at our weekly veterinary meeting, where we often discuss how recent literature findings might be incorporated into our daily activities.

Cardinal Rules for Communicating with and Enhancing Your Value to Your Supervisor

- Never blindside your supervisor.
- Provide accurate information.
- Don't let a problem get out of hand before discussing it.
- Be knowledgeable about the subject matter.
- Don't make up answers when you don't know an answer.
- Be prepared to document your comments if you are challenged.
- Take the initiative when it is reasonable to do so.
- Collaborate whenever possible with others in your organization.
- Stay current with the management and laboratory animal science literature, new products, and so forth.
- Adjust your communication style to fit your boss's management style.

Communicating and working with your boss is a two-way street. Your boss expects many things from you, including good management skills. In turn, you should expect your boss to be respectful of you, support you, give you constructive feedback, mentor you if necessary, and provide you with recognition when it is due. You also have a right to be included in certain decisions that affect you or the resources you manage. If you find that you are not getting the support and other items you need from your boss, I suggest you first perform an honest evaluation of your own performance to see if you are providing your boss with what he or she expects of you. This can include getting feedback from your work colleagues. Of course, there are some bosses who are easy to talk with and who are willing to try to resolve any perceived problem, but unfortunately, there are some with whom it's hard to have such a conversation. If the problem cannot be resolved by either a change on your part or a change on your boss's part, it may be time to think about getting a different position.

Accommodating Your Boss's Management Style

All managers have their own communication style that is a part of their overall management style (management styles are discussed in the next section). In the big picture, I would like to believe that all bosses, myself included, look at our subordinates as coworkers, not as *subordinates*, *direct reports*, or any other such term that suggests rank. (Sometimes in this book, I use the word *subordinate* simply to clarify my intent.) But, in the real world, this just isn't what happens all of the time. Some supervisors, for example, demand that their status be noticed, so we have to deal with it. This may not be easy, but we have to try. Some supervisors want you to do the job by whatever means will get the job done; others don't want you to act until you are told to do so. I'm one of the former: a "bottom-line, get the job done" person. Within reason, I'm more concerned about end results than the methods you used. My attitude is that I don't want to be bothered by knowing every detail about everything that is going on, but I do want to know the general situation. You get paid to do a job, and you have the training and experience to do it, so do it. I'll help you set goals, keep you informed, run interference, and support you, but if you have a problem, try to solve it yourself. If you can't, I'll try to help, but you can be sure I'll ask what steps you have already taken and why they didn't work. If you're the kind of person who needs all kinds of emotional support and managerial direction, you and I will have to strike some compromises.

Management Styles

Usually, a subordinate learns fairly quickly about a boss's communication style. Since different bosses may have different styles, the shrewd subordinate will not use a communication style that worked well in a previous position if it doesn't mesh with the communication style of his new boss. Here's an example of what can happen when conflicting management and communication styles come together:

> *Bill Williamson considered himself a no-nonsense type of manager. If you came to him with a problem, you had better be able to summarize the situation in a few sentences and tell him what you already did or wanted to do to solve it. His working*

philosophy was that you should do what you get paid to do. He would help you if you hit an absolute roadblock or if you needed his permission to proceed.

On the other hand, Kenny Roberts, who worked for Bill as the manager of his microbiology laboratory, hated to make decisions. He had much strength as a manager, but being decisive was not one of them. Williamson knew this, and he gave Kenny a little more leeway than he did with other people.

Kenny had to meet with Bill to discuss buying a new microscope. Before he walked into Bill's office, he knew that he had to have everything ready for Bill's bottom-line questions. He spent five minutes telling Bill why he needed a new microscope, and proceeded to give him a detailed breakdown of the comparative costs and features of different models. He then began describing the accessories that he could add. Kenny believed that this was the bottom-line information Bill would want. In reality, it was his standard approach to everything, and he probably wished that it was Bill's.

Bill was becoming somewhat agitated with what he considered to be superfluous detail, but he let Kenny go on. Finally, he stopped Kenny in midsentence and asked him which microscope he wanted. Kenny once again started speaking about cost versus value, but he would not commit himself. Williamson reached his limit. He told Kenny that he had no interest in making decisions for him, but because some sort of decision was required, he should consider repairing the old microscope rather than buying a new one. With that, the meeting ended.

Both Kenny and Bill knew the other's management style. Bill was a little more tolerant of Kenny than he would have been of some other people, but Kenny just used the time to give facts and figures that Bill wasn't interested in. Bill was right in not wanting to make the decision for Kenny. Whether he should have been so curt with Kenny is another issue. Kenny, on the other hand, should have known that eventually he would have to make a decision, and therefore, he should have considered all his options before going in to speak to Bill. Now he had two problems—he had to consider fixing the old microscope in addition to buying a new one. Nothing was accomplished at this meeting.

This example highlights Bill's bottom-line management and communication style. Other supervisors want you to keep them informed every step along the way, and they want direct involvement in every aspect of everything. This approach is often followed by new managers who have been promoted from the ranks. They want everything to be done their way, because whatever technical tasks they did before becoming a manager seemed to have worked. Now, they don't want anybody doing anything that might not be as good as their way, or might cast a shadow of doubt on their managerial ability. They will tell you what to do and how to do it. The word *trust* doesn't fit their managerial vocabulary. They waste valuable time constantly checking up on you [29]. This is what many people call micromanagement. Given my own prejudices, I don't like this management style. Still, you cannot dismiss it.

Some people want to speak about the weather or their family before they talk business. Then, there are some managers who would prefer that you talk very little; they want most everything in writing. You will simply have to know your boss's style of management and communication. You have to either work within the system to change it, accept it, or move on. If you understand how your supervisor communicates, it will be substantially easier for you to communicate with him. In the long run, your own job will be much easier.

Let me give you another absolutely true example (except for names) of a communications breakdown between a boss and a direct report. Both are people with professional degrees.

> *Sheila Roubideaux is a veterinarian having responsibility for many activities at Great Eastern University. She is friendly, articulate, and likes her job. Her boss, Larry Gant, is equally friendly and articulate and wants Sheila to succeed. Larry isn't a micromanager, but he wants to know what is happening in his department, so he frequently sends emails asking for brief written updates on specific activities. Sheila doesn't mind keeping Larry informed, but she hates communicating by email; she would rather Larry call her or come over to her office. She doesn't understand why this doesn't happen since it is such a simple thing to do.*
>
> *Communications begin to break down. Larry keeps sending emails, and Sheila doesn't respond. Nor does she give Larry an honest answer when he asks her about it. A small problem became magnified in Larry's mind, and eventually Sheila lost her job over another incident in which communications failed. It was the straw that broke the camel's back. Perhaps both Larry and Sheila should have shared some of the blame, but Sheila reported to Larry, not the other way around. You have to learn how to communicate with your boss.*

By this time, some of you must think that to communicate with everybody you work with, you have to be some sort of a chameleon. To a certain extent, that's true, as long as you are honest with what you say and just altering the manner in which you say it. Your style as a communicator, whether it is with your boss or others, may take some practice. You have to keep trying on different styles, like new clothes, until you find a good fit for yourself and your circumstances [30].

Recognize that you may have certain strengths that your supervisor lacks, and vice versa. You may be excellent with financial details, a trait appreciated by your supervisor, who is not a detail-oriented person. Your supervisor may be a whiz at time management, an area that is anything but your strong point. In most situations, you and your supervisor can complement each other. Open communication and a consistent managerial style can strengthen both of you. Similarly, you can benefit from the strengths of the people who work for you. Perhaps they are adapting to you in the same way you are adapting to them. That's fine; let them. It's a mutually beneficial arrangement, and it matters not one iota if everybody has to do a little adapting to everybody else.

Individual Management Styles

From the foregoing discussions, it should be apparent that everybody has his own management style. It's like people with different personalities. There are many published articles that expand on this subject and place management styles into classes, subclasses, and so on. I'm much more basic. The management styles that I have seen most often in laboratory animal facilities (and the workplace in general) are the *micromanager*, the *dictator*, the *mouse*, and the *compromiser*. I'm not about to do a psychoanalytical profile on each of these styles, nor could I, but I do want to describe them briefly so you are aware that they exist. These management styles describe how a manager interacts with people (primarily their subordinates) on a daily basis. They are personality styles as much as they are management styles. Although different people have different managerial styles, they all perform the basic managerial functions of planning, decision making, directing, controlling, and organizing.

I described the micromanager earlier and said I did not like this style because it signals a lack of trust. This is probably the worst management style for interacting with most other managers and veterinarians because much of their work depends on their own expertise and experience [31]. The micromanager may be a perfectly lovely person otherwise. There was an interesting published interview with Teresa Amabile (who is well known for her work on creativity in the workplace) in which she noted that subordinates may not want to be micromanaged, but they do want a certain amount of monitoring and support. They want their boss to monitor a project without becoming too involved, yet ask for their opinions about issues for which the boss might be able to provide them with help [32].

There may be times when micromanagement is appropriate. For example, some people micromanage when they first take a new position or are training a new person. When I do that, I tell people up front that I am simply trying to learn (or teach) what has to be done, and pretty soon I'll leave them alone.

If I don't care for the micromanager, I like the dictator even less. This person is the type who manages by raw power. This power can be derived from his position (e.g., a vice president or an animal care supervisor), his financial authority ("If we can't do it this way, I don't think I'll be able to fund it"), his physical ability ("Do it, or I'll pound you"), a combination thereof, or other aspects of power. The dictator doesn't always tell you *how* to do things (like a micromanager); rather, he tells you what is going to be done. There's no discussion, no compromise, no anything. There may be some good short-term results, but reaching long-term goals is probably out of the question because people resent being manipulated and controlled [33]. Authoritarian managers can lead to a greater likelihood of an employee leaving your organization [34]. If this person is your boss, and you object to this style, you will have to either work things out or move to another position. As one insightful writer noted, "Give someone monarch-like authority, and sooner or later there will be a royal screw-up" [35]. However, if this person reports to you, you're going to have to meet with him, tell him what others think of him, and describe how his behavior is (probably) affecting morale.

Once again, this type of person may not be inherently bad. He may truly care about the people he manages, and he might even socialize with them after hours. But, during the workday, it's another story. He thinks that you have to bulldoze people to get the work done.

Interestingly, there may be times when the dictator's managerial style (like any other managerial style) has its place. Consider a laboratory animal facility that is in trouble. The investigators are unhappy, upper management is unhappy, the U.S. Department of Agriculture (USDA) just cited it for multiple violations of the Animal Welfare Act regulations, the AAALAC International just revoked its accreditation, and the National Institutes of Health (NIH) is hinting at a not-so-pleasant site visit. This may be where you need a person with a firm hand to get things back on line.

The next example is the mouse. In my experience, the mouse is often a manager who has been moved up the ranks because of technical excellence, yet has limited or no managerial experience and is not or has not been willing to become a better manager. He is afraid to say most anything to you or those he supervises, because he might be saying the wrong thing or getting people mad at him. He always needs your approval and, in the process, can easily make his problems yours. When he does have to discuss a sensitive subject with those he supervises, he is prone to phrase it so that it seems he is little more than a messenger. This person may have excellent potential, but that potential should have been cultivated with managerial training programs before the person was promoted into a managerial position. His staff will quickly learn that the mouse is not much of a leader.

The compromiser (or consensus builder) is commonly found in managerial positions. This is the person who tries to find a middle ground that keeps everybody happy, yet can move work forward. This is not a bad way of doing things, and is often the best way, particularly if the manager is supportive and trustworthy [34]. Still, there are times when a manager has to take a firm, yet proper stand on an issue. Some compromisers refuse to do this and lose the respect of their staffs. Still, in most instances, building a consensus and compromising where necessary is a valuable management style. On the whole, I think it's the best one.

The importance of management style and the need to be able to change one's style when required by circumstances was underlined by a 2008 article in the *Boston Globe*. It concerned a physician who stepped down as chief of surgery at his institution because of his management style. The doctor said, "I agree I have a certain style." Surgeons "have very strong egos and by nature are not collaborative." And it requires a firm hand to get them to "fly in formation." "You have to switch styles when you go outside of the department and deal with nonsurgeons. That is very difficult to do, and I don't do it very well" [36]. It's refreshing to hear a person speak with such honesty. It's unfortunate that his lack of adaptability cost him his job.

Is there a management style that is uniformly to be avoided under most circumstances? Yes, and the answer is inconsistency. It's not even a style; it's a lack of style. Inconsistency characterizes the manager who wants to be involved in all the details one day, and then tells you to do it your own way the next day. After you do it your way, you get reprimanded for not keeping her informed every step of the way. Here's a concurring excerpt:

> The lack of day-to-day consistency in management expectations can make the most loyal employee hostile. Random responsibilities, conflicting instructions, and multiple supervisors with different measurements of success, all contribute to a workplace environment that undermines employee satisfaction and breeds frustration. [37]

Ronald Pickett describes a variation of the inconsistent manager, which he calls the psychobarbarian manager [38]. These are managers who instill self-doubt in employees by sending mixed messages. They like to keep people off balance. They fire employees quickly and often. They want to know everything that is going on outside of their own department, and they pay little attention to facts. In general, they exhibit paranoia. The psychobarbarian is one kind of manager from whom you want to stay far, far away.

If there is one lesson to learn relative to your own style, it is to be consistent as much as possible. There are certainly times when you have to change your management style to match the needs of different coworkers, but that doesn't happen every day. Your staff and your own supervisor want to know how to communicate with you and how you will likely react to a given situation. It's strongly complimentary to you if your staff says, "She's going to hate this," rather than "She'll hate it or love it, depending on which way the wind is blowing."

Be consistent in your management style.

Trust

Whether a person has a PhD or a high school education, it is crucial for us to learn to trust the people with whom we work and to have them trust us. I was distressed to read that about half of all managers don't trust their leaders [39], and so I have to question how many nonmanagers likewise distrust their managers? If it's also about half, that doesn't say much about the management of many organizations. One reason for engendering managerial trust is that without it, we cannot perform one of our most important managerial functions, the delegation of responsibility. You become the micromanager I described earlier. You work 12-hour days. You lose the respect of your staff. You become inefficient and ineffective. People don't want to talk to you and communications break down.

One possible reason for the occasional breakdown of trust between a manager and a subordinate may be that the concept of trust can differ between people. One study found that managers perceived trust as being related to the delegation of duties. Thus, managers trust a subordinate who makes herself available for delegated assignments, is receptive to the manager's needs, and shows good discretion. Subordinates viewed trust somewhat differently, putting more weight on a manager's competence, openness, and discretion [40]. The take-home message is that the wise manager understands that her needs and her coworker's needs may not be the same, and some tactful accommodations will help build trust between the two of them. For example, when delegating responsibility to a person, it can help if you, the manager, fully and clearly explain why the work is being delegated and demonstrate that you have a reasonable working knowledge about the delegated project.

In his fine essay, Gordon Shea notes, "Trust is the 'miracle ingredient' in organizational life—a lubricant that reduces friction, a bonding agent that glues together disparate parts, a catalyst that facilitates action. No substitute—neither threat nor promise—will do the job as well" [41].

Shea suggests a number of guidelines for people to use to build trust with their staffs. They make good sense; I use them, and I would like to pass some of them on to you in a modified form:

1. *Analyze the work you have, try to match the person to the task, and set high but realistic expectations for the work to be done.* I'll say more about realistic expectations (i.e., realistic goals) when we discuss employee motivation. At this time, though, I hope you can see the plain logic in matching the person to the task. Let people who enjoy working with primates work with primates.

If they do not have the required skills, work with them to develop a timetable for getting those skills. Don't put people in positions where you know there is a strong probability that they will fail. If and when they do fail, they will totally lose their trust in you.

2. *Train your staff to do a job in one acceptable way, define what a satisfactory outcome will be, and when they can perform the task in a competent manner, allow them to be innovative in performing the task.* This point is good general advice because, once you know what a satisfactory outcome is, you have the opportunity to be creative and find more efficient or effective ways of getting a job done. But, warning flags must be raised relative to the operation of laboratory animal facilities. Indeed, there may be more than one correct way to perform a task. Yet, in an effort to minimize any experimental variables, laboratory animal facilities frequently require that day after day, month after month, experiment after experiment, a particular task (such as blood sampling) be performed in the same way. And, as many of you know, we often codify these ways into SOPs. Therefore, you should check with the investigator before changing the way in which a task is performed. In addition, if you cannot change the manner of performing a particular procedure, let your employee know why. Keep that communication line open. I strongly recommend that you do not discourage employees from suggesting new ways of doing things (do just the opposite and encourage them), but they must understand that they cannot take it upon themselves to make changes without your concurrence. This is obviously a difficult concept because, on the one hand, we want to encourage employees to think for themselves and become creative team players, but on the other, we are inhibiting flexibility. The key is communicating why we don't do things in a different manner even if the end result is the same.

In recent years, some schools of managerial thought have advocated giving employees a good deal of freedom in making their own managerial decisions and then implementing those decisions. There may be room for this concept in laboratory animal science (and later on in this chapter, I'll discuss self-managed teams), but it is not quite as easy to implement as in other professions because of our need for consistency in the procedures we perform. For the present, I suggest that discretion should be the better part of valor.

3. *You should use your authority to help, motivate, train, and set good examples.* You should rarely use your authority to coerce people. Coercion is the hallmark of an authoritarian manager (the dictator I previously described). He orders you to do something rather than asking you. He makes it clear, directly or indirectly, that you are there to take orders, not to be a team player. Granted, there are times when coercion will be needed, but those times, I hope, will be few and far between.

If you have helped train someone, you should not be afraid to delegate responsibility to that person. In fact, delegation goes a long way toward building trust because you have essentially told the person that you trust his or her abilities. Of course, make sure that you are very clear about the limits of your delegation. You're delegating to someone a certain defined amount of responsibility, not your entire job. When you delegate, also delegate authority, and make sure others who work with you understand that you have given responsibility and authority to that other person. It can be hard to complete a task without the means to do it. This discussion on delegation will be expanded in Chapter 6.

4. *Mistakes will be made, especially if you want your staff to be innovative. To punish every mistake will obviously discourage innovation and lead to a lack of trust. Most important, future mistakes will be covered up.* In laboratory animal facilities, years of research and untold amounts of money can be lost if there is an unreported mistake, such as when the wrong treatment is administered to an animal. Even worse, the wrong results may be obtained without the investigators even realizing it. Can you imagine the consequences of such a cover-up? Don't be overly concerned with punishment. Your goal is to correct the mistake and make sure it doesn't happen again.

5. *Along these same lines, you should not waste undue amounts of time finding out who the guilty party is.* Rather, you should point out what the consequences of the error are, illustrate the desired outcome, and point the way toward correcting it and seeing that it does not happen again. Here's an example of a situation with which I was involved. A good amount of time and effort was properly expended finding a solution to a problem, rather than on finding a person to blame.

> Many years ago, a whole rack of cages containing mice fell over. This was one of those narrow racks that you rarely see in use anymore. A piece of string got caught in the wheels while a technician was pushing it, and down it went. There were nearly 200 white mice running around the floor. Even though the five mice in each cage were numbered from 1 to 5, there was no way of knowing which mouse came from which cage. As a result of this incident, we developed a system to identify not only the animal, but the cage it came from.

Although at that time we were not able to purchase wider racks, we were all alert to the possibility of the problem recurring. We were (of course) more careful, and we responded in the way we felt would best solve the problem. Putting the blame on the person who was pushing the rack, the person who bought the rack, or the company that designed the rack would have been without value. It is more important to prevent a problem from recurring than to try to find a person to blame. By doing this, you set an example of how to address a crisis the next time one occurs. You are opening the door to having people inform you of a problem, rather than hiding it from you.

Preventing a mistake from happening a second time is far more important than trying to find a person to blame for it.

Mistakes are inevitable and are made by people at all levels of an organization. It is better to plan ahead to establish systems that not only limit mistakes, but also limit the damage done if and when an error occurs. In the example of the fallen rack, we should have anticipated that their inherent instability could lead to a major problem if they were transported with animals on the shelves. We should have changed our work procedures. It was more of a management failure than an unintended and possibly unavoidable mistake by the technician.

Somewhere along the line, though, reality has to be faced. No matter how well you foster communications and trust, not everybody is going to want to tell you what's really on their mind or what's really happening. Suppose the rack that fell over had not led to any mix-up in animals. If you were the one who tipped it over, would you have told your supervisor? The answer is not that important. What is important is why you would or would not have said something. If you knew that your supervisor would raise havoc, the chances are you would say nothing. Why risk that kind of wrath? If you believed that your supervisor would work with you to ensure that the problem did not recur, you would be more apt to speak up because you wanted to, not because you had to. There will be some people who will say nothing no matter how supportive you are, but as managers, we try to build trust to limit the number of these people.

Just as a refresher about the use of resources by a manager, let's quickly review the actions of the manager in the fallen animal rack example. A problem was defined (poorly identified animals can get mixed up), a goal was set (creating a fail-safe animal identification system), and we worked hard to develop the system (a plan was made). In this case, we primarily used human resources (a manager and his staff) to come up with a solution, although we did use information resources (reading the literature and asking others for their techniques), and we were somewhat limited by time to come up with an identification system in case the problem recurred. The system we devised was one of special ear notches to identify the cage an animal came from.

Supporting Your Staff

Assume that the rack you were pushing fell over. The animals were mixed up and the researcher was fuming. He went steaming down to Dr. Peters, the animal facility director, and demanded you be fired for incompetence. You have been a fine employee and this was your first accident. What should Dr. Peters do? Among a multitude of possible responses, the bottom line is that he must support you. He has to gain your trust, and through you, he has to gain the trust of all his other employees. Even if you were clearly at fault, in general, he has to stand by you. Not lie for you, but support your past work, note that it was an accident, and try to calm the waters. You should never be hung out as shark bait. Trust has to be built up over time, and every step helps.

Should Dr. Peters continue to support you if you have repeated problems and seem to fit in like a square peg in a round hole? No. There comes a time when enough is enough, but that is not the thrust of this discussion. As a manager, you must care about and support subordinates to the extent that you reasonably can. Unfortunately, putting the blame on somebody seems to be a favorite pastime in some organizations. For some time now, I have facetiously told graduate students that the first rule of science is to "blame somebody else." The second rule is "take the credit for yourself." I suppose the third rule should be to disregard the first and second ones.

There are times when you have to support your staff not because something went as obviously wrong as a cage rack falling over, but rather because an employee tried her best to accomplish a goal, yet it simply didn't work out. That's something we've all experienced. In this context, and "assuming the failure is not attributable to something beyond her control," Marcus Buckingham suggests explaining failure as "a lack of effort, even if this is only partially accurate. This will put self-doubt in the background and give that person something to work on as she faces up to the next challenge" [42]. Although this may seem counterintuitive, Buckingham is simply suggesting that you *don't* say, "Look, Jean, I know you tried really hard and I'm sorry things didn't work out. I'm sure it will be better next time." This may leave Jean with some doubts about her own ability, potentially demoralize her, and perhaps make her fear starting a new project with a new goal. Rather, it would be better for you to show confidence in Jean and say, "Look, Jean, I know you tried your best and I'm sorry things didn't work out. I really think that with a little more effort, you'll make it work next time." By doing this, Jean's self-assurance is reinforced. And if Jean *does* reach her goal on the next project, the good manager praises her by telling her how good she used whatever special talents were needed on the project, not that she simply put in extra effort. We're focusing our praise on the specific areas of strength that Jean might have.

Building Trust through Listening

Earlier in this chapter, I discussed how important it is to listen to what a person has to say and, if appropriate, take subsequent actions. Listening is a critical part of providing feedback; we learn things by listening, and we show interest in other people by listening to them. But being a good listener has another important function: it can build trust by demonstrating interest in another person's opinions or expertise. As a manager who wants the trust of those you manage, report to, or provide with a service, you have to demonstrate honest interest in what they say. In other words, you have to be honestly interested in them. As an example, many of you know that researchers like nothing better than to talk about their research. Many researchers are very willing to give animal facility employees a nontechnical talk about their work and how laboratory animals fit into their research. What is one of the things *they* value most at the end of their presentation? The answer is a few questions that demonstrate people not only were listening, but also were interested in what was said. This can build trust toward the entire vivarium staff, not just those asking the questions. If you have any doubt about that, give it a try.

Motivation

As you know, part of your job as a manager is to do your best to ensure that your staff performs efficiently and effectively. Therefore, if an employee *wants* to perform efficiently and effectively, you can be somewhat assured that you and your company have taken some of the right steps toward successfully motivating that person. You have helped give that person the inner drive to reach a personal goal. You have provided a positive incentive to reach a goal. Let me put these thoughts together into a working definition of motivation in a business setting.

> **Motivation is any action taken by a manager (or an organization as a whole) that provides its employees with a desire to accomplish a goal.**

As you might guess, the incentives needed to accomplish a goal can be either tangible (e.g., money) or intangible (e.g., respect or trust). Hurley [39] reported that employees who worked in a low-trust environment described their work with words such as *stressful, threatening, divisive, unproductive*, and *tense*. In contrast, those working in a high-trust environment used words such as *fun, supportive, motivating, productive*, and *comfortable*. From this, we can draw an obvious conclusion: trust and motivation go hand in hand.

Your goal should be to keep motivation high. Since motivation is something the employee feels, not something that is forced upon him, the phrase "wants to perform" becomes that much more important.

> **Successful motivation of employees includes providing a work environment that engenders a desire to be efficient and effective.**

Unlike people, money, or information, motivation is not a resource, nor is it something you do while using a resource. Motivation, like communication and trust, is a core management and leadership function and is something that should permeate the workplace. It is an integral part of everything you do, of every resource that you use. To be formal about it, you should establish employee motivation as a continuing goal of management. Constantly consider ways to improve it. Use whatever resources you must. Plan, organize, direct, and do whatever it takes to reach that goal.

Motivating people is a continuing goal of management.

Motivation is not something that you can easily measure. It may show up as improved productivity or fewer headaches at the end of the day. (Appendix 1 describes some ways of measuring productivity.) You'll know motivation when you see it, and what you may see are actions such as people giving more than their job description requires, delivering extra effort when it's needed, and becoming truly focused on priority items [43].

Basic Requirements for Keeping Your Staff Motivated

Some jobs, such as cleaning dog cages (especially with the dogs in them) are downright difficult, repetitive, and sometimes dangerous. Pep talks will not change that. To begin to motivate that technician (or any person, for that matter), you yourself have to be motivated. If you are not, I hope you can give some Academy Award–winning performances, because it's going to be hard to motivate others.

If, in your opinion, your staff is completely unmotivated, start by being introspective and consider whether you are motivated. If you recognize that there may be factors that are limiting your own motivation, then you can develop personal goals and plans to possibly change those circumstances. Whether or not you realize it, you are a role model. People look to you for subtle clues on how to behave. If your work output is marginal and sloppy, why should you be surprised if your employees' work is marginal and sloppy? If your work is neat and of high quality, you have set the tone for expecting high-quality work from others, even if it requires change on their part. You should not assume that you or your staff is incapable of change. That's simply not true. When it is obvious to others that you are motivated, that will begin to influence those around you.

People will look at you as a role model. You must be motivated in order to motivate others.

Laboratory animal facilities, as previously noted, tend to be very hierarchal and do things in a set manner. Cages are changed the same way every day, we handle investigator complaints pretty much the same way every day, we prepare for inspections the same way, and so forth. These are the ways we accomplish our mission. In other words, even if we did not have written SOPs, we would function as if we did. Most of these procedures come from the top and filter down. The "top" can mean upper management or even middle management of the animal facility.

Think about where you work. Animal care technicians typically have repetitive work, and here is where a manager who is a strong motivator is worth his weight in gold. In larger animal facilities, there may be more than one full-time veterinarian. Although some veterinary work is repetitious, much of it is nonroutine and requires a self-directed and self-motivated person. Here is where an organization culture that encourages independence of thought and actions will help motivate a veterinarian [44]. If you are managing people who have repetitive work, how much input do people below you in the typical hierarchal vivarium structure have in setting certain their own work guidelines? If you are a supervisor of animal care, do you ask employees who work for you how they think certain tasks should be done? Or are you the type of person who defines the performance standards and then, and only then, asks for feedback? Yes, we do need standardization in laboratory animal science, but we also have to communicate honestly that we are enthusiastic about getting peoples' opinions and input *before* instituting a change. There's a good chance that animal technicians know more about how to do their job than the facility director does. Mundane work is mundane work, and whistles and bells will not change that too much. But when input has been seriously solicited from employees, discussed with them, and, when possible, incorporated into the daily work routine, motivation will follow. Your challenge is to minimize, not maximize, the mundane aspects of the job.

To keep our staffs motivated, we and our employers must be able to meet certain basic needs. Based on human research [45,46] and managerial experience, these needs are as follows.

Basic Needs for Motivating Employees

- **A reasonable salary and benefits**
- **Reasonably good working conditions**
- **Reasonably good supervision**
- **Respect and trust**
- **A feeling that the work being done is important**

These needs are not interchangeable. A high salary cannot compensate for poor supervision and respect, and trust alone doesn't put bread on the table [4,47]. How come job security isn't up on the list of basic needs? Job security is important, as is freedom to act independently (autonomy), the opportunity for recognition and promotion, and the ability to learn new things. But those items are not absolutely basic needs. They are items that fuel job motivation. In particular, I would argue that even if a promotion isn't a basic need, as managers, we must make it very clear to everybody just what promotional opportunities are available and what the requirements are for promotions. I'll talk about promotions as a reward in a little bit, but the *potential* for a promotion can be a significant motivator.

All employees should know about the opportunity for promotions and the requirements for promotions. The *potential* for a promotion can be a strong motivator.

Notice that I did not write that salaries must be very high for an employee to be motivated, or that working conditions must be superior. It is implied, however, that underpaid, underbenefited, overworked, and unrespected employees will not be motivated. Some people might argue that what is reasonable to me may not be reasonable to you. That's probably correct in some cases, but the concept of what is "reasonable" rests with the perception of the employee, not with me, you, or our organizations.

If an employee feels exploited, then that person's basic needs are not being met at that point in time. If just one person feels exploited, perhaps that person can be helped on an individual basis, using some of the suggestions offered below. If all of your employees feel the same way, there is a real problem. As Michael O'Malley pointed out, people may say they leave a job because of money, but in reality, they leave because of broken promises, abuses of power, and so on. People don't leave places where they love to work, not even for money [48]. Others have made the same point, noting that irrespective of what reasons people give for leaving a job, 88% actually leave because of workplace factors other than money, such as the job was not as expected, too few growth opportunities, stress from overwork, and too little feedback [49]. It is not always true that a well-paid person in a good working environment who is given respect, trust, and good supervision will be a well-motivated employee. However, the odds favor that this is true.

Let me elaborate a little more about basic needs. A *reasonable salary* is almost self-explanatory. It means that a person believes her salary is equitable based on her contributions to her organization's goals. Without a reasonable salary (and perhaps reasonable benefits), a person feels exploited, and an exploited person will not be motivated. I very much agree with what Rodd Wagner [50] wrote about wages:

> *A company that treats its people like widgets, that demands its people work as hard as they can for the lowest possible wages,*

> *gets the mirror response in return: people who want to work as little as they can for the highest wages they can get.*

By *reasonably good working conditions*, I am broadly referring to both the physical environment and nonfinancial support provided to employees. Impossibly long hours, hard work without a letup, intolerable noise without ear protection, extreme heat or cold, dangerous conditions without safety aids, a lack of adequate training, and a lack of learning opportunities are all examples of a work environment that is not conducive to keeping employees motivated. Still, the employee's perception has the final say.

> *Some years ago, there was a fire in Thailand that killed a large number of employees of a doll factory. Many doors were locked and many people could not escape the fire. Yet, the survivors wanted their jobs back, particularly if they could have safer working conditions. Their motivation came from economic necessity, not quality management, because there were very few other jobs available to them. By current American standards, we might argue that although the situation before the fire was better than nothing at all, it was still wrong, and some governmental agency should have stepped in and closed the factory until changes were made. But by the standards of the people working in the factory, they had jobs, and that was good. In other words, "better than nothing" was important to them. There was no union or enforced laws. The workers had to live and the factory was their livelihood.*

Reasonably good supervision is a key item because it is so variable and open to interpretation. It is not only dependent on the supervisor, but also on the expectations of the employee and the entire organizational culture. Some aspects of good supervision are clear-cut, and I have already discussed the most important ones (open communication and mutual trust). Good supervision also includes establishing clear goals and providing feedback as needed [51]. I have also discussed matching the person to the job, and in the pages that follow, I will review other aspects of good supervision, such as recognition of achievement through rewards and setting realistic goals. It's unfortunate but true that most people leave their job because they don't mesh with their manager [52].

Most people leave a job because they do not get along well with their manager.

We all want to feel that the work we do is important, and we can feel fulfilled when we accomplish one or more of our goals. This emotion may be the greatest basic motivator of all [53–55]. Not everybody has a job with the observable importance of an emergency room physician, but caring for animals that cannot care for themselves and being the voice of those animals is of tremendous importance. Just being part of an organization that develops drugs or new surgical techniques or studies the basic functioning of cells is important. Employees want to be proud of their institution's mission and their contributions to that mission. Unfortunately, not everybody can immediately grasp the importance of their work, and therefore, part of the responsibilities of a manager is to continually highlight the importance of animal husbandry and veterinary care and how the work that is performed by each person in the laboratory animal facility integrates with and contributes to the work being performed in the entire organization. This can be done, for example, at staff meetings, through newsletters, and by presentations made by researchers.

It seems to me that no matter what topic is discussed, *communication, respect,* and *trust* are involved. They are things we can all relate to. Nobody wants to be treated poorly—nor should they be—and we all desire the secure feeling we have when we know people trust us.

I'm intrigued by an aspect of trust that is promoted by Libby Sartain and Mark Schumann [56]. They write that an excellent way of motivating our staff is to consider a business (such as operating an animal facility) as a consumer brand. The same way consumers identify with and trust a brand such as Kleenex®, employees can identify and trust their own employer. For example, if we internally promote our animal facility as truly caring about our employees, and we deliver on that promise, we are establishing a culture of caring, which is one of those intangible rewards that keep motivation high. Sartain and Schumann point out that this can be a tremendous help in recruiting and retention. In practice, I've seen this work because some of our finest employees came to us through the recommendation of our other employees. Here are some thoughts on how a manager can demonstrate respect and trust [57].

How a Manager Can Demonstrate Respect and Trust

- Keep people informed about matters that can affect their jobs.
- Delegate responsibility.
- Involve staff in decision making.
- Establish mutually agreeable goals.
- Recognize achievement.
- Be attentive to people's needs.
- Don't make foolish rules.

You may have noticed that the above list does not include forming close personal friendships with employees. Sometimes, particularly in smaller animal facilities, it's hard for a manager *not* to form close friendships with the people she supervises. But friendships can lead to favoritism, or at least give people that perception. And, of course, it can blind your judgment when difficult problems (such as layoffs) occur. That's why, for years, I have been advising colleagues to keep an arm's-length relationship with people who report to them. There is a distinct difference between caring about your employees and forming personal relationships with them. I fully believe that you should ask about an employee's children, how their elderly grandmother is feeling, and the like. It's important to listen to what employees have to say, and I think it is important that you try to rearrange a person's work schedule if there is a pressing personal problem that he has to get resolved during the workday. That is not being a wimpy manager; it's being a caring manager. Nevertheless, I have yet to see an effective manager whose management style is, first and foremost, to be good friends with everybody. Similarly, don't get so involved with employees that you become the workplace psychologist. That's not what you are trained for [58].

Good managers care about people but focus on business.

What about not making stupid rules? This is a topic I'll discuss later in this chapter in the context of how to work with researchers, but as a generalization, it holds true for most people. Employees want simple and logical rules that apply equally to everybody, such as the way we teach people to grasp a rat or the way we gown up to go into a biocontainment area. Even if a person disagrees with a particular rule, he or she should know the reason behind it. Look around your own department and see if you have foolish rules that have been around forever and seem to irk people, but no one is taking the lead to question those rules. My suggestion to you is to take the lead in getting rid of those rules and let your staff know that they're gone. It will put you on a small pedestal and be appreciated by everyone.

After all is said and done, all employees must have their basic needs fulfilled, and that is the responsibility of management. If you and your organization cannot fulfill those needs, you are going to have an uphill battle trying to keep your staff motivated. Fortunately, most laboratory animal facilities are parts of organizations that adequately fulfill the basic needs of most of their employees. As managers, we have to go the extra step and ensure that the animal facility doesn't sit back and wait for the larger organization to act when we can make some changes by ourselves. We are the ones who set and perpetuate the management culture in the vivarium. It's our job to get it right.

Keeping Motivation High: Tangible and Intangible Rewards

Some years ago, I was talking to an employee about the marginal quality of his work. The conversation was not confrontational. After a while, the employee said that if he was paid more money, he would work harder. I countered that if he worked harder, I would pay him more money. This sort of conversation is not an unusual occurrence between a supervisor and an employee, although using 20–20 hindsight, I should not have been so curt with that person. But isn't it true that money is a motivator? It may motivate you to stay with your job a little longer (until another job that pays more comes along), but does it really make you a better employee? Using pay as the primary motivator can easily outweigh its benefits. Merit pay can undermine teamwork, encourage employees to focus on short-term results, and lead people to hone the political skills and personalities needed to impress their boss, rather than focusing on performance [59]. We can think about incentive pay in another way: the more incentive money one individual gets, the less there is for others, who are then demotivated. People think that if there are no clearly outstanding people, the better it is for them because there's going to be a bigger pot of money to share. Why would they want to help people get ahead in their job if, when they do, it hurts them financially [59]? The key is to have a workplace that isn't based on money alone, but pays a fair salary, has good supervision, does important work, and respects all employees for the contributions that they make.

Let's return to the employee who said he would work better if he got more money. Was the employee's request, and my counterrequest for him to do better work before paying him more, two sides of the same coin? Alfie Kohn has argued that there is no difference between punishing a person for doing the wrong thing or rewarding a person for doing the right thing [60]. Kohn believes that rewards, like punishment, lead to only temporary changes in behavior. Thus, not receiving an anticipated reward is like punishment. Kohn was focusing on financial incentive plans, and I personally experienced that feeling when a bonus I was promised did not materialize. I was depressed (no surprise here), and it actually had a negative effect on my performance. Kohn's viewpoint doesn't mean that people should never be rewarded. If that were the case, there would be no promotions and no salary increases. Kohn, like Pfeffer [59], is simply saying that motivation comes from treating people well and meeting their basic needs—not from hanging carrots in front of them. Thus, once again we find that meeting the basic needs of people (reasonable salary, adequate supervision, adequate working conditions, respect and trust, and emphasizing the importance of a person's work) is the foundation of motivation. If that foundation is not in place, all other means of motivating your staff probably will not have much impact. Once the basic needs of people are met, we can examine additional means of motivation.

Intangible Rewards

Earlier in this chapter, I wrote that motivation is typically based on two types of incentives, tangible ones such as money or extra vacation time and intangible ones such as respect and trust. You will quickly discover that most, if not all, people want some form of an intangible reward for performing their jobs. Three intangible rewards that are universal motivators are respect, trust, and feeling needed (recognition). They are far more important than any material reward, and are just an extension of the basic needs previously discussed. These three rewards are items that every good manager wants to give to a coworker whenever possible. When you think of your employees and superiors as coworkers or colleagues, rather than people who work for you, you will begin to appreciate this concept.

> **Intangible rewards that are universal motivators are respect, trust, and making a person feel needed.**

Positive reinforcement is another type of intangible reward. In his book *Other People's Habits* [61], Aubrey Daniels maintains that a key to keeping motivation high is to provide continual and strong positive reinforcement of desirable behaviors. He defines positive reinforcement as an action that follows a behavior and results in an increase of that behavior. (In fairness, there are those that believe intangible rewards are best given on an occasional basis [62], and even Daniels acknowledges that once a behavior begins to change for the better, it is appropriate to provide rewards less frequently [43].) Of course, you can reinforce undesired behaviors just as easily as wanted ones. If, every time your child throws a tantrum, you jump up and do whatever it takes (short of mayhem on your part) to stop him from yelling and stomping, you can be quite sure that you have just positively reinforced an unwanted behavior. That kind of attention, which is what tantrums seek to attract, just leads to more tantrums. On the other hand, if, as soon as the tantrum stops, you caress him, speak tenderly, and sit him on your lap (assuming that sitting on your lap is a desirable reward for him), you have positively reinforced a desired behavior. We have to ask why reinforcing positive behavior is a reward. The answer is that it is just another form of recognition, and perhaps we shouldn't even categorize it as a separate type of intangible reward. It gives people something they want for doing something right. When that happens, it is a reward. The keys to using positive behavior reinforcement are to know what type of a reward is important to the person and to implement that reward as soon as possible, preferably as soon as the positive behavior occurs.

> **A key to using positive behavior reinforcement is to know what type of a reward is important to the person.**

Most of us who work with animals know this, as it is a basic tenet in dog training. As soon as the dog properly responds to a command, we give her a biscuit, kiss, or pat on the head, or do whatever it is she likes. If we give her the biscuit 10 minutes later, she doesn't know why she got it, and it may even reinforce a bad behavior that she was doing at that moment. Understand, though, that people aren't dogs and a person's needs and perceptions may change over time, so a desired reward today may not be the same reward that is desired next year. This is where good general management skills are needed in order for you to know your direct reports well enough to discern what might be a motivator for a person at a particular point in time.

> **Reinforcing a desired behavior is a form of motivation.**

Changing a person's behavior through rewards can be difficult, and experience dictates that trying to change a behavior through punishments is even harder because everybody has a strong internal reason for behaving the way they do. Earlier in this chapter, there was a short discussion about finding the root cause of a problem. That's important because if we don't understand the true reason behind a person's behavior, it will be very difficult to develop methods to help change that behavior. For instance, if a person is routinely late for work, telling her to get to work on time will not help one bit if the problem is that her son's school bus doesn't arrive until 10 minutes before she's due to be at work. It's well worth your time to try to find the root cause of a negative behavior, at least by asking a person why she thinks the problem is occurring. We want to support a coworker, but we have no right to dig into her personal affairs.

It was already noted that not everybody wants money as a reward, including those who don't earn too much. To some people, having a champion to watch out for them is a reward, and they may adapt their behavior to be able to work with such a person. To another person, getting an office with a window is a desired reward, and one that person wants to maintain. You yourself can be a reward by managing in a manner that suits a person. Early on, when I discussed communication, I said that you have to adapt your communication style to the person you are talking to. That's an obvious part of management, and it helps people to feel comfortable with you as a manager. If being a little "chatty" with a person is a desired reward, you should do that, and do it reasonably often if it doesn't interfere with performance. Whatever it is, you will have to find out what's important to the individual. A simple, but often overlooked, way to do this is to ask the person directly or, if necessary, ask a friend of that person.

Now let me return to Alfie Kohn for a moment. Kohn observed that rewards, like punishment, lead to only temporary changes in behaviors. Aubrey Daniels agrees, and goes on to make a strong case that you have to constantly reinforce a positive behavior to have it remain. That's quite true, and I would add that along with using intangible rewards to help reinforce positive behaviors, it's helpful to review certain personnel policies (e.g., arriving on time) to see if they are clear and simple to understand and are not a contributing factor to negative behaviors.

Tangible Rewards

So far, this discussion has been about rewards that are largely intangible. Somewhere along the line, we all would like something a little more tangible, like a salary raise. We all appreciate a little something extra, whether it is money, tickets to a concert or sporting event, a dinner, a bigger office, a promotion, or most anything else. We still have to pay attention to the basics—respect, trust, adequate supervision, making people feel needed, and being positively rewarded (as with praise) for desirable behavior—but there are times when material rewards can help motivate people. Nevertheless, from my personal experience and that of others, I can tell you that giving more money to a basically dissatisfied person does no good at all. Therefore, I want to be very clear that the tangible rewards I will discuss have almost nothing to do with attempting to permanently change a person's behavior. In fact, it has been shown that once a person cannot qualify to get a reward (such as being out sick for a day and therefore not eligible to get a bonus for a month of perfect attendance), that person often reverts to his or her previous behaviors, which suggests that the reward did not increase that person's motivation to continue the initial positive behavior [63]. In contrast, the rewards I am referring to are for superior performances that add to the value of the service your animal facility is providing. Still, there are times when there is a fine line between tangible and nontangible rewards. I think we can all agree that reinforcing a positive behavior is a nontangible reward, but what about the praise I mentioned earlier? If we publicly praise a person's contribution, is that a tangible or nontangible reward? To me, it would be intangible; to you, it might be tangible. It really doesn't matter that much. What does matter is the effect on the person.

First, let me interject a word of caution. If you hand out tangible rewards too freely, they will soon lose their appeal. People begin to expect the reward, and soon more and larger rewards are desired. If, every year, irrespective of productivity, punctuality, or other parameters, your employees get a 4% raise in salary, the raise is no longer a reward. Rather, it's an expected part of the job. As a slightly off-base second example, think of a supervisor who always yells at her staff. Pretty soon, the employees get used to the yelling and pay no attention to it, so she has to yell even louder. Then they get used to that. After a while, the yelling means nothing to the employees because it has lost its meaning. The moral of the story is that, rather than having rewards continually handed out for most anything done, tangible rewards should become obtainable only through a special effort on the part of an employee. We see this in the business world in the form of commissions and bonuses. The salesperson is rewarded for what he contributes to the company, not for coming to work on time. In the laboratory animal facility, rewards can be given for reaching certain agreed-upon goals that add value to the facility's mission. An example of such a reward might be a trip to a professional meeting for an employee who initiated a program that cut costs and increased productivity.

Rewards that are given out too often lose effectiveness and are no longer special.

Although tangible rewards are important for motivation, we have to make the reward fit the action. That's because it can be demotivating to give someone a minor reward for a major accomplishment [64]. How do you think you would react if you just saved your animal facility $500,000 by negotiating a very favorable food and bedding contract, and in return you were given a $10 gift certificate as a thank you? Obviously, you would walk away scratching your head, trying to figure out the meaning of what just transpired. Maybe public praise and a personal thank-you note from the company's president would be more appropriate. The reverse is also true; we should not be giving a major reward for a minor accomplishment. That might even raise more questions than a minor reward for a major accomplishment.

But are there any practical limits to what can be expected of a person in order for that person to secure recognition or a tangible reward? If a person's extra effort is to lead to recognition, then the goal that spurs on that effort cannot be made too difficult. If it is, that person will probably not try to reach for an impossible goal.

Impossible goals can be of two types. The first is a performance goal that is so absurd that it simply cannot be achieved without lying, cheating, or something similar. For example, properly washing 4000 cages per day in a single rack-style washer is simply not going to happen, no matter how much of a tangible reward a manager is willing to give to a person who can reach that goal. The capacity of the washer and the time needed for each wash cycle will ensure that such a thoughtless goal cannot and will not be met. However, washing 400 cages per day may be a reasonable "stretch" goal for a person who currently can wash 325 cages per day using the rack washer. The second type of impossible goal is one that requires specific knowledge for the goal to be reached. A technician may not be able to wash even one cage a day in a rack washer if she doesn't know what a rack washer is or how to use it. As you can see, the challenge for the manager who wants to motivate a person to use extra effort to reach a goal is to make sure the person is reasonably capable of reaching the goal. "It is both foolish and immoral for organizations to assign 'stretch goals,' and then fail to give employees the means to succeed, yet punish them when they fail to attain the goals" [65].

Let's assume that a new technician has been taught how to change cages. Will this new technician be expected to change the same number of cages per day as a technician with 10 years of experience? In general, the answer is no. Research has shown that employees who have been challenged with goals fitted to their ability will proportionately outperform those coworkers in a "one-sized goal for everybody" condition [66]. Therefore, we should also consider rewards that are based on an individual's ability, not only "one-size-fits-all" goals that establish a reward for an entire group of people, regardless of ability differences within the group.

When we offer tangible rewards that are based on a person's ability, we help to avoid a potentially nasty problem. That is, if only our very top performers (the so-called A players) are able to earn rewards, then we might have the same person getting the "employee of the month" award every month, while other solid, but not superstar, employees get nothing. These latter people (the B players) are those who may get positive reviews from investigators, do a fine job of caring for animals, negotiate for better pricing of commodities, know how to get things done in the organization, and so forth, but at the same time, they may not want the pressure of trying to climb the workplace ladder. They are solid performers, but not superstars. Nevertheless, these are important people in your animal facility, and you should do your best to support and recognize their achievements and make sure they stay with you. Don't forget about them! They also need the periodic recognition and challenges that will motivate them to reach their potential. When people work at the upper limits of their ability and meet reasonable goals, they also should get a reward, even if it's just public recognition of effort, a personally written thank-you note, or a small gift certificate. If you recognize these solid employees along with your superstars, you have gone a long way toward establishing a positive organizational culture where morale and motivation are high among all employees, not just those at the very top.

Consider offering rewards that are based on an individual's ability rather than "one-size-fits-all" goals that establish a reward for an entire group of people regardless of their ability.

If we occasionally choose to use a carrot to motivate people toward a special reward, then the carrot must be obtainable. How obtainable? I don't know if anybody knows for sure, but the experience of many people suggests that there must be at least a 50–50 chance of success for a goal to be met in order for it to act as a motivator.

Sometimes, an incentive system may work to everybody's benefit, as when bonuses are provided based on overall business income or cost savings. At a practical level, this translates into giving people rewards, such as a salary bonus, for reaching a stated goal for the team. But a goal has to change over time, whether for a team or for an individual. For instance, you normally would not want to give a person with 10 years' experience a bonus for accomplishing the same goal as a person with 1 year's experience. As people progress in their careers or as teams become more productive, the bar has to be raised to obtain a bonus. You can think along the same lines for salary increases rather than bonuses.

> *Pete Smith received a 5% salary increase because his work was well above average, while an "average" employee in the animal facility received a 3% increase. Larry Lucas, Smith's supervisor, thought, "Good for Pete; he deserved it." But Larry knew that he had to work with Pete to set new and more challenging goals for the following year. If the goals were pretty much the same, and Pete was to meet those goals, then Larry knew Pete should get a 3% increase next year, not 5%. Larry kept raising the goal post as Pete built up more experience and skill, but he always kept the goals at a level where Pete had a reasonable chance of reaching them.*

A similar kind of goal setting is giving a salary increase for passing an American Association for Laboratory Animal Science (AALAS) examination. This type of incentive pay has been successfully used in many laboratory animal facilities. Of course, if a person doesn't pass the AALAS examination, the loss of the financial reward can be construed as a *de facto* penalty. Clearly, we are walking a fine line. On the one hand, we want to reward outstanding people by promotions, salary increases, recognition, and the like. On the other hand, we cannot continually hang out a carrot for everything, lest the carrots become meaningless and detrimental to motivation. There has to be a balance.

I do believe that managers have a right to demand that people work hard, and a little later on, I will discuss goal setting. At the same time, I also believe the people who work for you have a right to understand, without any ambiguity, what is expected of them (i.e., what defines satisfactory performance) and whether their goals can be reached with reasonable effort. The opposite is also true. People should understand what type of behavior and what type of output are not considered to be satisfactory.

Every person should understand, without ambiguity, what constitutes satisfactory performance of their job.

Quality Output Should Be Expected

There is a related issue that managers must consider when using rewards as motivators. To begin with, we understand that the potential for getting a reward helps define to an employee what constitutes good performance. There's nothing wrong with that part of the story because we all want to know what is expected of us. But what might happen if the reward, not the job, becomes the goal of a person's behavior? I've seen this happen at the service end of an automobile dealership. I was unashamedly asked to give the dealership the highest possible rating on the quality of their service because it meant a bonus for their employees and probably some kudos from the car manufacturer. That is, since the customer wielded the financial power, the dealership's service manager decided to tell the customer what to say, irrespective of the quality of the service. Obviously, the reward being offered to the dealership was too much of a motivator. Now, let's move away from the service department and instead assume you are selling cars. If your reward (maybe a bonus in addition to a commission) is based on the number of high-priced automobiles that you sell, will you pay much attention to the customer who wants a basic model? Of course not! If you get a financial bonus for making a presentation at an AALAS meeting, will you spend an inordinate amount of work time—perhaps too much time—planning your next AALAS project? You very well might do that because you're looking for a bonus. We all like rewards, but like all managers, we have to be on the alert to make sure that a person's value to our animal facility isn't negatively influenced by the reward itself.

With the above caveat in mind, and assuming that an employee's basic needs are being met, what else can you do to reward especially efficient and effective performance? You should be rewarding good performance that adds value to the animal facility's mission and provides value to the researcher or the animals. You do not want to give a tangible reward, such as incentive pay, to a person who accomplishes the basic work that he or she is being paid to do. That's a bribe, and in the long run, it will backfire.

> I'm reminded of a story I heard about Henry Kissinger, the former secretary of state. It emphasizes that employees are routinely expected to do the best that they can do. Dr. Kissinger had asked a young staff member to prepare a detailed report about a developing nation. After working hard for some time, the staffer placed the report in Kissinger's mail box. An hour later, it was returned, with a note from Kissinger asking for

more detail. Again, the staffer worked on the report, adding more detail, and again Kissinger quickly returned it, asking for still more information. Finally, after exhausting all resources, the staff member personally delivered the report to Kissinger. "Dr. Kissinger," he said, "this is the best I can do. There is no more information on the subject." "Are you sure?" asked Kissinger. "Yes," he responded. "Good," said Kissinger, "now I'll look at it."

We don't know if Kissinger had some past history with his aide and he was trying to make a point, whether this was his usual way of doing things (i.e., perhaps Dr. Kissinger was not particularly interested in motivating people), or if the story was embellished before it reached me. But the story does emphasize that a manager has a right to expect a person's best work.

In a situation that is similar to the Kissinger story, we had a conversation in my department about what was meant on a performance review form by the phrase "meets performance expectations." The discussion arose in the context of merit-based salary increases. We did not want to give a higher salary increase than was deserved for an employee's performance. On the surface, it would seem that no manager would want to give a higher salary increase than a person deserves, but we know it happens. Why? Because just as some teachers inflate grades to avoid arguments, be a nice guy, or make the school look better, some supervisors do the same (I'll discuss this a little more in the next section). They inflate performance ratings (which are often linked to salary increases) because it's easier for them to do that than to try to explain to a person why their work is good (i.e., meets performance expectations), but not exceptionally good. On the other side of the coin, there are some employees who do good basic work but believe that their work ethic is better than that of one or more of their coworkers. They figure that if they're better than somebody else, they're doing more than meeting their performance expectations. Therefore, they want a larger salary increase than would be expected for "meets performance expectations." In other words, they don't compare themselves to the expectations of the position; they compare themselves to other people. This can be a real problem because although a salary increase is usually seen as a tangible reward, an insufficient increase can be viewed as a punishment.

The astute manager doesn't fall into this trap. First, he will not use an annual or semiannual performance review as being the same as a salary review. Most people are waiting to hear about their salary, and they aren't paying too much attention to anything else you're saying. Therefore, at the most, you should only have to summarize for a person what you have previously discussed with them during the past year about their performance. Second, as I emphasized above, everybody must know what constitutes *satisfactory* performance because this provides a standard against which *superior* performance can be judged. In the example I used with Kissinger, he finally got the high level of performance he expected, not the superficial performance that the staff member thought was good enough. The last thing you want to do is to assume an average performer is superior, only because all of your other employees aren't even up to satisfactory performance. (This will be reemphasized in Appendix 1 in the discussion of Lean management.) If this scenario seems familiar to you, there are significant leadership and management problems in your organization that have to be corrected. Now let's look at some related issues.

Promotion as a Tangible Reward and Motivator

Promotions, another important form of a tangible reward and motivator, should be based on past performance and your assessment of the person's future potential, not solely on length of employment. What would you do if you had two employees, both equally skilled, applying for the same position, but one had been with the organization for 3 years and the other for 17 years? Wouldn't it be fair to give the job to the senior person who had worked so much longer for the company?

An animal facility supervisor asked me the above question. Her opinion was that the job should be given to the senior person if all other things (e.g., personality, job knowledge, leadership potential, and punctuality) were essentially equal. Not knowing anything more about either person or the job details, I asked her how it was possible that it took one person 17 years to get to the same level of skill that the other person had reached in 3 years. I was essentially asking, who really is the better person for the job? Who has the greater potential for adding value to your service mission? Seniority is important, but it isn't everything.

Sometimes, particularly in unionized organizations, promotions based on performance and potential are not possible due to contractual obligations. Nevertheless, and no matter what the criteria for promotion, there are still some managers who would rather give a promotion or a salary increase (or both) to an employee who really does not deserve it, rather than face a potentially disgruntled person. In essence, this practice is giving a reward for poor performance, and it was mentioned in the previous section. The rationale sounds something like this: "Why not? The money isn't coming out of my pocket" or "The promotion means nothing in terms of job responsibility." They argue that the change is just on paper. More often than not, this action creates, rather than solves, a problem.

To illustrate what problems can be created by giving an undeserved promotion, consider the following possibilities: Employees become agitated because they see an undeserving person being advanced, and now they want justice for themselves. Or, employees doing a job in one department of your university find out that they are at level A, but they do the same work as someone at level C in another department. The level C employee is paid more. Needless to say, the level A employees have a reasonable grievance. I hope that you will not be one of these managers, because when you are called on the line to justify your actions, all I can do is to wish you the best of luck.

In a worse situation, employees X and Y work side by side. Employee X is given a promotion but continues to perform the same work as employee Y. To justify the promotion, you give employee X the title of "supervisor" of employee Y. Employee Y isn't impressed. "Why," he argues, "should somebody who does the exact same work as I do be my supervisor?" Employee Y is right, especially if employee X does not act as a supervisor. Never give one person authority over another unless there are legitimate differences in responsibility.

Never give one person authority over another unless there are legitimate differences in responsibility.

In my opinion, you give promotions and salary increases when they are deserved and they meet the needs of your organization, not as a convenience. If you are a poor manager, incentive pay and promotions will simply act as bandages on a cancer. You cannot substitute rewards for poor management.

After you gain some managerial experience, you will find that some promotions have led to poor outcomes even though you thought the person had all the right qualifications. This may be due to emotional and not technical incompetence [67]. Some newly promoted managers are so overeager to demonstrate their managerial talents that they become overbearing to the people with whom they work. We want to give people the opportunity to make mistakes and learn from those mistakes, but when it comes to interpersonal relationships with coworkers, we have to be careful since good management is one of the key reasons that good people stay with good companies. With poor management, as you know, good people tend to leave. If you are a seasoned manager, you have to provide strong and rapid feedback to the new manager before the damage is out of control. If you have a new manager who pays little attention to his comanagers but tries to curry favor with you, you have to sit this person down and discuss the need to build relationships if he wants to be able to become a strong functioning manager in the future. Perhaps you can place him on a committee with managers from other departments so he can learn the importance of building relationships and compromising in order to accomplish a goal. As you advance in your career, you'll become more adept at detecting people who are nearly ready to be promoted, but need a little more emotional seasoning before the promotion becomes a reality. We can help these people by giving them the feedback and "clinical experience" they need before being rewarded with a promotion.

As I previously indicated, when it comes to rewards, not everybody is going to have the same priorities. One employee may want time-and-a-half pay for doing overtime work, another may want compensatory time off, while a third may not want either, but you are aware that that person is looking for a promotion. You should consider using the desires and ambitions of your employees to tailor their particular rewards for a job that was especially well done or when certain special goals are reached. Money is a very powerful motivator, especially when you are not making much to begin with. Nevertheless, for many people, it is not the only motivator. You must know your employee well enough to determine what reward or combination of rewards will be appropriate and not assume they will always stay the same [68].

Promotions can certainly be useful to an individual and a company, but we must be careful about who is promoted, even if a person is deserving of the promotion. Similar to the problem of using money as a reward, you also should not assume that everybody wants to be promoted. I know of more than one instance where a promotion was refused because the individual did not want the responsibilities that went with it. He or she was happy with his or her position and lifestyle and didn't want the change. There is nothing wrong with that. As I wrote a little earlier when discussing rewards, there are some very talented people who do not want to become known as superstars. They just want to have an internal feeling of job satisfaction, that they contributed unique knowledge to the design of the new vivarium, or that they have the unique experience and knowledge to keep the robotics working. They don't need or want lavish praise or promotions. That doesn't mean you can forget about them for salary increases or even promotions in the future. The fact that they are not self-promotors does not mean that they have no feelings or needs [69]. Therefore, if and when a promotion is desired, that person must understand the requirements of the position into which he or she will be promoted. It serves no useful purpose for him or her to be promoted into positions where he or she will be unhappy and possibly fail. Once again, this doesn't mean that these employees have no desire to succeed or no desire for recognition or tangible rewards. In fact, they may be very good at what they do, and "like any employees, no matter how secure and grounded, [they] need nurturing and recognition" [70].

Let's assume, however, that a supervisory position is open and you have one or more current employees who have applied for it. Superficially, all have the basic requirements for the position, but you still have to separate the wheat from the chaff. Who do you choose? This is analogous to the situation I described earlier wherein a supervisor wanted to promote the more senior of two employees. Rather than providing detailed criteria, which is all but impossible to do, I would prefer to provide you with some general managerial considerations to help guide you through the process.

To begin, you should define the qualities needed for the position. If it's a managerial position, it may require a good personality, job knowledge, resourcefulness, tact, and so on. For a nonmanagerial position, there may be a greater emphasis on punctuality, manual dexterity, ability to make rapid decisions, or other criteria. *Just because a person is an excellent technician, there is no guarantee that he or she will be a good manager.* Therefore, defining the necessary traits for the job is imperative. I hasten to add that you should be looking for a person's potential to succeed in the new position over the long run, not a guarantee that a person will be an overnight superstar. That rarely happens.

Once you have identified the important traits for the position, fill it with the person who most closely matches those criteria. There is always some political pressure to do this or that, but use your best managerial judgment, be ready to defend your choice, and do what has to be done. Promoting a person has many similarities to hiring a new person, so you may want to skim through Appendix 3 at this time.

If you are promoting from within your own organization (and, in particular, from within your own laboratory animal facility), I suggest that you do not simply say, "I'm sorry" to those candidates who were not selected. That's a sure way to break their morale and motivation. It is far better to talk with them and let them know the positive aspects of their applications, and that they are still highly valued members of the team. And although it may be somewhat controversial, I have had no problem telling an unsuccessful candidate who was not chosen why the chosen person had stronger qualifications. If you can't defend your choice, how *did* you make the decision?

Promoting Employees When Promotion Opportunities Are Limited

One problem many managers face relative to supervisory-level promotions is that there may be no more room at the top, or there is room at the top, but an excellent employee does not want a supervisory or managerial position. This can be a problem because some people, usually fine employees, may leave you if they have no room to grow. First, their motivation declines, and then they leave. Unfortunately, we can only have so many supervisors or managers before everybody becomes a captain and nobody is a sailor. You could, of course, create a completely new position, but this is not always easy, particularly in laboratory animal facilities where there frequently are rigid job categories. There is no easy answer to this problem, but here are some suggestions.

1. Make sure that new employees understand what opportunities for promotion are available before they are hired. Don't tell a new technician that if she works out well, she can become a research supervisor, when you know that the current supervisor has held the position for the past 10 years and is still going strong.

2. Consider using a "level" system for similar jobs. For example, there can be an animal technician level 1, level 2, and level 3, based on quality of service, education, experience, examinations passed, or any other criterion you feel is appropriate. Both authority and responsibility should be added to a new grade level. If a level 1 animal technician is responsible for changing cages, feeding, and observing animals, perhaps level 2 may do the same and also

be responsible for monitoring and ordering certain supplies. A level 3 animal technician may also be responsible for reviewing SOPs and overseeing the work of level 1 and 2 technicians. The level system need not be only for technicians. Managers who aspire for higher managerial positions can also progress upwards via a level system. Toward the end of Chapter 7, you will find a discussion of some methods a manager can use to move into a higher management or leadership position.

The significant concept is that you do not want the promoted employee to be responsible for doing the same things they did before, just more of them. Thus, if a level 1 technician changed cages most of the day, it's not much of a promotion if, as a level 2 technician, he or she is just expected to change more cages. If you add monitoring and ordering as responsibilities, the monitoring and ordering should be significant responsibilities, not just counting bags of food and calling in the next order. Consider giving the person the authority to evaluate the current supply ordering procedures and suggest means for improvement. By doing this, you are giving people the freedom to do something on their own without your dictating the way it is to be done. You have everything to gain and little to lose.

3. A variation of a level system is a dual ladder system. These are parallel career ladders that have equal institutional prestige and importance. One career ladder is the same as the level system in that it allows people to continue doing work that is in their current field, but they have an opportunity to take on more responsibilities and receive additional training, if needed. This path is often used for existing managers, but it is equally valid for veterinarians who enjoy clinical or other veterinary work, or for office workers (such as business managers) who desire to remain in that field. Part of any needed training can include mentoring, as described in Appendix 4.

The second ladder is for those employees who are in a nonmanagerial position but wish to progress into management. Is there a way for them to make this crossover, or does the pathway end at a high level of their current position? Not everybody wants to progress into management, but for those who do, there should be a clearly defined way to obtain a managerial position. If you require aspiring managers to have managerial experience, is there any reasonable way that current nonmanagerial employees can get that experience? Perhaps you can require a certain number of hours of formal managerial training, but if you do so, have you developed any in-house opportunities to obtain that training (e.g., lectures, hands-on training, or working alongside a manager)? Will your institution subsidize coursework at local colleges or online education? If not, maybe you should rethink the entire way people can get promoted in your animal facility.

4. On occasion, promotions can be made honorary, as long as it is a recognized procedure in your organization. A senior research biologist may not be paid more or have greater responsibilities than a research biologist. Nevertheless,

the title "senior" is given to recognize superior performance in lieu of other possible rewards. If you consider using this approach, also consider your organization's culture and how such a reward is commonly viewed. Some people appreciate honorary titles, as they can be used on résumés when they are seeking another position. I'll tell you something else about honorary titles that may not be obvious at first. That is, we live in a relatively egalitarian society in America, and those little honorary titles mean a lot to some people because to them, they actually symbolize a higher status level. I knew one PhD who received an honorary MD degree, and he wouldn't even sign a birthday card without writing MD (hc) after his name.

5. Finally, recognize that people rarely stay in the same position for their entire working lives. If positions are available in other divisions of your organization, help your employees to make the change if they want to move. If need be, help your employees find work outside of your organization. Use one of the best information resources you have: your friends. An unhappy employee is an unmotivated employee and is a reflection of you, the manager. In fact, some turnover in employees can breathe new life into stagnant organizations. You should try to keep a core of valued employees. Hire the best you can, with the understanding that some will leave. Even if you have a 20% or 25% turnover of noncore employees each year, you will probably not significantly impair your operations.

Keeping Your Star Performers

If an unhappy employee is an unmotivated employee, the obvious question is, how do I keep my star performers happy, motivated, and wanting to stay with me? We'll assume that you are a reasonably talented manager and a reasonable salary is being paid. But, perhaps the daily workload is falling off due to business setbacks for your company or because research grants are becoming more difficult to obtain. Your star performer wants more to do, or at least more stimulating work, but there isn't much more for you to give him. It's likely that he is smart, ambitious, and engaged in his work. He wants to be recognized for his output and talent, but what more can you do above and beyond the motivational suggestions already made in this book? I wish there were simple answers to these questions, but answers to such problems are never simple, and the reality is that in the twenty-first century, people rarely stay in the same job for their entire career. Still, there are things you can do to help retention, and I'll discuss some of them, but you have to brace yourself for the reality that you may lose some of your top people no matter how hard you try. For example, assume your institution has a functional career ladder in your department, business is booming, and your goal, along with your star performer's goal, is for her to be the next associate director. However, it may be obvious to everybody that the current associate director will not be retiring for another 15 or more years. Many star performers don't want to wait that long, and therefore they move on when another job opportunity presents itself. We also see situations in which a star employee receives a job offer from another company and asks you to match or exceed that offer. Can you match the offer or propose an alternate offer, or are you willing to lose your star? You really cannot fault a person for looking at a position that pays significantly more than her current position and has interesting new responsibilities, no matter how content she may be in her current position.

My advice, which I often repeat to the senior managers with whom I work, is to recognize talent as early as possible and start supporting and mentoring that person as soon as possible (see Appendix 4 for a discussion on mentoring). In my opinion, it's not a good idea to have your star's direct supervisor mentor him (although it may be unavoidable), but rather, a person above the direct supervisor should act as a mentor and advocate. It makes for an easier relationship and eliminates the likelihood of a direct supervisor undermining your star performer because they both have the same career aspirations. Not infrequently, you'll find that a star performer finds a mentor without your help.

Of course, do whatever is possible to keep your top people happy and engaged, including extra pay for that person if pay is an important consideration for retaining her. If she is a veterinarian with a research interest, work with her to develop small research projects that may lead to publications and future grant funding. If she is a technician who enjoys working with farm species, provide that opportunity and work with her to gain the knowledge needed to make a presentation at a national meeting, to teach others about caring for those animals, or to publish her observations. Do not just give a person a "project" and walk away from it. You have to show honest, active interest and give honest public praise when praise is warranted. Also, as part of the mentoring process, don't be afraid to include your star performer in the vivarium's business meetings or, when possible, invite her to be an observer at meetings of an institutional committee you serve on. When you do that, you demonstrate your interest in her career and support of her abilities. There will be those who complain that you are showing favoritism, and indeed you are. But you have to look down the road at the goal you have set with and for her.

After all is said and done from your end, the star performer has to be productive (and continue to be productive) in ways that are valuable to your animal facility. If your star can handle all assigned tasks, especially those that are challenging, and if that pattern continues over time, you have a win–win situation. But if the star is dimming after a while and if a lack of motivation and a feeling of entitlement begin to show, it's time to reevaluate the situation. You really have to work at keeping a star performer; it doesn't happen magically, and it takes time and effort. You want to make sure that the time and effort are being devoted to the right person.

Quick Summary about Motivation (and Demotivation)

Let's take a moment and review some thoughts about motivation before moving on to goal setting and its impact on motivation. We began by defining motivation as an action taken by a person or an organization that provides an employee with a desire to accomplish a goal. At a very basic level, employee motivation requires a reasonable salary, reasonably good working conditions, reasonably good supervision, imparting respect and trust to our coworkers, engendering a feeling of accomplishment, and having our coworkers feel that their work is important. Beyond these basic needs, there are tangible rewards (such as financial bonuses or promotions) that can add to the motivation of some (but not all) people. Intangible rewards (such as positive reinforcement for an accomplishment) are also important additions as motivators for many people.

Demotivation, which is the opposite of motivation, can occur when any of the above basic conditions are not met. There are daily workplace occurrences (all related to motivational needs) that exemplify the demotivating signals that may reach some employees. Here are just three examples [54]:

1. A manager who dismisses the importance of an employee's work by saying that he could get a kid off of the street to do the same job has just done a great job of demotivating a coworker. It shows a lack of respect for the person and a lack of respect for the importance of that person's work.
2. Motivation can unintentionally be destroyed by frequent and abrupt work reassignments by a manager who believes that all employees want to rotate their work duties. Although not all employees object to work rotations, there are those who do and would consider this an example of poor working conditions.
3. Demotivation can rear its ugly head through managerial actions that make a person's work seem to be of little or no value. Consider a group of employees who spent months developing an enrichment program for nonhuman primates, just to find out that a decision was made by upper management, without input from the vivarium staff, to no longer house nonhuman primates.

Motivation plays an important role in keeping star performers from leaving your institution. These top-notch employees need frequent recognition and interesting work challenges to maintain their engagement (i.e., their commitment to their work) and productivity. Something as basic as saying thank you means more to most employees than you might imagine. If you are a perceptive manager, you will identify these people early in their careers and nurture them with mentoring and tangible and intangible rewards, including recognition, pay, and projects that cultivate their interests.

Keeping Motivation High: Setting Goals with Individuals

Let's return to the basics for a minute. You may recall that managers follow the mission of their organization, and if the organization has a specific vision, managers are expected to partake in fulfilling that vision. Even if the vision does not require elaborate long-range goals and strategies, there are always short-range (operational) goals for managers. The latter goals are needed to accomplish the unique operational issues that arise every day. Likewise, all nonmanagerial employees also have to have goals to help them advance the needs of the animal facility and to help propel their own careers. Part of the skill of keeping employees motivated is to work with them to set realistic goals. Don't simply give a person a goal. You have to work with him or her to make sure it is mutually agreeable and clearly understood. Afterwards, periodically review with the person his or her progress toward the goal. Those of you who have a formal background in management theory will recognize that these suggestions are akin to the central theme of management by objectives (MBO). For the most part, I have only substituted the word *goal* for *objective*.

MBO sounds easier than it really is, and perhaps that's why you don't hear about it as much as you did years ago. You have to find the time to speak with the employee, you have to have some method of evaluating whether progress is being made toward reaching the goal, and most important, you have to make a commitment to be a good manager.

There are other problems as well, particularly for first- and middle-level managers. The one that is probably the most important, and occurs most often, is the question of who is actually setting the goals. Are you really communicating with the person, or are you dictating what *you* think the goal should be? If you think the employee should be changing at least 125 mouse cages per hour, and you tell that to the employee, whose goal is it? Is the employee afraid to speak up? Does he or she say okay, but you know that it is a meaningless statement? Sometimes there is a real communications problem when goals are being set.

In general, you should meet annually with each employee who reports directly to you to set goals for the coming year. Then, during that year, you should consider meeting periodically with each person (perhaps twice during the year) to assess the progress toward the goals and make any needed fine tuning. Needless to say, it also presents an opportunity for one-on-one communication that often cannot be fostered in hallways or lunch rooms.

Whatever the goal or goals, make sure they are clearly understood by both parties. I say this not to discourage you, but rather to encourage you to work even harder to cultivate open communication and mutual trust. I suggest that you put the agreed-upon goals in writing to eliminate any misunderstanding about what has been discussed.

Goals do not always have to be physical in nature, such as an increase in productivity. They can be behavioral as well, such as coming in on time or talking less to other employees. (In fact, I have been amazed how many employees have complained that other employees talk too much.) As a manager, your own goal might be to make sure that the talkative employee understands that constant talking is causing a problem, and why. That's common sense, but many employees who are creating a problem do not realize that one exists. You also have to make sure that this goal (less talking) does not have any obvious impediments that can inhibit the employee from reaching it. For instance, if it is mutually agreed to move the employee to a new area, it certainly would be counterproductive to place that person in an open office with a large number of other people.

It may be difficult to determine whether a behavioral goal has been achieved. Quantitative goals (such as an increase in production) are easier for you and an employee to judge. Quantifying goals at lower levels of management is often easier to do than at upper levels. In quantifying a behavioral goal, such as talking less to other employees, you might be able to set up some way of measuring if it is being reached, such as fewer complaints by employees who claimed that they were being distracted by "the talker."

You really have to think about what you are attempting to do when you try to quantify a goal. Let's say that you and an employee agree there should be an increase in efficiency and effectiveness. Increased efficiency is relatively easy to quantify. You can count the number of cages cleaned in a unit of time, for instance. But what did you mean when the goal you set with a person was to increase effectiveness? For you to have even asked for an increase in effectiveness, there must have been something that concerned you. What was it? Did you look at the cages and see they were dirty? That's asking for a debate. So how do you measure effectiveness? Perhaps you will use a pH measurement of water left in the cage, bacterial cultures, or an adenosine triphosphate (ATP) determination. As a manager of a laboratory animal facility, you should think about means of quantifying goals before you commit them to paper. In Chapter 1, there was a brief discussion about productivity (the sum of efficiency and effectiveness), and in Appendix 1, there is an expanded discussion about productivity and setting and measuring productivity goals.

Consider how you will measure the progress an employee has made toward fulfilling a goal.

You might also consider setting a time limit for reaching the goal. If you do, be sure that it is reasonable. You are asking for failure and the demotivation of an employee if you agree on unreasonably short times to accomplish a difficult goal.

You should not feel inhibited about praising a person for making steps toward the goal, even if the goal is not yet fully accomplished. Praise, and in particular praise that is personalized to the person's achievement, is a very real reward and a recognition that most of us seek. In fact, frequent praise, if deserved, is one of the best motivational rewards you can give an employee. You may praise an employee publicly or privately, but you should always use praise to reinforce something specific and positive (in other words, praise a specific action that is linked to the animal facility's or institutional values). You will get much further by using praise as a reward than by using criticism as a deterrent. While people remember criticism, they respond to praise [71].

People remember criticism but respond to praise.

Frequently, a sense of accomplishment and its recognition motivate people to reach a goal. Here's an example.

> *A laboratory animal facility had a policy, developed with the animal care staff, that stated that to be eligible for promotion, an employee, after a certain length of time, had to pass the appropriate examination given by AALAS. Even if no higher position was open, he or she would receive a $500 bonus for passing the test. The supervisor placed an attractive awards cabinet in the main hallway of the facility, and everybody who passed the examination had their certificate displayed in that cabinet. It became a matter of pride for the staff to have their certificate displayed with the others. There was peer pressure to pass the exam, and help was always offered in preparation for it.*

In this example, there were four motivating factors to reach the goal: a requirement for being promoted (which employees knew of before being hired), peer pressure, recognition of achievement, and a financial reward for an accomplishment that was beyond the basic responsibilities of their jobs. In addition, the organization itself demonstrated its desire to have its employees take and pass the examination by providing the display, financial incentive, and training. Most important, the employees themselves participated in establishing this policy.

Here is a summary of some key ideas that were discussed in this and earlier chapters that impact goal setting with employees.

Summary of Goal-Setting Concepts

- Set goals that are compatible with the mission and vision of your organization.
- Work with an employee to set a goal. Do not set your own goal and then force it upon another person.
- A goal should be neither too easy nor too hard to accomplish.
- A person must know what constitutes an acceptable completion of a goal.

- ▪ Try to set goals that can be measured and quantified.
- ▪ When possible, set a reasonable time limit for the completion of a goal.
- ▪ Put goals in writing.

Influencing People to Accomplish Goals

Influence is a management skill that every manager uses at one time or another because sometimes it is the most effective method to get a person to do something. In fact, influence is a large part of leadership (see Chapter 7). Influence should be used primarily when you're up against some resistance (or potential resistance) from a person or a group. But, you may ask, wouldn't that be the time for me to use my authority as a manager and just tell that person what to do? Granted, there are times when authority does have to be used to accomplish a goal, but it can quickly lead to a "me versus you" relationship, which is something you really don't want to have. Further, when you remove the pressure that emanates from using your authority, there is a good chance that a person will also stop trying to do what you would like him to do [72]. You are far better off having a person make his own choice to do what you want him to do, rather than by ordering him to do the same thing. That is the beauty of influence; you may be able to get the same result with fewer headaches. In a moment, I'll further clarify the difference between influence and authority. Also, if you want to read a short discussion on the types of authority that managers and leaders have, you will find that in Chapter 7.

True influence is more closely related to motivation than to authority because, at its best, influence can result in people *wanting* to do something for you rather than *having* to do something for you. You may not have any real authority over a person you want to influence (e.g., your boss), but you can still influence that person. Influence is usually used in a somewhat subtle manner, such as when a parent tries to influence a child to attend a particular college by telling her about the beautiful campus, the great social life, the closeness to a city, and maybe the quality of the education she will get (although that last reason might be low on the child's list of priorities). If she decides to apply to that college, the parent's influence has worked.

It's fair to wonder about the differences between *influence, persuasion, authority*, and *motivation* because all these words are interrelated. As you know, motivation implies that a person wants to do something. It is an inner drive that a person has to reach a goal, even if you gave the person a reason to be motivated. In contrast, influence is something one person says or does that helps induce another person to make a particular choice, whether or not the person is motivated to make that choice. This can be difficult at times because you are often trying to alter a person's behavior, and behaviors are not easy to change. Persuasion is quite similar to influence, and I will not try to split hairs by offering a separate definition. Authority refers to the power that a person has that compels (forces) another person to do something.

> **Influence is something one person says or does that helps induce another person to make a particular choice.**

If I suggest that you should see a particular movie because I saw that movie and liked it, I have tried to influence your choice. You may be willing to give the movie a try, but there may be no internal drive (motivation) to see that particular movie. Another movie might have been just as good. If I told you that you must watch a particular movie as part of a class I teach, then I have used my authority as a teacher to compel you to watch a specific movie. I'm not influencing or truly motivating you, I'm just making you do something even if you would rather not do it.

One of the keys to using influence is to be very clear about what you want to happen and what the other person may want in return (this is also a key to good negotiating, which is discussed later in this chapter). Your goal may be very clear to you, but you still have to make sure the other person understands what you want or need. Knowing the other person's goal may be obvious in some situations, sometimes the other person may not have specific goals or needs, and there will be times when you have to find out what a person wants.

> **A key to using influence is to be very clear about what you want to happen and what the other person may want in return.**

For example, let's say you want to influence a person to see a movie that you saw and liked because the person said he wants to see a good movie (in this case, the other person had a need, which was to see a good movie). There was probably a short discussion about movies before you suggested to your friend the particular movie that you saw. You can strengthen your influence attempt by adding to your recommendation the reasons why the movie was good, such as excellent acting and a great plot. As another example, let's assume you want to build up some political capital with your boss, so you tell her of a potential problem you heard about in the hallway. She gets information, and you may have influenced her to give you some positive recognition, even if it's only a sincere thank you. Sometimes influence can work like a negotiation between people. For example, let's say that you need a person to work late tonight and Joe wants to come in late tomorrow. You influence Joe to work late today by letting him come in late tomorrow. At the same time, Joe has influenced you to let him come in late tomorrow by agreeing to work late today. You have influenced each other to reach goals that are satisfactory to both of you. It's also a win–win negotiation.

Related to knowing what the other person may want is the concept of reciprocity. That is, people are more likely to want to help you if they feel they owe you something from a favor you did for them in the past [73,74]. Something basic, like lending a person a book or emailing them an interesting article, can become a major benefit for you in the long run.

What do we do if we don't know what might influence a person? One possibility, of course, is to ask the person, but a more subtle and effective way is to listen to the person. See what you can learn by hearing what the person has to say while you are asking some mild questions. This can help to develop a positive rapport with the person well before there is any need to influence that person. That way, you know the person and how she thinks and what might be needed to influence her. You can build upon this rapport by being honest, having open communications, showing respect for her work, standing up for her when it's appropriate to do so, and even by doing little things like stopping her in the hallway and asking about her kid who got hurt last week when he fell off his bike.

Another concept to consider when attempting to influence a person is to understand the person's background and training. For those of us who work in the academic research arena, we know that the scientists we interact with are trained to be analytical problem solvers, and therefore we should be prepared to first present the principles behind what we would like to do, then present data to back up the principles (if at all possible), and finally, conclude our presentation by explaining how it might benefit them (again, if possible).

Yet another way, and always one of the best ways of influencing a person, is to think or even act a little like the person you are trying to influence. In other words, sharing a common trait increases your chances of influencing a person because there's a better chance he or she will like you. Therefore, if you both drive the same kind of car or like the same kind of Italian food or work out in a gym before going to work, there's a better chance for you to be able to influence that person than if you had nothing at all in common. If you are trying to influence a veterinary technician to stay late to monitor an animal and you both are frustrated that surgery began late, don't be afraid to express your frustration, because it is sharing a common problem. If you really need that new animal rack but your boss has other managers who have competing needs, try finding a similarity between you and your boss. It can only help.

I'll tell you a story about me and my former boss. I come to work every day neatly dressed with a tie, jacket, ironed shirt, and so forth. I would much rather dress casually, especially without a tie (if Barak Obama can make speeches not wearing a tie, I think I can come to work without a tie). I was chatting casually with my boss (who almost never wore a tie), and he asked me why I wore one. My answer was that the dean, the chancellor, and some of the department heads wore ties, and since I frequently interacted with them, I wanted to project the same image they presented. In other words, I wanted to build up some influence reserves by having a common bond with them, and that bond was the way we dressed.

When trying to influence a person, it helps to try to find areas of similarity with that person.

Another important recommendation for helping to influence a person is to display strength or warmth as the situation requires [75,76]. Little things, such as tilting your head to the side, speaking with a low voice, or keeping your hands and arms near your body, project warmth. The opposite, keeping your head and body straight and repeated arm gesturing, projects strength. Whether you should use strength or warmth may very well depend on the reactions of the person you are trying to influence. If that person projects warmth to you, then reciprocate with warmth. If that person projects strength, then you should consider projecting strength to help influence that person. If you are not sure which way to go, it's always safer to begin by projecting warmth [75].

Here's my final suggestion on how to influence a person or people: get peer pressure to work for you. Every organization, every animal facility, has one or more people who are opinion leaders. As I'll describe in more detail later in this chapter, under the heading of "Resistance to Change," an opinion leader does not have to be a person high up on the corporate ladder; rather, an opinion leader can be any person whose opinion is listened to and sought out by others in the group. If you feel you need help to influence more than one person, use many of the tactics I just described to influence the opinion leader, because the opinion leader will be able to help influence other people.

A little earlier in this chapter, I alluded to the fact that managers sometimes have to be chameleons and adapt to the person or people they're working with. This certainly rings true when you are trying to influence people. But trying to influence a person can have a benefit that most people don't consider: maybe that person can influence you. As managers, we have to be open-minded and consider that our view of a situation is not the only viewpoint. When trying to influence a person, there is usually a conversation in which a person presents his or her viewpoint. Are you listening to that person? If that person presents an opinion or data that makes perfectly good sense to you, even if it is contrary to your current belief, then accept that the person has influenced you to change your mind. We're not playing a win–lose game, and it's a rare day that you will lose face by agreeing with another person's opinion.

It may seem strange to have only a brief discussion about using influence to get people to accomplish a goal, because a good argument can be made that managerial leadership is all about motivating and influencing people to be efficient and effective in reaching goals. However, one of the intents of this book is to focus on the basics of management so that motivation is enhanced and the need for using tactics such as influence is reduced. If you would like an easy-to-understand and thoughtful discussion on influence, you should read *Influence without Authority* [77] by Allan Cohen and David Bradford.

Motivation and Entitlement

Now let's turn our attention to entitlement and see how it affects motivation. What can you do when your organization meets the basic needs of its employees, but people on your staff are beginning to take things for granted? This is the so-called entitlement paradigm. I was quite interested in Judith Bardwick's book [78], in which she discussed the entitlement attitude that many employees have. We've all seen people who believe you should be honored to have them show up for work and do little more than accept their paycheck. These are often the employees who are the least motivated and cause you the most headaches. Are you willing to accept this attitude, or should you do something to motivate these people? Rather than jump in with a bag of magical managerial tricks, Bardwick first challenges us to be introspective. She tells us to look at the culture of our parent organization (and our laboratory animal facility) and see if we can get some understanding of why these attitudes may exist. For example, does your organizational culture support a sense of entitlement by giving salary increases that are based on titles, seniority, and performance relative to budgets, rather than for value created [79]? If entitlement is a problem with one employee, we might address the problem one way, but if everyone feels they have a job for life and there is no need to be motivated, then the organization's entitlement system has to change concurrently with changes employees will be expected to make. I'm willing to guess that in some laboratory animal facilities, entitlement is more the norm than the exception.

Don't make the error of trying to overcome this problem by becoming a tyrannical manager who attempts to motivate people through fear of being fired, transferred, and so forth. Both efficiency and effectiveness will suffer, because, under tyranny, few people want to move forward by taking risks. Everybody wants to protect their own little bit of turf. People become polite, follow the rules, and look to you to tell them what to do. You really can't blame them if making a mistake will cost them their jobs.

What can we do if we look around and find that our staff is not motivated because of a feeling of entitlement? Bardwick [78] focuses on three needs.

Help Employees Out of a Feeling of Entitlement By

1. Challenging people to take risks, learn, and fulfill their potential
2. Empowering people to be free to be creative and take actions (i.e., giving people autonomy)
3. Giving people a feeling that their work is significant

None of these ideas is new or revolutionary. I've talked about them as part of goal setting and the need for recognition. The purpose of repeating them at this juncture is to emphasize their importance in many different managerial situations. Most people want to be challenged to succeed, but there must be a reasonable chance of success. Likewise, most people want to be free to do things their own way without being castigated. As you know, sometimes this is hard to accomplish due to the necessary consistency required with some laboratory animal science jobs. And, as important as anything else, if employees do not think their work is significant, they will never reach their full potential. I always go back to the example of the technicians who work in the cage cleaning area. They are a vital link, perhaps the most vital link in the day-to-day operations of many animal facilities. Within the usual constraints of maintaining research consistency, how much leeway do they have in determining how the cage wash area should be managed? How often have you told those people how significant their work is, and why? But of greater importance, the cage wash team, like all other coworkers, should be reminded that their efforts are important contributions to the science or education that is the "end product" of your institution's mission.

Morale

General Comments

I don't remember how or why, but one day, while working in New York, I decided that it would be excellent for both morale and job performance if we had our animal care and animal research technicians meet with investigators before a study started. I know many of you routinely do that, but I was still a neophyte in laboratory animal science at that time. In any case, it worked wonders. The investigators briefly described their research, the laboratory animal facility technicians asked questions that were important to them, and everybody was happy. We continued this program, and instituted a second one in which we invited all the research technicians in the building to regularly scheduled seminars that the investigators presented. They discussed their research at a level that was understandable by technicians from the laboratory animal facility, chemistry, pathology, and other departments of the institute. Once again, it was a success.

In both instances, the success was not only a result of the quality of the meetings and seminars, but also because we provided a continuing education opportunity for the staff and showed how much we really cared about them. Their importance in understanding and contributing to the goals of our organization was recognized, and by doing so, we increased their job satisfaction. A little earlier in this chapter, I noted that recognition was one of those intangible rewards that everybody seeks.

One of our functions as managers is to try to ensure that our employees have a high level of job satisfaction, that is, that their morale is high. This tenet is another broad concept in the management of human resources.

Motivation versus Morale

Motivation is providing people with the *incentive* to accomplish a goal. Morale is a person's or group's *willingness and desire* to accomplish that goal.

We know that to keep motivation and morale high, we must ensure that basic needs are being met (money, respect, good working conditions, and good supervision) and that a reward is available for unusually good performance. The reward should be important to the person who is receiving it. If there is no appropriate motivation to do a good job, you are probably safe in assuming that morale is not going to be high.

When I started to think about the importance of morale in the workplace, my mind wandered back to my time in veterinary school, where one of the most likeable professors I had always kept the class mesmerized with humorous anecdotes. It was a fun class. Unfortunately, I learned almost nothing of medical value. The enjoyment was there, but I was totally unengaged. That means that although the class made me happy, I was never motivated to excel or even try a little harder to understand the subject matter. The class gave me the same satisfaction I can get now by listening to a good comedian. It's interesting, it's fun, but there's no motivation for me to become a better listener, a comedian, or anything else. It's just a nice way of passing a little time. Branham and Hirschfeld [80] provide a short but important statement about morale that in many ways mimics my experience in veterinary school. They wrote that enjoyment doesn't necessarily equate to engagement, the latter being an outcome that can be directly linked to the overall productivity and retention of employees. That is, engagement in our work includes an outcome measurement, which is to become more productive and to have less employee turnover. To gain productivity, the basic needs described in the previous paragraph must be met. In addition, there is a need for people to be motivated and have high morale to see true workplace engagement. As managers, we have a reasonable expectation that our employees will try their best to perform satisfactorily in order to give us the impetus to work with them and help advance their careers. But there is also an expectation on us to help increase morale by paying attention to employees' needs, providing them with recognition and growth opportunities, and ensuring that their day-to-day experiences in the animal facility are satisfying.

When a coworker does a job well, give that person recognition for the achievement. That's important because it's a severe blow to morale to know that you have done a job (reached a goal), but someone else has taken the credit, or you received no recognition. Publicly (and gracefully) acknowledge the work of others who contributed to the success of a project. Give full credit to those who made it work. You will never lose respect doing this because everybody still knows who the boss is, and indirectly, you will get your share of the credit. In fact, the only way you can lose face and break morale in this type of a situation is to steal the spotlight from someone else.

Likewise, you have to support your staff by taking or sharing the blame when something goes wrong. I wrote earlier: Do what you can to ensure that a mistake will not recur. Do not waste your time looking for a scapegoat. Your coworkers' morale and your own image will both suffer.

There will be times when an employee has to be reprimanded. Every manager knows that, and just about every manager wishes it wouldn't happen. But it does. In fact, there was a long discussion earlier in this chapter on how to approach a difficult conversation. When a reprimand is appropriate, you should criticize the person's action, not the person. For example, it's better to say, "Putting male mice together can lead to their fighting" than "You should have known better than to put male mice together." The former comment criticizes the action, while the latter criticizes the person.

Never publicly criticize an employee. That's a sure way of injuring that person's morale and pride, and it leads to an antagonistic situation. It also sends a dangerous message to everybody else: do things the boss's way or you're going to be publicly embarrassed. No company should ever want to have that kind of a reputation. Therefore, if you must criticize, do it in private and do it quickly and, if possible, inconspicuously. Reprimands should be short, clear, and leave the door open for future improvement. To the extent possible, use praise as a reward rather than criticism as a means of control.

Praise a person's accomplishments in public. Criticize a person's actions (not the person) in private.

If criticizing people in public is one fine way of diminishing morale, another is having somebody on your staff that doesn't pull his or her own weight. (You may recall that this was used as an example earlier in this chapter.) This problem is compounded by having a manager who doesn't have the courage to do anything about it [81]. All of us have probably seen this happen, and all of us probably have been upset. When managers do nothing about freeloaders, then other employees start to loaf as well, and there are even other employees who spend a good deal of their own work time trying to figure out ways to get even with the loafer [81]. All in all, it's bad for morale and bad for business. The answer for managers, as you might have guessed, is twofold. First, we have to generate an organizational culture that doesn't tolerate employees who specialize in doing nothing. Second, when the problem rears its head, we have to do something and do it quickly. You can make sure the person has enough work to do, you can make sure that the person has the needed tools to get his or her work done, and you can talk to him or her, using some of the techniques described in this chapter (see "Difficult Conversations"). But you have to do something.

> *In one very poorly handled situation, there was a veterinarian who simply wasn't competent. To compound the problem, he was handing off much of his work to the veterinary technicians, who in turn were becoming frustrated. The other two veterinarians who worked with him were getting livid and complained frequently, but their department chair did nothing because she was the ultimate nonconfrontational person. She hoped that by closing her eyes and waiting long enough, the problem would resolve itself. Well, in a way it did. Two very fine veterinarians quit and the department chair was left with one ineffective veterinarian and one practical option: finding two new veterinarians to work with the slacker. It was a lose–lose situation for everybody but the slacker.*

Fitting the Person to the Job

One aspect of motivation and morale is to make sure you give the right job to the right person. Your goal is to increase morale by using people's strengths, not their weaknesses. In an earlier example, I referred to letting a person work with primates if he or she likes to work with primates. This assumes, of course, that the person has the qualifications to do the job. Sometimes, it's not possible to match the person with the job. Many laboratory animal facilities are small, and the animal technicians have to do almost everything. Under those circumstances, it becomes difficult to pick and choose the tasks that are to be assigned. There are also those times when we all have to compromise and perform certain tasks that we would prefer not to do. Grading essay exams used to be one of my responsibilities that I would gladly have passed on to someone else.

In a small number of instances, a person's expression of dissatisfaction about taking on a new responsibility is really a fear of the unknown or a fear of failure. To help that person, you must be reasonably sure that he knows what is expected of him. You also must be reasonably confident that he has the technical, emotional, and behavioral capabilities to perform the new task. For example, decapitation is used to euthanize certain laboratory rodents. We all can readily understand why some employees may not desire to perform this procedure. You can give that responsibility to another person, do it yourself, or explain to employees why decapitation is performed. For the last option, you should concurrently discuss the humane considerations and note that all staff members are expected to learn the technique. This discussion can be difficult because it involves altering people's perceptions.

Nevertheless, I have often heard laboratory animal technicians say that although they don't like to perform certain procedures, they do them because they understand the importance of the study. That's realistic, and it is complimentary to the communication, motivation, and morale in that animal facility.

There is a dark side to this. Some managers have intentionally given undesirable tasks to certain employees to encourage them to resign. While this is a well-known management tactic, it must be used with great care, as an unhappy, unmotivated employee can spread this feeling to many other employees. Our job is to keep morale up, not find ways to pull it down.

One additional thought about morale. At the beginning of this chapter, I stated the importance of trust and communication. These factors are critically important for keeping morale high. Ask for feedback, and do everything in your power to get it. Find out what people think. Remember that perceptions, whether right or wrong, are what people react to. Be honest with your staff, give them advance notice of proposed changes or other significant events, ask for their opinions, care about them, and you will have gone a long way toward keeping morale high.

Morale and Organizational Conflict

There will be times in any organization when conflicts occur. Most managers do everything possible to avoid conflict, and as a general statement, I can say that much of this chapter is focused on communicating with people and establishing a culture of trust in order to avoid conflict. But, just as there are good stressors and bad stressors that can affect laboratory animals, there are good and bad conflicts that can affect an animal facility. To put this into perspective, think about some business executive or politician who was surrounded by "yes men," meaning that the executive didn't want to hear any bad news and his or her cronies were perfectly content not to challenge the boss. Common sense tells us that this type of an action can create far more problems than it cures. If you're losing a war and every one of your advisors tells you you're winning, just to save their jobs, you're not much of a leader. This same problem can affect an entire animal facility. Sometimes, a reality check is needed, and this is where some conflict can be good. It provides an opportunity to develop solutions to problems. But to do this, there have to be some rules, the first of which is to act civilly and have a formal means of handling conflict (such as SOPs for difficult conversations, as was previously discussed in this chapter).

Let me give you an example. Let's assume that your animal facility has a chronic problem with dirt embedded into the floor and walls of the locker room's shower stalls. It's been going on long enough that the technicians are hounding their supervisors to do "something," and eventually the problem reaches you, the facility's director. By this time, tempers are starting to flare. What do you do to defuse the situation and maintain morale?

Let's not discuss the actual solution to the problem because that's not as important as the process I would like to describe. I suspect that you would get on the phone or send an email to the director of building maintenance and ask for that person to find a solution or come down for a meeting. Eventually, the problem would be resolved. In the interim, employees have learned that if they make enough of an issue with their supervisors, the problem will go up the ladder and they'll get a response. If it isn't the response they want, they'll make more noise. This indirect employee–supervisor conflict is hardly a way to build morale. It helps establish walls rather than bridges, so let's look at another way to resolve intradepartmental conflict, a way that has been reasonably successful at my animal facility.

The Department of Animal Medicine, building on the experience of another school department, found that the establishment of a focus group helped to resolve many conflicts in a positive manner. The group is comprised of about eight people, representing animal care technicians, veterinary service technicians, office personnel, and managerial personnel. One senior person, not a member of the group itself, serves as a facilitator. The facilitator is a person who has the respect of the group and functions to keep the discussion focused on the root causes of the problem (to see if it can be prevented in the future) and, of course, seek a resolution for the immediate problem. The problem is embedded dirt in shower stalls. The discussion might go something like this:

Dave (the facilitator): We all know what the issue is, but can somebody tell me what was already tried to resolve it?

Angela (a veterinary technician): Well, before this was a real problem, Stacy [a technician not in the focus group] saw Kim cleaning the shower stalls and asked her if she could try to get the dirt out. Kim said she tried, but she couldn't. I'm not sure she tried hard enough.

Heather (an animal care technician): Then Kim brought some sort of a scrubbing machine and that didn't work either. She said it was a stain, not dirt. She tried bleach, and it didn't work. I think she's right; it's just badly stained. But it's really gross!

Dave: Did anybody tell Andy (Kim's boss) about this?

Heather: I don't know.

Angela: I don't know either.

Mike (a veterinary technician): If bleach and a scrubbing machine can't get rid of the stain, then what can Andy do?

Karen (a purchasing specialist): If he doesn't have any cleaning tricks up his sleeve, he can tell us if the tiles can be replaced or if we can put a shower enclosure over the existing tiles. I did that with my bathtub at home and it looks fine.

Dave: Is that OK with everybody? We'll get Andy down here, but we'll ask him to talk first to Kim or even bring Kim with him. Then we'll sit down with them and hear what they have to say. OK? Heather, at next week's staff meeting, will you report back to the department what happened and what's going to happen?

Heather: Sure, no problem.
Dave: Karen, are you comfortable with calling Andy, explaining the situation, and seeing if we can get him to the meeting?
Karen: No problem.

Let's end the scenario. The main accomplishments were that people talked to each other in a civil manner, there was no animosity, there were no labor versus management issues, and everybody learned that the problem wasn't just that Kim wasn't cleaning well enough, but that she had actually tried to solve the problem. Perhaps Kim should have told Andy, her supervisor, that she couldn't get the stains out, but she didn't. That's not the issue for right now. The issue is that the focus group learned the full extent of the problem, and Heather was able to report to everybody in the department the steps that the focus group was taking on behalf of the entire department. This approach may not be applicable to every conflict that might arise, but it does work with many situations.

There are three additional points to be made. First, it's important that there is a formal structure for resolving the conflict. The focus group was that structure. It was similar to a self-managed team, which will be described in the next section. Second, Dave asked specific people to do specific things for the next meeting. He wanted to make sure there was going to be some accountability by a person, not a generic "somebody." I think most of us already know that if you don't ask a specific person to do a specific thing, there's more than a good chance that nothing will get done. Lastly, and most important, Dave approached the problem as an outsider looking in. He didn't approach it as the boss who was going to resolve the problem immediately and then go on with his business (had he done that, it may have led to lasting resentment from the "losing" party). Rather, he talked with an unbiased point of view and helped the focus group resolve the problem. Earlier in this chapter, when discussing how to have a difficult conversation between individuals, one of the points made was that you should approach a difficult conversation from an unbiased point of view, gathering information and listening, rather than accusing. You try to determine if there's an underlying problem. In this case, the real problem was that the stain grossed people out, even if it wasn't dirt. The similarities between resolving intrapersonal problems and group problems are noteworthy, and you might want to revisit the earlier part of this chapter. When handled properly, the resolution of conflict can lead to enhanced morale.

Bridging the Generational Gap

When we think about fitting the person to the job, we have to give consideration to the age differences that are so often found in animal facilities. Where I work, at the University of Massachusetts Medical School, we have department members whose ages range from new high school graduates to those getting ready to retire.

The facts are simple: people of significantly different ages may think and react differently when exposed to the same set of circumstances. This is discussed briefly at the end of Appendix 3, in terms of what you might anticipate when interviewing and hiring people from a generation different than your own.

The newest generation is commonly called Generation Y or the Millennial Generation. These are people born anywhere from about 1985 to 2005, and some are beginning to enter the workforce. My daughter, born in 1997, fits into this generation. If you know any Millennials, you know that they are so attached to their smartphones that you rarely see one without the other. They readily accept racial and cultural diversity, are ambitious, crave feedback and recognition, are independent (yet tolerant of micromanagement), and are willing to work hard to achieve personal goals (including money) but also want the time to play hard. These descriptors apply to Millennials who are technicians, managers, office workers, veterinarians, and so forth. They all share similar characteristics that are not defined by their jobs in the animal facility.

What does this mean to you if you are their manager but not a Millennial? It means that you have to do much of what was described earlier in this chapter and provide the same feedback and recognition that everybody wants but perhaps, at first, a little more than you might give to older coworkers. We (who are often older) want to help Millennials find the flexibility and satisfaction they want from their work, but we have to be honest with them and point out that many of the daily tasks we do in laboratory animal science are repetitive and are usually done the same way to help ensure research consistency, such as the way we change cages or care for mice with dermatitis. Because Millennials value their free time, we might find that a desire to have flexible work hours is more of the norm rather than the exception, and we should plan for implementing the same. We should also plan on incorporating Millennials' opinions into many of our decisions, as they want to feel like their opinions count [82]. If we can provide Millennials with exposure to all the different animals and responsibilities that are found in many animal facilities, we should consider doing so. Further, Millennials often want detailed instructions on what is expected of them. so be explicit about how to change a cage, hold a mouse, and even when to arrive and when to leave.

We should be promoting continuing education wherever possible and never forget the that Millennials (and others) may want us to help by becoming their mentor. Understand, of course, that Millennials will often change jobs even with the best of supervision, and this has given them the reputation for job hopping. However, I have not seen any statistics indicating that Millennials change jobs any faster than the generation that came before them (the so-called Gen X) when they were of the same age or, for that matter, the baby boomers who came before Gen X. If the Millennial Generation does not seem to be loyal to their employers, ask yourself how loyal most employers are to their current employees. For better or worse, we now live in a nation where you have to take charge of your own career.

Eventually, Millennials will get older and their outlook on life will likely change somewhat. They will get married, have families, and look for more job security. Yet, no matter how old you or I or Millennials get, and although we all change over time, we also retain certain traits from our youth, which is what keeps managers on their toes, always facing new and sometimes stimulating personnel challenges.

Allowing Employees to Solve Problems: Empowerment through Self-Managed Work Teams

Now that the third edition of this book is being published, it seems fair to ask if my earlier disappointment with the lack of team empowerment in laboratory animal facilities (as stated in the second edition) has dissipated. With men and women from the Millennium Generation already in or entering into the workforce, should we expect a greater desire to work in teams emanating from a generation that appears to enjoy teamwork? It's hard to give a firm answer because there have been no published surveys or other documents indicating the extent that employee-empowered teams are being used in laboratory animal facilities, although word of mouth suggests that they are starting to be used more than in the past. Perhaps this slowness in developing and using self-managed teams is a reflection of the extent that a typical hierarchal management pyramid is still the norm in most animal facilities, and it may take more time for the general culture to change to the point where employees feel more comfortable working in self-managed teams and a newer generation of managers feels comfortable advocating for self-managed teams.

Employee-empowered teams (such as the focus group mentioned earlier) are groups of employees who have been given the freedom to act independently, yet within the boundaries of the consistency that is needed for biomedical research. By becoming part of the decision-making process, rather than the recipients of another person's decision, there is greater "buy-in" from the entire group, resulting in greater motivation to reach the team's goal. There also are some potential benefits to individual team members. For instance, individual team members can be positively impacted by team-level empowerment [83], and it has been shown (in women) that the motivation of the team members with the least ability can be increased by the strong social ties that often form between members of a team [84].

The concept behind team empowerment is to define the parameters of a goal, make sure the team understands why the goal is important, and then let the members of the team decide how to accomplish the goal. If the goal is reached, there should be tangible rewards whenever possible (e.g., a bonus or other form of recognition) for *all* members of that team. This does not mean that we throw consistency in animal care out the window and let people do as they choose; to suggest that is research suicide. Rather, the need for consistency is simply a "given" in establishing a new model of managing laboratory animal facilities. There are other key benefits to empowering employees that are listed below [85].

Benefits of Empowering Employees

- **Employees are less afraid to take risks and may produce better solutions to problems.**
- **Employees are encouraged to be innovative and creative.**
- **Well-functioning teams lead to motivated employees.**
- **There is teamwork among managers and employees.**
- **They work together as adults, not in an adult-to-child relationship.**
- **Managers have more time available to handle other tasks.**

The example that follows uses a crisis as the impetus to initiate the process of empowering employees. That is not at all necessary. Look around you. Is entitlement a problem? Could you be more efficient and effective than you are now in taking an animal census? Do you think that your environmental enrichment program needs to be improved? Is there wasted time or effort during cage changing or cage washing? If the answer to any of these or similar questions is yes, then this is a good time to start thinking about using employee teams rather than depending solely on supervisors and managers to come up with solutions. Sometimes, there is no clear answer to the question, and if that occurs, an employee team, working with other involved employees, can investigate a process to see if there is wasted time or supplies. Very few people know more about where waste is occurring than those technicians who work every day on the floor of the animal facility.

> *McGee Pharmaceuticals has just been purchased by a larger pharmaceutical company. Both companies have laboratory animal facilities. Knowing that the new owners will keep only the most financially efficient animal facility open, you must figure out how to become the surviving facility. In other words, the problem is, how can we decrease costs, eliminate errors, and give the same or a higher level of service to our two customers—the animals and the investigators?*

This is not an easy problem because if it were, you would have solved it long ago. One way of approaching the problem is to take it upon yourself, alone or with other managers, and develop goals, plans, and strategies that will increase efficiency while maintaining effectiveness. If you like this option, you might want to jump to Appendix 1 and quickly read through it, paying particular attention to the discussion on Lean management. Another option is to have one or more teams of frontline employees have nearly full say over how this should be done. Sometimes, there will be a simple working group (similar to what we all know as a "committee"), and at times, there will be true teams (shared leadership, shared and individual accountability, and a shared desire to have the team succeed). Either way, your job is to be a facilitator, and only make decisions when it is absolutely necessary to do so. In fact, the more I work with empowered teams, the more I have recognized that a team leader (i.e., a person who takes the leadership role from within the team) or a good team facilitator (a person who is not formally part of the team but helps clarify tricky concepts and keeps the team going in the right direction) is vitally important to the success of the team. I'll have more to say about this in just a moment.

Your job as a manager or team facilitator is also to be a cheerleader, and continually encourage the team members and remind them about the importance of their work. Indeed, without your support and the support of the other leadership of your animal facility, there is a good chance that the team will not be successful. At all other times, the team should have the authority and responsibility for achieving its goals. Your job is to do the following:

- *Define the problem to all the team members* (e.g., we are in competition with another laboratory animal facility to see which one remains open). We all may lose our jobs if we can't substantially cut back on our expenses and keep up at least the same level of service. All people involved in the change must clearly understand the need for change and what it means to them personally. Do not be afraid to share this type of information; it establishes the urgency and importance of the project. One of the main reasons some empowered teams fail is that the team does not clearly understand their overall task. For example, if the problem is that an animal census takes every animal care technician about four hours a month, and the accuracy of the census is questionable, how will you present that problem to the team? Will you define the problem as "We have to do a better job with taking the census" or "We have 10 animal care technicians and together they spend 40 hours a month taking an animal census that the investigators complain about as being inaccurate. Our goal is to increase accuracy while decreasing the time it takes to do the census."
- *Define the working rules for the team.* The facilitator has to make it very clear that team members are to be honest and open with each other. Like a sports team, they have to be able to depend on each other, support each other, and be accountable to each other. Nobody can have a personal agenda; rather, there has to be a team agenda. It's most important that they are all goal focused and willing to work as a unit to reach the goal.
- *Define the goals.* Employees may know what the problem is and why it's important, but everybody must understand the goals to be reached. Very

often, the employees themselves will define the goals, but you should be ready to help. In the McGee example, one goal may be to decrease overtime by 50% within a four-week period, or perhaps to increase the census accuracy to 98% (of course, you will first have to determine the current accuracy of the census). Goals must be reasonably objective so that they can be readily evaluated and quantified (i.e., you must be able to easily determine if you are reaching your goal). Once the goal is established, the team will have to develop the strategies (plans) to accomplish the goal.

Be cautious about a team member who is enthusiastic about the team's goal but is subtly (or sometimes not so subtly) advocating for his own goals. Sometimes the team can handle this problem by itself, but especially for newly formed teams, the team facilitator may have to intervene to provide the direction and demeanor needed to keep the team working as a cohesive unit.

▪ *Guide the team into establishing specific responsibilities for different people on the team.* "Collaboration improves when the roles of individual team members are clearly defined and well understood" [86]. Nevertheless, don't forget that the team is a team, not a committee. There will be times when people have to work alone or with another person and report back to the team, but if it becomes common to have everybody going off in different directions, it can undermine the cohesiveness of the team.

▪ *Define the limiting parameters for the team* (e.g., the amount of money available for the team, the available time, and whether the team has authority to change work hours for some people or, for example, meet with vendors who have bar-coded or similar census systems).

▪ *Define how success will be evaluated.* For instance, both animal facilities will house primarily rodents and rabbits. At the end of each month, our total expenses will be measured against the total number of animals in our facility. This process will go on for six months, during which time investigators will be asked by management about their level of satisfaction with our work. Then, a final decision will be made by management.

If it is not clear how success will be evaluated, let the team make this decision. The bottom line is that there have to be results.

▪ *Give people the opportunity to succeed and to fail.* This is critical. Once the team understands its goal, it has to develop the strategies needed to accomplish the goal. Your job as a manager or team facilitator is to run interference for the team when necessary. Give the team the resources it needs. Let them consult with specialists. Let them consult with investigators. Don't let a supervisor keep a good person from joining the team because she wants him to spend all of his time caring for the animals. In the example given about McGee Pharmaceuticals, there is no need to assume that all the team members have to be efficiency experts. It may be sufficient that, if desired, they have the competency and ability to hire or consult with such a person. Nevertheless, you don't want to be such a protective mother that people on the team aren't given the opportunity to make readily correctable (noncritical) errors that might provide them with a learning experience about what not to do if a similar situation arises. In other words, in addition to giving people the opportunity to succeed, there are times when you have to give them the opportunity to fail. You should not read this as being a license to fail under any conditions. It simply means that there are times when we can learn as much from failures as from successes, particularly when *failure* indicates that a best effort was made, but it was not successful. As a manager, part of your job is to help analyze a failure to see if a lesson can be learned from it.

In addition to giving people the opportunity to succeed, there are times when you have to give them the opportunity to fail.

■ *Show your own leadership as a facilitator* by gently guiding the team into developing a plan of action, even if it isn't the one you would choose. Don't let the team waste 90% of its time on theoretical considerations or allow one person to dominate the discussion. Explain potentially difficult concepts in plain language that everybody on the team can understand. Remember that serving on the team, as important as it may be, is just a part of a person's job, and sometimes, when the going gets tough, people are willing to take the path of least resistance (i.e., the easiest path) rather than the best path. The facilitator cannot demand that the team go in a particular direction, but you can explain the pitfalls of taking a path that clearly is not the best path. Keep the team focused!

■ *Make sure that all the players, not just the team members, know what is happening and are willing to accept change.* This includes you and your own boss. All of your actions, and your boss's actions, are being closely followed by your employees. If you say one thing but do another, you just undermined your credibility. Communication is more than words. Body language and other actions count heavily. As always, provide feedback. Open communications within the team is absolutely critical for its success.

■ *Use small, functional teams.* You are better off with small, functional teams, concentrating on one issue, rather than one large team concentrating on all issues. If appropriate, a smaller team can report to the entire group. Even in small laboratory animal facilities, groups of employees can focus on a limited number of issues because the same person can be on more than one team. How small should a team be? I would suggest 5–10 people. Others have suggested as many as 20 people, but I think you will find that about 7 people is the most you can expect to work together as a strong team. It helps if some of the team members know each other, as the establishment of trusting relationships will occur in less time [86].

▪ You may find it useful to assemble a team with members that have some diversity in their backgrounds and personalities. Some people tend to summarize discussions, some act as devil's advocates, some are analytical, and others may be more emotional. Taken as a whole, teams that are comprised of people with different viewpoints can offer a more balanced discussion than people who act and think alike. Having disagreements that are approached from a common-goal viewpoint rather than a selfish viewpoint can actually strengthen teams because across-the-board agreement suggests that the group doesn't have very many ideas, or they value agreement over the quality of a suggestion [87]. Those of you with any IACUC experience recognize that one of the strengths of that committee lies in the diversity of the backgrounds of its members.

▪ *Put words into action.* All the plans in the world mean nothing if you do not act on them. As a manager, you should be ready to accept a team's plan and help implement that plan. One of the only circumstances where you might have to deny support to a team's plan is if you are quite certain that the plan will fail and subsequently create problems for the animal facility. In that unusual circumstance, it behooves you to meet with the team, clearly explain your concern, and work with the team to find a satisfactory resolution.

▪ *Measure the working group's or team's productivity.* The success or failure of a team or working group should be monitored on an ongoing basis, even if there is a single goal. I suggest going back to Chapter 1 and rereading the discussion on productivity evaluation, and then turn to Appendix 1 for a much more detailed discussion on the subject.

▪ *Reward success.* Whether it is a small party, a bonus for all members, or something else, a little tangible reward is always appreciated when a group succeeds.

▪ *Keep group membership fresh.* The longer the same group of people works together, the greater is the possibility of "same think" or a developing bureaucracy [88]. On the other hand, Huckman and Staats [89] have made a strong argument for keeping teams intact, noting that team member familiarity leads to better functional results in many areas. I can appreciate both points of view. For example, a cage wash team that has been together for years is likely to work more efficiently than one that rotates its technicians every year or even more often than that. An environmental enrichment committee may do better work by using staggered terms of service, so there is a turnover of about one-third of the group every year. That gives people three years of service before they rotate off and give others an opportunity to serve.

What I just described is your job as a facilitator for either a true team or a working group. Whether a team or simple working group is better for self-empowered employees is likely to be a function of the organizational culture and available leadership. In either instance, the single most critical aspect of group dynamics in an empowered group is to have trust and open communications between the group's members. Without the feeling that the group members are there to support each other and do what is best for the group, self-empowerment will fail because the members will think that others are looking out only for their own interests. Additionally, the group must be willing to accept all opinions brought forth, accept disagreements, and look at the benefit to the group, not any one individual within the group. Lastly, for a group to be truly empowered, it has to look to the group itself for accountability, not to any one person within or outside of the group [90].

To sum things up, your managerial function with self-directed groups is to help direct, organize, facilitate, give support, untangle problems, and reinforce and promote the team's achievements. If this sounds suspiciously like the basic roles of a manager, you're right, of course. The employees decide on work schedules, needs, discipline, and so forth. This is a difficult pill to swallow for some managers because it goes against the old "command and control" managerial thinking. Not only do you have to give up power, but also you have to bite the bullet if the team decides to do something you don't entirely agree with. Still, it fits the definition of management, which is to use resources efficiently and effectively to reach a goal. It does not interfere with the need for consistency in animal care or research because you, the manager, have set the limits of the group's authority. Research integrity is not being compromised. At some future time, once the team concept sinks in and is working, you may be able to expand the responsibility of your teams, and therein further empower the teams and their members.

"That won't work in my facility," says the cynic, "we're unionized." So what? Unionization should make no difference in motivation or team empowerment. Does joining a union make a person's brain deteriorate into putty? Does your union representative routinely tell your staff not to be motivated or empowered and to walk around like zombies? Unions have a stake in the success of their members and the success of the entire organization. Unions want fairness, not aggravation. I have worked with unions and without unions, and I have been in a union myself. The presence of a union should have little to do with your daily interactions with people. In most organizations (unfortunately, not all), you can almost invariably implement everything we are discussing with or without a union.

Two problems occasionally arise. First, although you may have willingly given up some of your power and counseled and trained your supervisors to do the same, the real world may not be following the textbook. There may be a supervisor who doesn't want to give up any power. When your back is turned, the supervisor is conducting business as usual because she believes that exhibiting power leads to order, and order will lead to peace within her group. That's unfortunate because a command and control management style is no longer viable. (We know what would happen to most dictators if the people being suppressed had their way.) It's outside of the intent of this book to delve into specific tactics for handling the recalcitrant supervisor, other than the communication needs discussed earlier in this chapter. Certainly, this person isn't a team player and may have to leave if she doesn't make some rapid attitudinal changes. But as a manager, you have the basic tools to help prevent this problem from happening. These include knowing the personality of that supervisor (and therefore how to communicate with her), clearly defining why something is being done, providing training as necessary, and remembering to listen and get feedback from the supervisor before moving ahead. Most people are willing to change, and if empowered, they are usually willing to help be the cheerleaders for change. Yet, try as you may, some people are simply resistant to change. It's not going to do you much good to insist upon doing things your way or trying to compromise, because it's easy for a person to sabotage your efforts. If you can, find out from the person why that behavior is occurring and try to address the underlying problem. This is often easier said than done, but you have nothing to lose by trying. If nothing works, somebody may have to move on to a new job.

Although the majority of employees are willing to work in a team, not all are so helpful. That's the second problem. You may come across a person (as I did) whose primary concern was, "What do we get out of it?" I asked him if he was speaking for himself or the rest of the team, but he dodged the question. He was implying he wanted more money for being on the team since he would be doing "supervisor's work." I told him about the benefits of self-determination in terms of the team being able to put its own ideas into action; that it would mean less, rather than more, supervision; that there is a feeling of accomplishment; that it will give him status among his coworkers; and so on. I pointed out that, within the confines of research needs, there is more than one way of getting the job done, and that employee empowerment is meant to give people both self-determination and better use of the skills they've developed over the years. Those arguments didn't hold any water, and even some of the team members were unable to convince him of the advantages of self-management. In retrospect, I should have asked the team what they would want to get out of an employee empowerment program and what it would take for them to buy into it. Unfortunately, I didn't do that. Not everybody is cut out to be a team member. You will just have to accept that fact and work with the team to find a new member who shares the team's all-for-one and one-for-all philosophy.

I think there may be good reasons to provide rewards to people working on teams, if the teams have successfully accomplished their goals. Parker et al. [91], based on their real-world experience, suggest six different reward structures, focusing primarily on group, rather than individual, rewards. Nevertheless, rewards have to be handled carefully, lest those teams that do not get a reward simply throw in the towel and stop trying.

As I'll note below, the concept of teams didn't work for me at first, largely because of my own inability to describe and implement the basics of an employee empowerment program. It seems to me that before anybody wades into the waters of management teams, they should first learn a lot more than I did.

Is this idea of team empowerment realistic or simplistic? I really don't know yet because, to the best of my knowledge, the use of self-managed teams has not been tried on a large scale in laboratory animal science. Some organizations, particularly in the pharmaceutical industry, have used quality circles, total quality management, and similar management tools, such as Lean management, but I know of none that has taken the leap into totally self-managed teams.

Self-managed teams are being tried in many industries, at different levels, including with production line workers [92]. Their efforts and successes are bolstered by financial and other rewards, as well as true recognition from managers throughout the organization. The only reminder I would add is that rewards should be for all or for none. For this teamwork approach to be successful, there must be group, not individual, accountability.

For the teamwork approach to be successful, there must be group, not individual, accountability.

Resistance to Change

Throughout this chapter, there has been frequent use of the word *change*. There were discussions on changes in behavior, understanding the need for change, the need to inform employees of upcoming changes, and so forth. Nevertheless, we all know from our personal experiences that at one time or another, people are reluctant to initiate a change. Some people just have a low tolerance for change, some do not understand what the change means and therefore prefer to keep the status quo, a few may think that they will lose something of value from the change (such as vacation time), and some may simply think that the proposed change makes no sense at all [93]. Fear of change is not limited to managers. Indeed, one of the most difficult problems I have faced over the years is a reluctance of our managerial staff to initiate changes. Take this another level higher, and you will find that even college deans often have trouble selling a proposed change to department chairpersons. If a proposed change does not have the support of supervisors and managers, you can be reasonably sure it will not get very far with other employees. This statement is related to the earlier discussion about motivation, where I noted that if managers themselves are not motivated, it's going to be hard for them to motivate others. Therefore, we have to ask how a manager can begin to motivate and influence people to overcome resistance to change.

Once again, we fall back on the two foundations of human resources management: communication and trust. A manager should openly and honestly describe the proposed change and why it is important, while at the same time listening to and evaluating any concerns, including opposing viewpoints. Many times, proposed changes can be very broad and similar to a vision statement, and as with vision statements, it is important to frequently discuss and promote the proposed change with all of your coworkers. This is part of the process of trying to influence people to accept the change. Some employees tend to have more influence or credibility among coworkers, and it is important to spend additional time with these influential people to make very sure that they fully understand the significance of the change, why it is needed, and that they will support the change.

Here's a question for you that has been raised by others [94]: Do you think that an early and open discussion about initiating a change should be made? Or, do you think it is better for a manager to delay announcing a planned changed, and in the interim try to influence selected people to accept the change, while at the same time making small initial steps toward effecting the change? There's no clear answer to this question, even though both methods rely on communicating the change. It's just the timing that differs. The first method may be more typically used in the United States, where early and honest communication is often advocated. In other countries, such as Japan, it is often believed that a gradual change paradigm allows a manager greater flexibility in fine-tuning a planned change and allowing people to get used to some aspects of the change before the entire change is rolled out [94].

Whether you roll out the change little by little or all at once, getting support for the change from the opinion leaders of your group, department, or institution as a whole is often a critical part of initiating a change. If these people don't buy into your idea, it may become an uphill battle to get others to go along. Ted Wright [95] states that if we want these influential people to agree with our reason for a change, the story about the change must be relevant, interesting, and authentic. That is, it should have meaning to that opinion leading individual (e.g., the initiation of flexible working hours), it should be interesting (e.g., the advantages and disadvantages of flexible hours), and it should be authentic (it involves something that the person has heard of and may know a little about it). If your story can meet those three conditions, you should tell it over and over and over again. Those influential people, assuming you have influenced them, will carry the ball for you by repeating it to others.

When telling your story to influential people in your organization, make sure that it is relevant, interesting, and authentic.

There will be times when a manager has to make some compromises to move a change forward (see the discussion below on negotiations), but before doing so, you should look for a win–win outcome. For example, if your animal facility will begin to use a computerized animal census process, you may find some resistance based on one or more of the factors previously mentioned. However, if the training required to use the new system will provide some people with an opportunity for career advancement, a pay increase, job diversity, and so forth, you may have a win–win situation staring you in the face.

It is also a responsibility of managers to make absolutely sure that if a change requires new processes or new equipment, the equipment will be available and people will be trained in the processes. Here's what happened at the University of Massachusetts Medical School.

> The University of Massachusetts Medical School was preparing to build a new vivarium, and there were early discussions about the use of washroom robotics to help clean mouse cages. Before any definitive decisions were made, the entire staff was told that robotics was being considered; what did they think about that idea? Although some of the staff had prior experience with robotics, nobody had experience with the type of robots that were being considered for the new building. Most, but not all, of the technicians favored the change. A small number preferred to keep the status quo, which was the use of a tunnel washer with people at one end dumping cages and people at the other end removing them from the washer belt after they were cleaned. They felt that the manual method contributed to full employment. In response, managers made it clear that there would be no layoffs, animal care technicians would still be needed to work in the cage cleaning areas, and there would be a need for even more animal care technicians because there would be additional housing rooms for animals. To add to the win–win situation that was developing, the technicians were reminded that those people working in the cage cleaning area who needed and received robotics training would be in a higher salary grade level if they were able to satisfactorily implement their training. That sealed the deal.
>
> As the vivarium was being built, tours were provided for all employees so everybody could see how the pieces were coming together. The technicians who volunteered to work in the new cage cleaning area were provided with training and retraining as the new equipment was being installed. Before the vivarium had its official opening, the technicians had already identified numerous "bugs" in the system, and they played an active part in helping the vendors correct the problems. After the official opening, the technicians continued to find small operational problems, but they capably handled almost every incident that arose.

This is a nice example of a change that was well planned, communicated, and executed. Not every change will be as smooth as this one, but communications, trust, planning, and training were the tools used to facilitate the change.

Basic Requirements for Implementing a Change

1. Have a strong and well-considered reason for implementing the change.
2. Communicate often about the need for the change and the benefits that are anticipated.
3. Be honest. Do not promise what reasonably cannot be delivered.
4. Solicit the support of the opinion leaders of your group.
5. Understand that different people will have different reasons for accepting or rejecting a change.
6. Accept and honestly evaluate feedback about the change.
7. Keep people informed about progress in implementing the change. Mention successes and failures and what will be done to try to rectify failures.

Negotiation

In the previous sections, there were discussions about influence, motivation, and resistance to change. I said that influence is often used when we are faced with resistance to change because we are trying to encourage a person to make a particular choice, whether or not the person is motivated to make that choice. Influence often leads to the desired result, but not always. When it does not succeed, there may be room for negotiation. A negotiation is a discussion between people who are trying to avoid a dispute and reach a mutually acceptable outcome. Managers may find themselves negotiating many things on a daily basis, from the starting salary of a new employee to the time and place of a meeting. As with most management topics, negotiating is a skill that can be learned. There may be some people who are natural negotiators, but most of us are not. Most of us have a certain amount of anxiety when entering into a negotiation (think about how you felt if you ever had to negotiate to increase your own salary), and anxiety can lead to suboptimal negotiation outcomes (you may make a weak first offer, respond too quickly to a counteroffer, or try to end the negotiation too early) [96]. One good recommendation is to warm yourself up for the negotiation by trying to establish a particular mindset [97]. For example, if you want to come across as a pleasant but forceful person, but that's not your true persona, you might try to picture yourself as having the temperament of former Secretary of State Colin Powell, or perhaps football quarterback Peyton Manning. You have to prepare yourself emotionally for a negotiation [97]. Practice, and then practice some more.

> **A negotiation is a discussion between people who are trying to avoid a dispute and reach a mutually acceptable outcome.**

The best outcome of a negotiation is when all of the participants feel they have reached a *reasonably good* outcome, even if it is not the *best* outcome for them. This is the desired win–win outcome. It's far too easy to fall into the trap of thinking that your side of the negotiation has to get everything it wants as long as the opposing side has a minor gain of some sort. Let me give you an example:

> *While I was updating this chapter, U.S. Secretary of State John Kerry had finished negotiating a framework agreement with his Iranian counterpart that would limit the use of nuclear materials in Iran to processes that would not lead to the manufacture of nuclear weapons. As soon as the agreement was announced, there was an outcry from some members of the U.S. Congress, claiming that it wasn't the best deal the United States could have negotiated. Perhaps that was true, but if it was the best deal for the United States, it would not have been the best deal for Iran, and there may not have been any agreement at all.*

We have to ask, is it better to have a win–win framework for future negotiations, or should you obtain an "I win–you don't win" outcome as the starting point for future bargaining? "In the long run, win-lose is really *lose-lose*. One of the negotiators is bound to go into the next negotiations with a jaundiced eye and ready for combat instead of intelligent exchanges" [98].

Before entering any negotiation, you have to make a basic decision: Should I accept the status quo, or should I negotiate? Some valuable rules of thumb are [99]

1. *Do not negotiate when you might easily lose something you truly want.* For example, if you find that good, fully reconditioned monkey cages are available to you at a very good price, don't waste much time negotiating. You may lose the cages to another person who quickly agrees to pay the asking price for the cages.
2. *Do not negotiate if little or no benefit will come from negotiating.* Let's look at the caging example I just used from the viewpoint of the seller. If the seller is confident that she will be able to get her asking price for the cages, then there is no need for her to waste her time negotiating with a potential buyer who wants a lower price.

3. *You should negotiate when you have a cushion to fall back on.* If you need monkey cages and there are a few vendors who have good reconditioned cages available at a good price, try negotiating with the vendor of your choice for an even better price. Even if the negotiation fails, you have alternative vendors with the same or a similar product.

The basics behind being a good negotiator don't differ much from the basics that were discussed earlier under the headings of "Difficult Conversations" and "Influencing People to Accomplish Goals." First and foremost, you have to put yourself in the shoes of the other person. That is, you have to be empathetic and try to understand *why* the opposing person is taking a particular position. Any reader who has been to an IACUC meeting knows that investigators are often passionate about their research and just want to get it started as fast as possible. The IACUC has certain legal constraints that affect the study approval process, and the committee may not be able to approve the research protocol as quickly as an investigator wants it to be approved. Both parties have to show empathy toward each other. Ideally, they also have to find the root causes of the problem and not waste too much time dancing around unimportant topics unless they need at least one small mutual "win" to get the negotiation process into full swing.

Negotiators, like people trying to exert influence on others, will find it useful to seek common ground with their opponent. If you are dealing with a person who likes to use examples, then you should use examples in your negotiations. If your opponent seems to be a bottom-line, "just get the job done" person, then you should be very clear and direct about what you say and come prepared with documented facts. If your opponent seems to be conflict averse, then don't start a fight. Indeed, when negotiating, it might be tempting to threaten your counterpart or intentionally lose your temper when making a point. In some instances, this has worked, but in general, it is far better to remain calm and collected and act civilly. Does that mean that you should be quiet and indecisive? Not at all! A good negotiator is not afraid to raise his voice to make a point and always acts decisively, but that is not the same as acting aggressively.

Think about negotiating the price of a new car with a salesman. You have done your homework and you enter the negotiation with a price in mind. The salesman also has a price in mind, but neither of you know what the other is thinking until you ask for the price and the salesman says $25,000 plus the $850 transportation fee and $125 for needed documents and the car's state-required inspection. That's a total of $25,975. You really want the car, but your goal is to pay $25,500 for everything. You look at the salesman and say that you are willing to pay a total of $25,000 and nothing more. The salesman says he doesn't think that's possible, but he'll check with the sales manager. After leaving you to look at the wall for about 10 minutes, he returns and says the sales manager is willing to drop the $125 document and inspection fee, but he can't go any lower; the final price will be $25,850.

That's still $350 over the maximum you wanted to pay. Will you go a little higher and hope the sales manager will continue to bargain with you, or will you walk away from the negotiation? Well, you have nothing to lose by trying because by simply assuming the sales manager will not go any lower, you've lost any further negotiation potential. But maybe there's something else you can do other than negotiate for a lower price. What would happen if you said you would pay $25,850 as long as the dealership will include routine oil changes and tire rotations for the first year of ownership? That might save you about $175 but only cost the dealership about $95. It might be a win–win negotiation for you and the dealership. You saved more money, and not only did the dealership make a profit, but also, more important to the dealership, you may become used to coming to the dealership and turn out to be a regular customer in its service area.

Sometimes you have to negotiate for your own needs. Perhaps for a salary adjustment, a new position, a need to work from home, or a quieter office. The same principles that I just described for general negotiations hold for your personal negotiations. By entering into a negotiation about something that is important to you, especially if you are being given new or additional responsibilities, you are probably showing more initiative than most people would do. Don't be shy about negotiating, but don't negotiate unless it's important to you and you can show how a successful agreement can be of benefit to your organization.

There are five important points to any negotiation.

1. *Know what you want to accomplish.* In the car example I just used, you knew what car you wanted and how much you wanted to pay. It's appropriate (and a good idea) to justify your offer [100]. For example, why are you willing to pay $25,000 but not $25,850? Is it because during your online price search you found that a good number of people payed $25,000 for the same car having the same options that you want?

2. *Know what your opponent wants.* You can use many online services to determine what the dealership paid for the car and what a fair markup would be. The dealership wants at least the fair markup and possibly a little more.

3. *Don't rush into the negotiation.* It is better to assess the personality of your opponent and try to find areas of commonality.

4. *Begin the negotiation.* If possible, have your opponents state first what they want (that will give you a feel for your negotiating limits). If you, and not the salesman, make the first offer, it should be the most extreme offer that you think your counterpart will seriously consider [99].

5. *Finally, know when to walk away from the negotiation.* Either you or the dealership has the right to say no to each other's offers or to take a time-out if it seems that tempers are getting hot. If you can't reach a win–win outcome, and if the product or service is not critical to you, walk away from the negotiation with a handshake and expression of regrets. Perhaps you can renew the negotiation at another time.

Interacting with Highly Educated People: Managing the Unmanageable?

Having read an entire chapter on how to interact with and manage most people with whom we work, it seems appropriate to add a few additional words about managing a unique group of people. These are investigators and other highly educated people. Some people claim they are unmanageable, but that's not so. Everything written earlier in this chapter holds true, but because many highly educated people are very analytical in their thinking and driven in their work, some additional understanding is needed.

As you know, laboratory animal facilities are somewhat unique because of the diversity of people that work in and around them. There are talented people with little formal education working side by side with veterinarians, PhDs, physicians, and other highly educated people. Working with, or managing highly educated people, can be a challenge for the unwary manager. Some years ago, Peter Drucker said that the best way to manage knowledge workers was to leave them alone. I'm not sure if those were his actual words, but it does drive home the fact that highly educated people may require management techniques that differ a little from the traditional ways of management. What's the reason for this? It was nicely condensed into one sentence by Goffee and Jones [101], who wrote, "If clever people have one defining characteristic, it is that they do not want to be led." I can only agree. Scientists want independence and not to be managed. It's one of the reasons they like being principal investigators. I'm sympathetic to that desire, even though at times it can be frustrating to me as an animal facility director. My coworkers have heard me say for many years now that I don't like authority. I don't like being bossed, and I get mad at myself when I remember that I'm somebody else's boss. I don't think of myself as being particularly clever, but I do have to admit to having a lot of education, and after all is said and done, there is a basic truth that managing and interacting with highly educated people can be a challenge. When I wrote my final report for my master's degree, the topic was managing scientists. One point from that paper that has remained with me over the years is that if anybody has to lead a scientist, it should preferably be another scientist in the same discipline. That's why the chairperson of a pathology department is likely to be a pathologist and the head of a veterinary group is likely to be a veterinarian. How, then, does one manage (or at least interact with) the scientists we often encounter in laboratory animal science? The first thing we have to understand is that they are as passionate about their work as we are about ours. Scientists don't like to be bogged down in what they consider to be administrative nonsense. Many who work with animals consider the IACUC to be a prime example of administrative nonsense. The scientist wakes up in the morning with a "Eureka!" idea and doesn't want excuses about why she has to get approval from the IACUC to do an experiment. In fact, she expects you to drop everything you're doing to expedite her research. It's not as simple as saying scientists are self-centered; rather, it's part of their passion for science. Second, like all of us, scientists want, and often demand, respect. The only difference is that the scientist is often not shy about her demand. You may hear her say, "I'm Dr. Smith, not Joan." Don't get upset; just accept this as a typical personality trait of some, but certainly not all, scientists. Finally, don't expect too many thank-you statements; such "frivolities" are not a high-priority item for many scientists.

Of course, not all highly educated people are scientists or veterinarians. There are many different highly educated people with whom we interact (such as physicians, chief financial officers, chief operating officers, and hospital administrators), but the fact remains that the research scientist and veterinarian are typically the ones with whom we most often work in animal facilities. Now that we know some key personality traits of highly educated people, let's return to answer my question about how we interact with and manage them. The answer is that we have to treat them somewhat differently from others. This is usually not a big problem. For example, don't begin by addressing the scientist as Joan until she gives you the opportunity to do so. If she says, "Please call me Joan," then do just that. Otherwise, she's Dr. Smith. If she addresses you by your first name, there's no reason not to use her first name. Because most scientists are frustrated with bureaucracy and want quick access to you and quick results in general, you should try to provide that access and service. Can you work with this person to find a solution to their problem? Are you going into the conversation with a "Yes, I'm going to try to help" frame of mind? If not, you should be. These are people we have to work with day in and day out. They are our customers. They help pay our rent. We want them to think positively of the vivarium and its staff and convey these positive opinions to other investigators. Go back to Chapter 2, where I emphasized that we usually have two customers, animals and researchers. We cannot view the scientist as an inconvenience to be dealt with, because if we do, they're going to return the favor and make our lives miserable.

What are additional ways to work with scientists? From here, I'll draw on some of the key points made by Goffee and Jones [101]. We should understand that scientists are organizationally savvy and know how to work the system to get what they need. For us, as managers, that means they will go over our heads to our bosses to get supplies, money, or animal housing space. In fact, there's a good chance they know our bosses quite well because they are typically well connected within the organization. But, if part of our job is to try to expedite science, then the best way to interact with the scientist is to become an ally, not a hurdle for her to traverse. Try your best to have a "can do" attitude if you want to become a person's ally. That doesn't mean flaunting federal regulations, becoming subservient, or doing something foolish; it just means that we have to try to be as helpful as possible within the context of good animal facility management. It's as important to approach the situation with a positive attitude as it is to be able to satisfy a scientist's needs.

There will be times when you, the manager, are a scientist who is leading other scientists, or a veterinarian leading other veterinarians. In this situation, your job is not to prove that you're smarter than the others, but to be able to converse on the same level as the others. In plain language, that means it will help you quite a bit if you're up on the literature and new trends in your field. You have to be a champion and do what you can to remove barriers to intellectual endeavors. You also have to make sure that there is intellectual stimulation available. For example, do you seize upon interesting clinical problems for veterinarians to work up in detail, even if they aren't immediately pertinent to your daily operations? Do you suggest research projects when opportunities and time permit? Can you provide start-up funds for small research projects? Do you have regular clinical rounds or related meetings to discuss veterinary medicine? If the answer is no, because you feel that they know their jobs and they're paid well to do them, then I wouldn't be surprised if the best and brightest of your staff move on because of a lack of intellectual stimulation. If the answer is yes, then you're doing your job, but as I wrote earlier, don't expect a thank-you note. You're only doing what highly educated people expect of you.

Final Thoughts

Thanks for the Advice, but I've Tried It All and Nothing Seems to Work

Throughout this chapter, I've attempted to address the most important aspects of providing good management and a good working environment for our coworkers. A good manager knows his coworkers' strengths and weakness and tries to work within those parameters. But sometimes, as I've noted elsewhere, a person just doesn't seem to fit in and your first thought is, "How can I get rid of her?" That should be your last thought, not your first one. Let me provide a brief summary of thought processes before you lower the boom.

1. Make sure that all basic needs have been satisfied; don't just assume they are. Is this person being fairly paid, is her supervision satisfactory, is she recognized for her contributions, and so forth?
2. Does she clearly and unambiguously know what her job is? This is not a simple matter in some circumstances. When a manager comes to me complaining that so-and-so isn't doing his or her job, the first question I always ask is, "Have you clearly explained what is expected and what constitutes satisfactory performance?" If the answer is something like "She's an adult and I shouldn't have to hand-feed her," then I wonder if there's a communication problem between the manager and the employee. Let me repeat yet again: People have to know what's expected of them. In fact, this should have been made very clear during the interviewing and hiring process. If there is doubt in your mind, ask a person what she thinks her job is and compare the answer to your own perception. You may be surprised.

3. Does this person have to be retrained or given additional training? Almost every person who has worked in a laboratory animal facility has run across a technician or manager who doesn't seem as adept as his or her peers. However, if that person has potential (and that must be the case; otherwise, why was he given the position?), see if you can provide supplemental training. You're helping to train yourself by reading this book. Perhaps you can take that person under your wing and mentor him, or pair him up for a while with a seasoned partner.

4. Use discipline only if you are near the end of your rope. Sometimes, a little tough love helps a person, and sometimes not. Before you reach this point, I hope you have reread the section on difficult conversations.

5. If nothing else works, and before terminating a person's employment, think about another position that might be a better fit for the person [42]. This assumes, of course, that another position is available within or outside of the animal facility. A very friendly person who can't stop chatting and socializing all day long may not be the best fit as a veterinary technician in a busy animal facility, but he or she may make a wonderful receptionist or recruiter. A person who can't cut it as an animal care supervisor may have the skills and temperament to be a fine veterinary technician.

6. Only as a last resort should a person be let go, and even if that happens, it is in everybody's interest to do that without animosity.

People Are Our Most Important Resource

In essentially every laboratory animal facility, people are our most important financial and functional resource. Because of this importance, some managers may think that managing people is their only job. However, managers have to consistently manage all of their resources (finances, people, information, capital equipment, and time). Focusing on human resources management to the near exclusion of the management of other resources is a formula for failure. If you are interested in developing a scorecard to help gauge your animal facility's strengths and weaknesses in human resources management, a good deal of quality information can be found in a fine article by Bassi and McMurrer [102].

The day-to-day management of human resources revolves around good communication and trust. To that, we add basic human needs such as a fair salary, good working conditions, and good supervision. People want to feel their work is important, and they are respected for their contributions. Without these basic factors in place, rewards and other forms of motivation will have minimal benefit.

> People have to feel that their contribution is important—that they are taking on responsibility for a significant thing. Typically, in a structured environment, people don't feel that the responsibility they're assuming is really important, so they ask, why bother?

We in leadership positions have to recognize and appreciate that when we ask other people to change from past practices, we at the top have got to change first. [103]

References

1. Brown, S.P., and Lam, S.K. 2008. A meta-analysis of relationships linking employee satisfaction to customer responses. *J. Retailing* 84: 243.
2. Myers, K.K., and Sadaghiani, K. 2010. Millennials in the workplace: A communication perspective on millennials' organizational relationships and performance. *J. Bus. Psychol.* 25: 225.
3. Bennis, W., Goleman, D., O'Toole, J., with Biederman, P. 2008. *Transparency: How Leaders Create a Culture of Candor.* Hoboken, NJ: John Wiley & Sons.
4. Sirota, D., Mischkind, L.A., and Meltzer, M.I. 2006. Stop demotivating your employees! *Harvard Manage. Update.* 11(1): 1.
5. Latson, A. 2014. Does your staff respect you … or do they fear you? *Lab Manager Magazine*, February 3. http://www.labmanager.com/management-tips/2014/01/does-your-staff-respect-you-or-do-they-fear-you- (accessed February 27, 2014).
6. Meyer, E. 2014. *The Culture Map: Breaking through the Invisible Boundaries of Global Business.* New York: Perseus Books.
7. Casciaro, T., and Lobo, M.G. 2005. Competent jerks, lovable fools, and the formation of social networks. *Harvard Bus. Rev.* 83(6): 92.
8. Gilmour, D. 2003. How to fix knowledge management. *Harvard Bus. Rev.* 81(10): 16.
9. Detert, J.R., and Burris, E.R. 2015. Can your employees really speak freely? *Harvard Bus. Rev.* 93(1–2): 81.
10. Porath, C., and Pearson, C. 2013. The price of incivility: Lack of respect hurts morale—and the bottom line. *Harvard Bus. Rev.* 91(1–2): 115.
11. Pearson, C.M., and Porath, C.L. 2005. On the nature, consequences and remedies of workplace incivility: No time for "nice"? Think again. *Acad. Manage. Exec.* 19: 7.
12. Bartolomé, F., and Weeks, J. 2007. Find the gold in toxic feedback. *Harvard Bus. Rev.* 85(4): 24.
14. Stone, D., Patton, B., and Heen, S. 2010. *Difficult Conversations: How to Discuss What Matters Most.* New York: Penguin Group.
15. Patterson, K., Grenny, J., McMillan, R., and Switzler, A. 2004. *Crucial Confrontations: Tools for Resolving Broken Promises, Violated Expectations, and Bad Behavior.* New York: McGraw-Hill.
16. Lencioni, P. 2012. *The Advantage: Why Organizational Health Trumps Everything Else in Business.* Hoboken, NJ: John Wiley & Sons.
17. Blanchard, K., and Johnson, S. 2000. *The One Minute Manager.* Revised ed. New York: HarperCollins Business.
18. Heen, S., and Stone, D. 2014. Finding the coaching in criticism. *Harvard Bus. Rev.* 92(1–2): 109.
19. Culbert, S.A. 2008. Get rid of the performance review! *Wall Street Journal*, October 20. http://online.wsj.com/article/SB122426318874844933.html (accessed March 1, 2016).
20. Lemov, D., Woolway, E., and Yezzi, K. 2012. *Practice Perfect: 42 Rules for Getting Better at Getting Better.* Hoboken, NJ: John Wiley & Sons.

21. DeNisi, A.S. 2011. Managing performance to change behavior. *J. Organ. Behav. Manage.* 31: 262.

22. Cullen, K.L., Fan, J., and Liu, C. 2014. Employee popularity mediates the relationship between political skill and workplace interpersonal mistreatment. *J. Manage.* 40: 1760.

23. Lazare, A., and Levy, R.S. 2011. Apologizing for humiliations in medical practice. *Chest* 139: 746.

24. Goldsmith, M., and Reiter, M. 2007. *What Got You Here Won't Get You There: How Successful People Become Even More Successful!* New York: Hyperion.

25. Hasson, G. 2015. *How to Deal with Difficult People: Smart Tactics for Overcoming the Problem People in Your Life.* Hoboken, NJ: John Wiley & Sons.

26. Goldberg, J.H., Lerner, J.L., and Tetlock, P.E. 1999. Rage and reason: The psychology of the intuitive prosecutor. *Eur. J. Soc. Psychol.* 29: 781.

27. Bossidy, L. 2007. What your leader expects of you. *Harvard Bus. Rev.* 85(4): 58.

28. Sull, D., Homkes, R., and Sull, C. 2015. Why strategy execution unravels and what to do about it. *Harvard Bus. Rev.* 93(3): 58.

29. Bernstein, A.J., and Rozen, S.C. 1994. *Sacred Bull: The Inner Obstacles That Hold You Back at Work and How to Overcome Them.* Hoboken, NJ: John Wiley & Sons.

30. Ibarra, H. 2015. The authenticity paradox. *Harvard Bus. Rev.* 93(1–2): 52.

31. Davenport, T.H. 2005. *Thinking for a Living: How to Get Better Performance and Results from Knowledge Workers.* Boston: Harvard Business School Press.

32. Anonymous. 2005. Five questions about … how leaders influence creativity. *Harvard Manage. Update.* 10(7): 5.

33. Hunsaker, P., and Alessandra, T. 2008. *The New Art of Managing People: Updated and Revised Person-to-Person Skills, Guidelines, and Techniques Every Manager Needs to Guide, Direct, and Motivate the Team.* New York: Free Press.

34. Chou, W.-J., Sibley, C.G., Liu, J.H., Lin, T.-T., and Cheng, B.-S. 2015. Paternalistic leadership profiles: A person-centered approach. *Group Organ. Manage.* 40: 685.

35. Hamel, G. 2011. Management is the least efficient activity in your organization. *Harvard Bus. Rev.* 89(12): 50.

36. Kowalczyk, L. 2008. Doctor in bias suit was asked to resign. *Boston Globe,* July 26, p. B.1.

37. Luntz, F.I. 2009. *What Americans Really Want … Really: The Truth about Our Hopes, Dreams, and Fears.* New York: Hyperion Books.

38. Pickett, R.B. 2011. The psychobarbarian manager. *Lab Manager Magazine,* October 21. http://www.ourdigitalmags.com/display_article.php?id=847151 (accessed March 3, 2014).

39. Hurley, R.F. 2006. The decision to trust. *Harvard Bus. Rev.* 84(9): 55.

40. Werbel, J.D., and Henriques, P.L. 2009. Different views of trust and relational leadership: Supervisor and subordinate perspectives. *J. Manage. Psychol.* 24: 780.

41. Shea, G. 1984. *Building Trust in the Workplace.* AMA Management Briefing. New York: American Management Association.

42. Buckingham, M. 2006. What great managers do. *Harvard Bus. Rev.* 83(3): 70.

43. Daniels, A.C., and Daniels, J.E. 2005. *Measure of a Leader: An Actionable Formula for Legendary Leadership.* New York: McGraw-Hill.

44. Pink, D.H. 2009. *Drive: The Surprising Truth about What Motivates Us.* New York: Penguin Group.

45. Maslow, A.H. 1943. A theory of human motivation. *Psychol. Rev.* 50: 370.

46. Alderfer, D.P. 1969. An empirical test of a new theory of human needs. *Org. Behav. Human Perform.* 4: 142.
47. Jawahar, I.M., and Stone, T.H. 2011. Fairness perceptions and satisfaction with components of pay satisfaction. *J. Manage. Psychol.* 26: 297.
48. O'Malley, M.N. 2000. *Creating Commitment: How to Attract and Retain Talented Employees by Building Relationships That Last.* Hoboken, NJ: John Wiley & Sons.
49. Ross, J. 2005. Dealing with the real reason people leave. *Harvard Manage. Update.* 10(8): 1. The author referenced Branham, L. 2005. *The 7 Hidden Reasons Employees Leave: How to Recognize the Subtle Signs and Act before It's Too Late.* New York: Amacom.
50. Wagner, R. 2015. *Widgets: The 12 New Rules for Managing Your Employees as if They're Real People.* Columbus, OH: McGraw-Hill Education.
51. Locke, E.A., and Latham, G.P. 1984. *Goal Setting: A Motivation Technique That Works!* New York: Prentice Hall.
52. Ryan, K. 2012. Gilt group's CEO on building a team of A players. *Harvard Bus. Rev.* 90 (1–2): 43.
53. Katzenbach, J.R., and Khan, A. 2010. *Leading Outside the Lines: How to Mobilize the (In)formal Organization, Energize Your Team, and Get Better Results.* Hoboken, NJ: Jossey-Bass.
54. Amabile, T.M., and Kramer, S.J. 2010. What really motivates workers. *Harvard Bus. Rev.* 88(1–2): 44.
55. Amabile, T.M., and Kramer, S.J. 2011. The power of small wins. *Harvard Bus Rev.* 89(5): 70.
56. Sartain, L., and Schumann, M. 2006. *Brand from the Inside: Eight Essentials to Emotionally Connect Your Employees to Your Business.* Hoboken, NJ: John Wiley & Sons.
57. Goffee, R., and Jones, G. 2013. Creating the best workplace on earth. *Harvard Bus. Rev.* 91(5): 99.
58. Taguri, R. 1995. Managing people: Ten essential behaviors. *Harvard Bus. Rev.* 73(1): 10.
59. Pfeffer, J. 1998. Six dangerous myths about pay. *Harvard Bus. Rev.* 5–6: 109.
60. Kohn, A. 1993. Why incentive plans cannot work. *Harvard Bus. Rev.* 71(5): 54.
61. Daniels, A.C. 2001. *Other People's Habits: How to Use Positive Reinforcement to Bring Out the Best in People around You.* New York: McGraw-Hill.
62. Howell, J.P., and Costley, D.L. 2005. *Understanding Behaviors for Effective Leadership.* 2nd ed. Upper Saddle River, NJ: Prentice Hall.
63. Gubler, T., Larkin, I., and Pierce, L. 2015. Motivational spillovers from awards: Crowding out in a multitasking environment. Harvard Business School NOM Unit Working Paper No. 13-069. July 14. http://papers.ssrn.com/sol3/papers.cfm?abstract_id=2215922 (accessed March 1, 2016).
64. Bielaszka-DuVernay, C. 2007. Are you using recognition effectively? *Harvard Manage. Update.* 12(5): 4. Quoting Adrian Gostick.
65. Seijts, G.H., and Latham, G.P. 2005. Learning versus performance goals: When should each be used? *Acad. Manage. Exec.* 19: 124.
66. Jeffrey, S.A., Schulz, D., and Webb, A. 2012. The performance effects of an ability-based approach to goal assignment. *J. Organ. Behav. Manage.* 32: 221.
67. Bunker, K.A., Kram, K.D., and Ting, S. 2002. The young and the clueless. *Harvard Bus. Rev.* 80(12): 80.

68. Wine, B., Gilroy, S., and Hantula, D.A. 2012. Temporal (in)stability of employee preferences for rewards. *J. Organ. Behav. Manage.* 32: 58.
69. Zweig, D. 2014. Managing the "invisibles." *Harvard Bus. Rev.* 92(5): 97.
70. Delong, T.J., and Vijayaraghavan, V. 2003. Let's hear it for B players. *Harvard Bus. Rev.* 81(6): 96.
71. Roberts, L.M., Spreitzer, G., Dutton, J.E., Quinn, R.E., Heaphy, E., and Barker, B. 2005. How to play to your strengths. *Harvard Bus. Rev.* 83(1): 75.
72. Patterson, K., Grenny, J., Maxfield, D., McMillan, R., and Switzler, A. 2008. *Influencer: The Power to Change Anything.* New York: McGraw-Hill.
73. Goldstein, N.J., Martin, S.J., and Cialdini, R.B. 2009. *Yes! 50 Scientifically Proven Ways to Be Persuasive.* New York: Free Press.
74. Anonymous. 2013. Spotlight: Interview with Robert Cialdini. The uses (and abuses) of influence. *Harv. Bus. Rev.* 91: 77.
75. Cuddy, A.J.C., Kohut, M., and Neffinger, J. 2013. Connect, then lead. *Harvard Bus. Rev.* 91: 56.
76. Neffinger, J., and Kohut, M. 2013. *Compelling People: The Hidden Qualities That Make Us Influential.* New York: Hudson Street Press.
77. Cohen, A.R., and Bradford, D.L. 2005. *Influence without Authority.* 2nd ed. of the classic work. Hoboken, NJ: John Wiley & Sons.
78. Bardwick, J. 1995. *Danger in the Comfort Zone.* New York: AMACOM Books.
79. Koch, C.G. 2007. *The Science of Success: How Market-Based Management Built the World's Largest Private Company.* Hoboken, NJ: John Wiley & Sons.
80. Branham, L., and Hirschfeld, M. 2010. *Re-Engage: How America's Best Places to Work Inspire Extra Effort in Extraordinary Times.* New York: McGraw-Hill.
81. Wagner, R., and Harter, J.K. 2007. When there's a freeloader on your team. *Harvard Manage. Update.* 12(1): 5.
82. Tapscott, D. 2008. *Grown Up Digital: How the Net Generation Is Changing Your World.* New York: McGraw-Hill.
83. Kukenberger, M.R., and Mathieu, J.E. 2015. A cross-level test of empowerment and process influences on members' informal learning and team commitment. *J. Manage.* 41: 957.
84. Kern, N.L., and Seok, D.-H. 2011. "… with a little help from my friends": Friendship, effort norms, and group motivation gain. *J. Manage. Psychol.* 26: 205.
85. Brymer, R. 1991. Employee empowerment: A guest-driven leadership strategy. *Cornell Hotel Rest. A* 32(1): 58.
86. Gratton, L., and Erickson, T. 2007. Ways to build collaborative teams. *Harvard Bus. Rev.* 85(11): 101.
87. Burkus, D. 2013. *The Myths of Creativity: The Truth about How Innovative Companies and People Generate Great Ideas.* Hoboken, NJ: John Wiley & Sons.
88. Girard, B. 2009. *The Google Way: How One Company Is Revolutionizing Management as We Know It.* San Francisco: No Starch Press.
89. Huckman, R., and Staats, B. 2013. The hidden benefits of keeping teams intact. *Harvard Bus. Rev.* 91(12): 27.
90. Lencioni, P. 2005. *Overcoming the Five Dysfunctions of a Team: A Field Guide for Leaders, Managers, and Facilitators.* San Francisco: Jossey-Bass.
91. Parker, G., McAdams, J., and Zielinski, D. 2000. *Rewarding Teams: Lessons from the Trenches.* San Francisco: Jossey-Bass.

92. Waterman, R., Jr. 1994. *What America Does Right: Learning from Companies That Put People First.* New York: W.W. Norton and Co.
93. Kotter, J.P., and Schlesinger, L.A. 2008. Choosing strategies for change. *Harvard Bus. Rev.* 7: 49.
94. Pascale, R.T. 1979. Zen and the art of management. In *Harvard Business Review on Human Relations*, 125. New York: Harper & Row.
95. Wright, T. 2014. *Fizz: Harness the Power of Word of Mouth Marketing to Drive Brand Growth.* New York: McGraw-Hill.
96. Brooks, A.W. 2015. Emotion and the art of negotiation. *Harvard Bus. Rev.* 93(12): 57.
97. Leary, K., Pillemer, J., and Wheeler, M. 2013. Negotiating with emotion. *Harvard Bus. Rev.* 91 (1–2): 96.
98. Hornickel, J. 2013. *Negotiating Success: Tips and Tools for Building Rapport and Dissolving Conflict While Still Getting What You Want.* Hoboken, NJ: John Wiley & Sons.
99. Neale, M.A., and Lys, T.Z. 2015. *Getting (More of) What You Want: How the Secrets of Economics and Psychology Can Help You Negotiate Anything, in Business and in Life.* New York: Basic Books.
100. Malhotra, D. 2015. Control the negotiation before it begins. *Harvard Bus. Rev.* 93(12): 67.
101. Goffee, R., and Jones, G. 2007. Leading clever people. *Harvard Bus. Rev.* 85(3): 72.
102. Bassi, L., and McMurrer, D. 2007. Maximizing your return on people. *Harvard Bus. Rev.* 85(3): 115.
103. Pascarella, P., DiBianca, V., and Gioja, L. 1988. The power of being responsible. *Independent Weekly*, December 5, p. 41. Quoting Richard Schiavo.

Chapter 4

Managing Financial Resources

> Money and mission are inextricably linked and woe betides the leader
> who ignores either one.
>
> **Jeremy Lim**

It sometimes appears that first-level managers spend an inordinate amount of time managing people. As the management ladder is climbed and you become a middle-level manager, you become progressively more concerned about the management of money. Once that happens, it seems that there is a black financial cloud constantly hanging over your head. College presidents have been called glorified fund-raisers. Deans calmly tell their faculty that new equipment cannot be purchased and that open positions will remain open. Animal facility managers get caught in the bind of balancing their budgets by decreasing services or increasing fees. And it's not much different in the corporate world. Somebody is always telling you that money is tight; you have to cut back.

Although money is certainly a resource that managers may have at their disposal and must manage, on occasion we become so involved with the management of money that we make it the be-all and end-all of our job. Money is important, of course. Profit-making organizations such as pharmaceutical companies or contract research organizations have to make a profit to stay in business. These companies may have to put being profitable above any important mission they may have, such as discovering and producing life-saving drugs. On the other hand, in not-for-profit organizations, such as most colleges and universities, their service and educational missions may be more important than their income. Still, even not-for-profit institutions need money to carry out their mission, leading to the often-heard dictum "no money, no mission." My advice to you is to be very careful, especially if you work in the not-for-profit world, and make sure that you do not lose sight of your organization's mission. If you stay alive financially, but are not fulfilling your mission, you should begin to question the purpose of your organization's existence. With that caveat, it still is the responsibility of the manager to make sound financial decisions. In larger animal facilities with an accounting or financial management division, it's all too easy to take the easy road and let the "bean counters" develop the budget or set financial goals. I suggest to you that this isn't a good idea. We should be working together with our financial people to develop financial goals and monitor them. However, there is something I do like to delegate to them, and that is the interpretation of financial reports. In my opinion, that is a critical function of any financial manager. If the best your financial team can do, or is expected to do, is give you raw data without any interpretation, then you might consider revising their job description. They should be more fully integrated into the leadership of the animal facility.

This chapter will primarily discuss managing the funds that you have or potentially have available for your use. In fact, I think that it is fair to say that the basic purpose of financial management is to make sure that money is available when you need it, where you need it, and in the amount that you need. "When you need it" means that money must be available in a timely manner. If you need money today or tomorrow, it doesn't help if you can't get it until next year. "Where you need it" means it has to be accessible to you. Consider what can happen if you find out that John Johnson, who you have been courting off and on for a year, is now ready to accept your job offer. Although your university has sufficient money to pay his salary, there isn't enough left in your budget to hire him this year. Two weeks from now he'll have a new job and you'll have the same budget without him. Sorry, Charlie. It's unfortunate that most laboratory animal facilities do not have the flexibility to grab at opportunities when they arise. Our budgets are often carved in granite, and they remain like that for a year. I yearn for the day when at least some of the decisions about allocating money will be made in real time, based on value to the organization. But I accept reality, and like everybody else, I do my best to prepare budgets that have a chance of providing the money I need when I need it.

The basic purpose of financial management is to make money available when you need it, where you need it, and in the amount that you need.

It would be more accurate to include items such as your inventory of animal cages as part of your financial resources, because you can sell the cages. But, as a practical matter, laboratory animal facility managers are mostly concerned with using funds that have already been allocated as part of the animal facility's budget.

You don't have to go to a bank for loans. Your job is to work with your staff and superiors to develop a budget that will eventually provide the funds you need.

How Money Is Allocated for Animal Purchases

Profit-Making Organizations

It would be nice if we could pick our salary off a money tree, but we still have to earn it the hard way. Organizations have to do the same. In a profit-making organization such as a pharmaceutical company, money can come from the sale of various products or services, the sale of shares of the company (i.e., stocks), borrowing, and investments that the company makes. A portion of this income will eventually be allocated to the animal facility. Some of it will be used for salaries, some for supplies and equipment, and some to purchase animals.

The money needed to purchase animals is generally provided for in one of two ways. In the first instance, each department of the company that uses the animal facility estimates the number and types of animals it will need for a certain time period (usually a year in advance). Then, either the investigator's department or the laboratory animal facility estimates how much money it will take to purchase those animals. This figure becomes part of the budget request of the research (or teaching) department. When the budget is approved, the company allocates that amount of money to the department's operating budget. The department is expected to work within that budget. Therefore, if the toxicology department estimates that it will need 10,000 rats during the budget year, and each rat costs $10.00, toxicology would have to request $100,000 in its budget for rat purchases.

In the second instance, the laboratory animal facility estimates the amount of money it will require to purchase animals for all users in the coming year, and that amount of money is requested in the animal facility's budget. When the budget is approved and the money is actually being used, the animal facility becomes responsible for monitoring its animal purchase budget and not allowing it to be overspent.

Not-for-Profit Organizations

In not-for-profit organizations, such as most universities, government, or some private research laboratories, money for the organization may come from government and private grants, special endowment funds, borrowing, and different types of gifts the organization might receive. Additionally, income may come from the sale of services, products, or various investments.

Money used to purchase animals is usually part of the research funds of a particular investigator or the investigator's department. On occasion, the animal facility is given the responsibility of monitoring the amount of money in an investigator's animal purchase account and advising him or her when it is low. Sometimes, the laboratory animal facility has its own budget for purchasing animals for investigators. In fact, that's how we do it where I work. When this occurs, the laboratory animal facility estimates how much money it will need to purchase animals for all the investigators in the upcoming budget year. It requests those funds as part of its own budget. When the facility purchases an animal for an investigator, the funds to purchase the animal come from the animal facility's budget. Then, the animal facility transfers money from the investigator's account back into its own account. The net effect is that it's an even trade financially, but it's a cumbersome way of doing things, and in general, it's easier for the animal facility to purchase the animals using the investigator's money. One good reason for doing the latter is that the animal facility doesn't have to worry about an investigator running out of money and not being able to repay the cost of the animals.

How Money Is Obtained for Animal Care

A laboratory animal facility must have money to pay its staff, purchase cages and cleaning supplies, get pens and pencils for the office, and for everything else needed to run the facility. In many profit-making organizations, the laboratory animal facility requests money directly from the company, based on either past history or the development of an annual animal care budget. In a few profit-making organizations, and in most not-for-profit organizations, the income from *per diem* charges is used to fully or partially cover the expense of operating the laboratory animal facility.

A *per diem* charge (or a *per diem* rate, as it is sometimes called) is the amount of money an investigator has to pay the animal facility to maintain one animal for one day. Each species you house will have its own *per diem* charge. There are some variations possible. For example, a *per diem* charge in your institution may refer to the cost of keeping one cage of animals in the facility for one day, even though the cage may have more than one animal in it. In vivariums with many thousands of cages of animals, such as mice or rats, it is much easier to count cages than it is to count individual animals, so a *per diem* rate in these institutions is typically based on cages, not individual rodents. If your animal facility receives money from federal grants or contracts, you may be required by federal policy to charge a *per diem* fee to investigators who use the vivarium. I will discuss below some of the components of the *per diem* charge, and how they are formulated.

> A *per diem* charge is the amount of money you charge an investigator to house one animal (or one cage of animals) for one day.

In general, a laboratory animal facility receives its income from *per diem* charges in one of two ways. In the first, the facility sends monthly animal housing bills to the investigators, and it is the investigators' responsibility to pay those bills using their research funds. A variation of this method occurs if the animal facility sends the bill to the institution's accounting office, which then transfers money from investigators' accounts into the animal facility's account. In the second method, the organization puts a certain amount of the investigators' money in a special account that is exclusively for the use of the laboratory animal facility. The animal facility requests money from that account once a month. When this second method is used, investigators must be informed of the amount of the monthly deductions from their accounts so they can keep accurate records. Although these are typical ways of funding the operations of a laboratory animal facility, there are other variations in use.

In any organization, but particularly in not-for-profits, investigators should have sufficient money to cover all anticipated animal facility charges. Their funds should cover not only *per diem* charges, but also any anticipated extra charges, such as for radiographs, special treatments, special caging, special diets, and animal transportation. If money is not available, either the study should not begin or the parent organization should provide a mechanism for reimbursing (supporting) the expenses of the laboratory animal facility. If, as the manager of the facility, you know that there will not be sufficient money, it is your responsibility to bring it to the attention of the appropriate people in your organization. In other words, my advice is that when possible you estimate the animal facility cost for the entire research project as accurately as you can, and make sure there is money to cover it. I grant you that sometimes, especially in larger organizations, this is not possible or perhaps not practical. But when you can do it, even though accounting may not be part of your job description, it is part of managing. You are doing your best to control the use of a resource, which, in this case, happens to be money. Politically speaking, it is also to your advantage to point out that a potential problem exists before it becomes a reality.

Before moving on to a discussion about budgets, I want to be sure that I have not given you the impression that *per diem* income is "extra" income that supplements the money that has already been allocated to fund your budget. That is not the case. Your organization uses *per diem* income to pay all or part of the vivarium's current expenses unless your organization just approves your expense budget and doesn't expect your animal facility to generate any income. If there is insufficient *per diem* income to cover your expenses, your organization has to make up the difference with other funds, or work with you to come up with another solution. *Per diem* income is not extra income, it's *the* main source of income that is typically used to balance all or most of your expenses.

Other Sources of Income for Animal Facilities

It's not unusual for *per diem* income to account for 60%–70% of a not-for-profit animal facility's overall income. But where does the remaining 30%–40% come from? In many instances, the parent organization (e.g., your university) provides a financial subsidy to the animal facility to cover costs that are not recovered by *per diem* charges (and I think that this is more of the rule rather than the exception). Nevertheless, there are additional potential sources of income. For example, most animal facilities charge investigators a separate fee for providing services that are not part of the *per diem* rate. Some examples of this might be charges for helping in surgery, treating a health problem that resulted from the research being performed (such as treating surgical wound infections), staying late to care for animals that haven't yet awakened from anesthesia, and so forth. Some animal facilities are even more entrepreneurial and charge a fee for training research laboratories in animal handling and use. Others charge for providing specific surgical services to their research or teaching community. In order to consider performing some of these extra services, you have to have some way of knowing if the service is wanted. Often, researchers will simply ask you if a service is available. Other times, you may have to talk to the research or teaching teams at your institution to discern if a need is present, and if there is such a need, you will want to know how often the need arises and how much you might charge for providing the service. It really doesn't make a lot of sense to train people on your staff to perform ovariectomies on mice if the research community would use such a service once or twice a year. On the other hand, you may find that catheter placement and maintenance is a wanted service, and your current staff can fit it into their existing schedule while bringing additional income and appreciation to the vivarium. Of course, you can perform all of these services for free if you want to be kindhearted to the research community, but money in not-for-profit organizations is usually quite tight, and it's unlikely that your budget can carry such a financial load. One thing to keep in mind is that if you directly bill an investigator for supplemental services provided by your animal facility and its staff, you cannot include the cost of those same services as part of your *per diem* rate because that would be charging twice for the same service.

As a final example, some vivariums increase their income by renting out space to research organizations that do not have their own vivariums. Doing this can be a blessing or a curse, depending on specific circumstances, and any animal facility considering this should first speak to colleagues who are experienced with such an arrangement.

Budgets

From what you've just read, it should be obvious that money management is an important part of the responsibilities of many managers. Such responsibility can transform even quiet, withdrawn managers into fire-breathing dragons. Planning, making decisions, organizing, controlling, and directing are all used in the management of financial resources. Whether you like it or not, the budget often will be the center of your attention.

A budget is a financial plan, and as I described above, it is often considered by some institutions as a completely rigid and carved-in-granite plan. That's unfortunate because most every plan should allow for some flexibility to meet unforeseen needs, be they business threats or opportunities. That issue aside, budgets are prepared for a set period of time, which is usually one year. Generally, a budget requires you to plan for both income and expenses, although, in some organizations (such as some pharmaceutical companies), the animal facility may plan only for its expenses. Assuming there is no *per diem* charge, the pharmaceutical company will provide the entire operating income for the animal facility. In most not-for-profit institutions, you must estimate your operating expenses as well as your anticipated income from *per diem* and other sources. *Your goal in preparing your budget is to have your income equalize your expenses.*

A budget is a financial plan.

Types of Budgets

From conversations with colleagues, it appears that one of the most commonly used budgets in laboratory animal facilities is the *incremental budget.* In this type of budget, the expenses of the previous year are assumed to have been reasonable. For the coming year, you simply add or subtract a percentage to the old budget. A similar percentage is added to the *per diem* charge and other income sources so that income and expenses are balanced. For example, if your expenses totaled $100,000 this year, you might estimate that next year your expenses will be $105,000, assuming an inflation rate of 5%. You would then increase your *per diem* rates and your charges for other sources of income (if any) by 5%, to make your income balance with your expenses. If you anticipate housing fewer animals in the coming year, you may decrease your budget by a small percentage or perhaps keep it the same to compensate for inflation.

Incremental budgets are often used because they are satisfactory for many animal facilities and they are the easiest to develop. Unfortunately, they do not require any vigorous assessment of the actual requirements of the laboratory animal facility, and they do not fulfill the requirements for establishing a *per diem* rate if your institution is required to document the methods used to develop your rate. At most, with an incremental budget you may be asked to justify increased spending above a given percentage. For example, assume you are asked to justify any increases in supplies that are 20% more than last year's budget. So if you paid $1000 for cleaning chemicals last year, you would have to justify any increase greater than $1200 ($1000 + 20% = $1200). Most managers can do that, and it often becomes an exercise in thinking up excuses for requesting budget increases. It can be argued that, if you have too much money left over at the end of the current budget year, upper management will not approve a substantial increase for the next budget year without detailed explanations. True, perhaps, but then, most managers are aware of this common problem and simply make sure they spend most of their budgets.

All of this makes little managerial sense. If you want to be a second-rate manager, you can get away with all kinds of tricks. But you really have to look at the entire organization, not just your area of immediate responsibility. All the managers in an organization, on all levels, have to budget as honestly as possible. I truly empathize with a manager who submits an honest, bare-bones budget and has it approved, only to be asked to cut it by 10% once the budget year begins. If the request was due to unforeseen problems, that may be understandable, but if it was meant to be an across-the-board cost-cutting gimmick, the savvy manager will not put in an honest budget the next year. Trust, trust, and more trust is needed to grease the wheels of financial management.

A second type of budget, used less often, is the *zero-based budget*. Zero-based budgeting assumes that you start with a budget of $0 and you justify every item in the budget until you get your entire budget completed. Therefore, if you claim you need $3750 for diagnostic serology for mice, you should be ready to justify that amount by, for example, explaining that you will be doing 50 serology panels at $75 per panel. It is far more difficult to make unsubstantiated guesses with a zero-based budget, since it requires that every year you justify all your planned expenses. It also requires that you justify why each *per diem* charge is what it is. That is, you have to justify why the mouse *per diem* is $0.60 per mouse cage per day, rather than $0.55 or $0.63. Although this procedure requires greater time and effort on your part, it also gives you an opportunity to study your needs for the coming year in depth.

If you really want to get a handle on your financial status and understand what's going on, prepare a zero-based budget. I did that when I first went to The Ohio State University. It was a chore, but I learned the ins and outs of where money came from, where it went, what records we had, what records were needed, what we could eliminate, and so on. I was able to speak intelligently to the university's account managers, and I was able to clearly articulate to my staff where we were financially and how we got there. Fortunately, there are a number of commercially available computer programs that can greatly simplify—but not replace—the development of zero-based budgets for animal facility managers.

A combination of incremental and zero-based budgets can be used. A percentage of the last year's budget, such as 80%, might be approved for the coming year. The manager then must justify further deletions or additions by zero-based budgeting.

You might want to consider yet another type of budget, although I don't know how often it is used in laboratory animal facilities. This is a *continuous budget*. In this system, you develop a part of your next year's budget at the end of each month of the current year. For example, assume that at the end of November 2015 your records indicate that you spent $2400 on travel (because of the national American

Association for Laboratory Animal Science [AALAS] meeting), $750 on office supplies, $950 on equipment repairs, and so on. You then develop your November 2016 budget using the same figures, perhaps adding a small percentage for inflation. As you probably can see, this is similar to developing an incremental budget. The advantage with the continuous budget is the month-by-month visualization of how you anticipate your money will be spent. Even if your organization's business processes do not accommodate a continuous budget, it's a good idea to have it for your own needs, so you can readily see what expenses are forthcoming. I tried continuous budgeting for a while, and it actually had some value, but not so much as to warrant the additional time it took. That doesn't mean it will not be good for you, so give it full consideration.

A sample budget for a laboratory animal facility is shown in Table 4.1, for the period of July 1, 2015, to June 30, 2016. You will notice that, by the end of the 2016 budget year, the Laboratory Animal Research Center of Great Eastern University anticipates a deficit of $21,410. When this budget is submitted, the university will have to agree to subsidize the animal facility for that amount, or the facility will have to revise its budget. Certainly, the large expense for office supplies should be questioned.

The 20% benefits that are included with salaries was not a figure determined by the animal facility. Rather, it was designated by Great Eastern University's upper-level financial managers and then used by the entire university. Different groups of employees might have a different benefits percentage, but if that occurs, it also is determined by upper management.

Table 4.2 subdivides the salaries from Table 4.1 and also indicates where the effort for those salaries will go. For example, in Table 4.1, the total for all salaries is shown to be $135,120, and in Table 4.2, this is detailed to describe the base salaries (i.e., salaries without benefits) of the individual staff members.

In my calculations for Table 4.1, I did not include the animal facility's share of the cost for heat, electricity, janitorial services, and the like. In some organizations, these expenses (the so-called indirect or overhead expenses) are put into the budget, whereas in other organizations, the parent organization pays for these items with separate funds. You must be aware of this information before you begin your budget-making process. Usually, you will receive this information well in advance.

Information Needed to Plan a Budget

Assuming that you have been asked to prepare a budget, what information will you need? First and foremost, relate your budget to your goals. We know we have to care for the animals and their environment, but what about your goal of replacing all your monkey cages within three years? What kind of financial commitment is needed? Will you need an increase in *per diem* rates? If yes, will it be everybody's *per diem* or just those who use monkeys? What did you say you would do when you developed the strategies to implement your long-range goals? What financial

Table 4.1 General Budget of a Laboratory Animal Facility

Great Eastern University

Laboratory Animal Research Center

2015–2016 Budget

Expenses	Total	General and Administrative	Animal Health Care	Animal Husbandry	Research Services
Salary (with 20% benefits)	$135,120	$10,200	$11,400	$108,120	$5,400
Office supplies	$90,350	$785	$12,515	$45,235	$31,815
Food	$8,300			$8,300	
Bedding	$12,900			$12,900	
Laboratory services	$4,500		$4,500		
Equipment	$3,800			$3,800	
Equipment repair	$1,000			$500	$500
Travel	$800			$800	
Dues/ subscriptions	$350	$350			
Freight	$200			$200	
Service contracts	$425			$425	
Animal purchases	$150				$150
Animal maintenance	$500				$500
Postage	$100	$100			
Messenger service	$65	$65			
Total expenses	$258,560	$11,500	$28,415	$180,280	$38,365
Revenues					
Per diem billing	$205,150			$205,150	
Research services	$32,000				$32,000
Total revenues	$237,150			$205,150	$32,000
Surplus (deficit)	($21,410)	($11,500)	($28,415)	$24,870	($6,365)

Table 4.2 Salary Breakout of a Laboratory Animal Facility Budget

Great Eastern University					
Laboratory Animal Research Center					
2015–2016 Administrative and Technical Salaries					
Item	Total Less Benefits	General and Administrative	Animal Health Care	Animal Husbandry	Research Services
Secretary 50%	$8,500	$8,500			
Vet tech 50%	$9,500		$9,500		
Senior animal tech	$22,850			$18,350	$4,500
Animal tech II	$16,150			$16,150	
Animal tech I	$13,900			$13,900	
Animal tech I	$13,900			$13,900	
Animal tech I	$13,900			$13,900	
Animal tech I	$13,900			$13,900	
Total salaries	**$112,600**	**$8,500**	**$9,500**	**$90,100**	**$4,500**

resources were needed? Now, at budget time, you have the opportunity to implement those strategies.

> *One of my favorite stories concerns a woman who had been the executive director of a research-related organization. We were having lunch together, and she was bemoaning the fact that her organization needed more money. I asked her why it needed more money. She said, "What do you mean by 'why'?" to which I responded, "If your organization had all the money it needed, what would you do with it?" This threw her for a loop, and after thinking for a few seconds, she answered, "I guess we would do what we're doing now, but more of it."*
>
> *In other words, she really didn't have any noteworthy plan for using extra money. She's probably still wishing for the extra cash, and probably still doesn't know what she will do with it if she ever gets it, because to date I haven't heard anything about her organization's plans for the future. Your budget, of course, is a somewhat more immediate document, but it is still a financial plan, and you have to know where you want to be when*

> *you are preparing it. If a manager can't figure that out, the best he can hope for are no changes from past years because there are no plans for the future.*

If your budget will only be a percentage increase from the previous year's budget (an incremental budget), then very little information may be needed. Even here, however, you might have to provide a general explanation about the previous year's expenses and income to justify your new requests. Certainly, if you had responsibility for approving someone else's budget, you would want to know at least that much. Simply approving a new budget by basing that approval on an increase or decrease in the Consumer Price Index or some similar index is questionable financial management.

Whether you have short- and long-range goals and strategies or none, there are certain basic items that you must have to plan a budget for an animal facility. The specifics, of course, will vary with the organization. As a rule of thumb, if you don't know what to include in your budget, try to obtain a copy of a recent budget from your department, from another department, or even from a colleague in another institution. Use whatever information resources you have. Simply asking experienced people what goes into their budgets is an effective but often overlooked use of a resource. Even if you are an old pro at making budgets, its format and what is included in it may be quite different in a new organization. Some commonly included budget items are shown in Table 4.1.

If you are working with a zero-based budget, you will need an estimate of the average number of animals that will be housed each day, by species. I'll discuss this necessity further in just a moment. You will want to know the salaries of your staff and what percentage of their salaries should be added to the budget for fringe benefits (such as life or medical insurance and social security). What percentage should be added to current salaries for raises? Usually, you will have a reasonable idea of what constitutes an average raise in your department. In some organizations, you will be told what an average raise will be for the coming year.

With any type of budget, you will want to know the cost of all the items and services you purchased in the current budget year so that you can plan for next year. You will want to know about these expenses by category—dog food, mouse food, rat bedding, uniform rental, and the like. Many laboratory animal facilities have this information readily available, whereas in others, you must do a little work to obtain it. If your organization doesn't have the information readily available, you should take the time to organize your records so the information will be available for next year.

There are commercially available computer programs that make this work a little easier, although it's time-consuming but not overly difficult to do this manually. The problem with most of the computer programs that are designed for animal facility management is that they often do not have a direct communication with the accounting programs used by most institutions. Therefore, it is not unusual to have to enter financial and related information into your institution's financial system, and if you also want to use a separate animal facility management system, you may find yourself reentering the same information into the latter system. Another potential problem with some computerized systems (particularly those *not* made for an animal facility) is that you may not be able to subdivide expense categories. For example, if you feed rats, mice, and hamsters the same food, you have to categorize this as "rodent food," rather than rat, mouse, or hamster food. Do not forget to add possible price increases or decreases when you prepare your budget.

In making a budget, you should

- **Relate the budget to your financial and nonfinancial goals**
- **Know what items have to be included**

   Remember to take into account any salary or other cost changes

---

Reviewing Programs and Operations

While preparing your budget, I recommend that you review your ongoing programs to see if they are satisfactory as is, or if they should be modified. The purpose of this review is to make sure your facility is financially running efficiently and effectively. There is a legitimate desire on the part of upper management to save money. At the same time, your job is not only to save money (i.e., to be efficient), but also to keep your operation effective.

The concept of budget review is a touchy subject. To some people, it's a catch-phrase that means cut the budget by any means, whether by letting people go or making fewer long-distance phone calls. That's not what I'm referring to. I am suggesting you review major categories of what you do, such as the services (e.g., animal health monitoring and radiography) or benefits (e.g., continuing education) you provide. If, in your opinion, programs should be dropped, revised, or expanded, these matters should be discussed with your staff, your boss, and when appropriate, the users of the facility. Get their feedback, because there are times when efficiency has to take a back seat to effectiveness.

Most seasoned animal facility managers know there is a limit to the number of services they can offer. It sounds nice on paper to want to provide animal husbandry, surgical help, radiography, ultrasonography, veterinary service, account balances, a diagnostic laboratory, and everything else imaginable, but reality tells us otherwise. The typical laboratory animal facility is either not big enough, does not have the budget, does not have the people, or does not have a need to provide all possible services. Think about what you have and what you really need to do, as opposed to what would be nice to do. Obviously, you must provide for animal husbandry and veterinary oversight. Beyond that, you have to evaluate the need versus the cost. There may be times when you choose to continue a service that loses money because it is necessary to your organization. There will be other times when you have to give up financially profitable services because they take the space or time that can be better used for other endeavors that more closely match your organization's needs. If you have the financial and other resources, and you can provide both what is needed and what we would generally call "icing on the cake," then so much the better.

When preparing your budget, review current programs to determine if they should be continued, modified, or deleted.

Let me again emphasize that discussing your budgetary and other views with your staff and receiving their input can be invaluable in the long run and is essential to good management. This feedback is part of the concept of keeping communications open and making everybody have a stake in the organization. This is also part of your responsibility to determine where you are, where you want to be, and how you are going to get there.

Working with a Zero-Based Budget

Let's return to zero-based budgets and use the subject as an introduction to calculating *per diem* rates. You will quickly find that listing your planned expenses and income is more important to you when you work with a zero-based budget than when you work with an incremental budget. The zero-based budget requires that you will be able to estimate in advance the costs for most of the individual items that will comprise your budget (e.g., mouse food, rat bedding, equipment repairs, and cleaning chemicals), whereas with the incremental-based budget, it's good to have that information available, but you can do without it.

For the moment, let's concentrate on estimating your expenses for the coming year using a zero-based budget. First, you must have a reasonably accurate estimate of the average number of animals, by species, which will be housed in your facility. In addition, the length of time that those animals will be housed must also be estimated. The calculation is fairly simple. If you think you will be housing 1000 rats for 60 days, 300 rats for 30 days and 150 rats for 365 days, the average number of rats housed is 339 per day:

$$1000 \text{ rats} \times 60 \text{ days} = \quad 60{,}000 \text{ rat-days}$$

$$300 \text{ rats} \times 30 \text{ days} = \quad 9{,}000 \text{ rat-days}$$

$$150 \text{ rats} \times 365 \text{ days} = \quad 54{,}750 \text{ rat-days}$$

$$\text{Total} = 123{,}750 \text{ rat-days}$$

$$123{,}750 \text{ rat} - \text{days} \div 365 \text{ days in a year} = 339 \text{ rats housed per day}$$

Using these calculations, if an average of 339 rats are to be housed every day of the year, how many people are needed to care for them? How much food will be needed to feed 339 rats every day? How many brooms, mops, and shoe covers are needed? How much will all of this cost? Past experience will help answer this question because the needs of each facility differ. As the manager, you are expected to know how many people are needed to care for an average of 339 rats per day. In some facilities, this may be less than one person, but this can vary, depending on the level of care required. Nude rats will undoubtedly require more care time than conventional rats. You may also have to plan to use short-term help, because, at times, you will be housing more than the average number of animals. You have to plan for sickness, vacations, and so forth. The important point is that, through the zero-budgeting process, you begin to understand what your needs will be. Although this example is quite simplistic, it serves to illustrate that zero-based budgets almost force you to reexamine your operations to ensure that both are efficient and effective.

Who tells you how many animals to expect in the coming year? Usually, investigators, your animal facility business office, or the research administration office will be able to estimate the number. In some organizations, the number of animals housed per day rarely varies over time, and therefore, it is easy to estimate what will happen next year. In others, the number varies greatly each year and getting an accurate estimate is quite difficult. It's not impossible—just difficult and sometimes quite inaccurate. You may recall the discussion in Chapter 2 about the internal and external environments that can affect animal facilities. For example, at the time of writing this chapter (2016), there still is a major concern with bioterrorism and your animal facility may want to gear up for related animal-based studies. Managers are planning to house certain types of animals for certain types of bioterrorism-related studies, and these animal facility managers have to budget appropriately. Fortunately, in most instances, investigators know how much money they have or will have available and can make an educated guess as to what animal studies will be done in the coming year. In addition, your business office can estimate the number of grant or contract applications that are likely to be approved and funded in the coming year. By putting this information together, you can get a rough idea of your estimated facility usage. Be advised, though, that wide fluctuations can occur.

Once you have an idea of what you will be spending on animal care, salaries, overhead, and so forth, you can begin to calculate a *per diem* charge based on more than a simple percentage increase over last year's rate.

How to Calculate Per Diem Rates

Here's another true story. In fact, a colleague had this happen to her on two separate occasions.

> The director of an animal facility at a major university asked his operations manager to calculate per diem rates for the facility. Since it had not been done accurately in past years, the operations manager spent a great amount of time methodically gathering information for the calculations, did them, and presented the suggested new rates to the director. The numbers indicated that the vivarium's per diem rates had been far too low, which helped explain why the facility was losing substantial amounts of money each year. The director thanked his manager for the information and proceeded to simply raise the previous year's per diem by 5%, ignoring all the efforts of the operations manager.

This probably sounds familiar to those of you who have had experience calculating *per diem* rates. To paraphrase a song, "Don't nobody give me no bad news." I'm relating the above story to help prepare you for some of the realities of business, not to discourage you. Many people in laboratory animal science have not had to calculate *per diem* rates, and therefore, they are afraid to try to do so. Part of the fear is that they will discover they should have been doing those calculations all along.

At this point, I will state that I cannot provide you with all the details that are needed to establish an accurate *per diem* charge for each species of animal. Room simply does not permit us to do so; nevertheless, I will describe the major concepts here and then I suggest you turn to Appendix 2, where more of the nitty-gritty details are provided. For your information, the National Center for Research Resources has published a manual that goes into great detail on how to establish *per diem* rates [1], and work sessions on the subject are occasionally presented at professional meetings. I've read the manual a few times, and I think the method used below is substantially easier and provides about the same information.

There are some commercially available software programs that claim to be able to compute *per diem* rates. Still, if you do not understand what the computer is doing for you, you will not know what raw material is needed and how to best use the computer program for your own facility's needs.

I am going to describe the process of establishing the *per diem* rate in a series of steps, but there is one overall point that should be kept in mind: the main concept is to have all budgeted animal facility expenses covered (i.e., offset) by income-generating activities. For example, if your facility has an administrative assistant, his or her salary is part of the animal facility's budget, and is an expense for the facility. The cost of the animal care that the animal facility provides is also an expense to you, but animal care will also generate income when an investigator is charged a *per diem* for that service. Therefore, we will have to make sure that the income from animal care will cover both the administrative assistant's salary and the cost of taking care of the animals. In effect, we include the salary expense for the administrative assistant (and often everybody else who works in the animal facility) in the *per diem* charge. At the end of the year, we would like to have all our revenues balance all our expenses. That is, we want to break even. Nevertheless, it is entirely appropriate to have a little surplus revenue to cover unanticipated expenses or evolving needs.

The concept of covering (transferring) expense is used in most businesses. The cost you pay for an apple is more than what the grocer paid for it. That's because part of what he charges you has to cover his expenditures for heat, electricity, telephone, repairs to his store, and taxes, and of course, it includes his profit. The grocer has transferred a part of his cost of doing business into what he charges you for an apple. Because the grocer sells more products than just apples, part of his cost for doing business will also be transferred to pears, plums, and anything else that is sold.

The main concept in setting *per diem* rates is to have all antici-pated expenses covered by activities that will generate income.

Unlike the grocer, you may not be in a profit-making business, but, at the very least, you would like to break even. Therefore, you too have to transfer your expenses from areas that do not make money (such as disposable outerwear costs or equipment repair costs) into areas that do make money (such as the *per diem* charges for animal care). In fact, the income recognized from the *per diem* charges is the biggest moneymaker in most not-for-profit animal facilities. So, instead of figuring out how much to charge for an apple or a pear, you have to figure out how much to charge to maintain a mouse or a rat (or a mouse or rat cage) for one day.

If we house just one species, and the only service provided is animal husbandry, then calculating the *per diem* charge is quite easy. Assume we anticipate caring for an average of 1000 mice per day. The formula for calculating the *per diem* rate for one mouse is

$$\frac{\text{Total mouse expenses for the year}}{365} \div \text{Average number of mice housed per day}$$

$$= \textit{per diem} \text{ rate}$$

Assume that you estimate from your budget that the total cost of caring for these animals for one year will be $50,000. This figure includes salaries, food, bedding, cleaning supplies, repairs, and everything else you want to include. Then,

$$\$50,000 \div 365 \text{ days in a year} = \$136.99 \text{ (the cost to house 1000 mice for 1 day)}$$

$$\$136.99 \div 1000 \text{ mice} = \$0.137 \text{ (the cost to house one mouse for one day;}$$

$$\text{the } per \ diem \ \text{rate)}$$

This $0.137 per day is the *per diem* charge. It is the amount you will charge an investigator to house one mouse for one day. This charge is the major way you earn revenue for your animal facility. You are anticipating expenditures of $50,000 for the coming year, and with a *per diem* of $0.137 per mouse per day, your revenues and expenses will be about the same at the end of the year.

This is a very simple example of the concept behind *per diem* rates. Since these rates influence managerial decisions and are quite important in a laboratory animal facility, a more thorough explanation of how *per diem* rates are calculated is presented in Appendix 2.

There are a few more points that I must mention relative to *per diem* rates. The first is that most institutions do not set *per diem* rates by planning *ahead* with a zero-based budget. Rather, they look *back* at the past year's expenditures and, using the calculations just described, determine what the actual *per diem* cost *had been*. Then, as I said earlier, they add or subtract an estimated percentage to cover their financial needs for the coming year. This process is similar to how some managers develop an incremental budget. If their budget increases by 5%, they increase their *per diem* rate by 4% or 5%. This strategy can work if you anticipate no major changes in your operations, such as significant increases or decreases in animal numbers or the number of animals of each species to be housed, no major changes in salary structure, or no major construction and equipment purchases. The 4% or 5% is just an estimation of what the general escalation of costs will be for the coming year.

The unspoken attitude is that, if nobody complains, it certainly saves a lot of work to use an incremental *per diem*. You can rest easy with this approach. That's true until somebody who understands how *per diem* charges are calculated (such as an internal or federal auditor) starts questioning your rates. That's true until a major animal user leaves your organization. That's true until you are asked to justify an increase in *per diem* charges. And finally, that's true if you are not interested in changes to the component parts of a *per diem* charge that occur over time. In other words, you cannot effectively manage your financial resources (and sometimes your human resources) if you do not understand the component parts of your *per diem* charge.

I must emphasize that, no matter how well you plan your budget and how carefully you formulate your *per diem* charges, they mean very little unless upper management approves them. Most organizations having a laboratory animal facility also have an Institutional Animal Care and Use Committee (IACUC). The IACUC interacts with the facility director and also represents various divisions of the organization. In some instances, the committee will help plan the budget and *per diem* charges. You may find it useful to have the IACUC endorse your budget and *per diem* charges before they are sent to upper management for final approvals. The budget will then have the indirect endorsement of many research divisions of your institution, which will also help deflect the inevitable complaints that accompany any increase in *per diem* charges. Of course, although you may have the best of intentions by asking the IACUC to become involved with your budget, some IACUCs are not going to want to take on that responsibility. You'll just have to learn the culture of your IACUC.

An occasional downside to involving the IACUC is that it might refuse to support a new *per diem* rate. This might happen when the researchers on the IACUC try to protect their own budgets. The most useful way of preventing this lack of support is to justify every component of your *per diem* and have an open dialogue with the IACUC. Don't spring this request on the committee just before the new fiscal year begins. Keep the IACUC apprised of your financial and other activities throughout the year, so that trust is built. Once again, communication and trust are used to grease the wheels of management.

Finally, there is an operational tactic that may be of value to you. Even with the best of intentions, the best documentation, and the best institutional support, investigators are not going to be happy with any increase in *per diem* charges. This is just human nature. Some, perhaps many, will want to know how your charges compare with those of other institutions. This information is available through the informal network of laboratory animal facility directors and managers and an annual survey coordinated by Yale University (you have to join the survey to get the results). It's not specifically important if your rates are at the top, middle, or bottom of the list, as long as the rates are based on sound management and a consistent method of calculating the *per diems*. It helps if you have an idea of what other institutions include in their *per diem* charges. For example, if your animal care staff is paid through your budget, and that is offset by collecting *per diem* charges, you obviously cannot make a comparison with an organization that uses student labor paid by investigators. It's a comparison between apples and oranges. No matter what the reason, if your rates are way out of line with all other institutions in your survey, or even nationally, you'd better have some strong and convincing documentation behind you.

Management Control of Budgets

In addition to serving as a planning document, the budget can serve as a controlling document. You cannot make a budget, have it approved, and then never monitor its progress. Simply stated, you should periodically compare your actual revenues and expenses to your planned (budgeted) revenues and expenses. If a significant deviation occurs, you should determine the reasons for it and, if appropriate, make adjustments to get it back on target. In this way, the budget helps you to monitor your financial progress and control unanticipated changes.

In most animal facilities, there is one person who does this monitoring. This can be the same person who had the major responsibility for the budget's preparation. Even if that person is not you, you may be the one who ultimately has responsibility for explaining and correcting any variances between what was budgeted and what is actually happening. When you study the budget in this manner, especially if you take steps to correct (or at least understand) any variances, you are using the budget as a financial control document.

Look for significant discrepancies between budgeted and actual revenues and expenses. Try to prevent significant discrepancies from recurring.

To enable the detection of any budgetary variances, there must be sufficient information available to you from your organization's accounting department. Usually, the actual revenue and expense amounts are made available to you on a monthly basis. With that information, you can compare the monthly revenue and expenses against the budgeted revenue and expenses for the same month. The accounting department generally takes your yearly budget and divides it into 12 equal monthly parts, although, in some organizations, the accounting program allows you to budget different amounts for each month of the budget year.

Unfortunately, the most frustrating documents I have ever worked with are the monthly financial reports sent by an institution's accounting office. I'm convinced they are prepared by sadistic gnomes who spend most of their working hours devising statements that no living being, themselves included, can possibly understand. My pleas are simple. Please keep my income and expenses separate. Do not redistribute my income or expenses to categories I never heard of, particularly without telling me. Be reasonably up to date and give me enough detail to figure out what I'm paying for, how much it costs, and how much money I have left. Of course, none of these wishes will come true for many of you, but my lament is not unique. Over the years, I have heard the same tale of woe from scores of managers. What amazes me is that accounting departments think those statements are excellent—or maybe they only need a little fine tuning. There is a big difference in perception. Nevertheless, we all must try to adapt to those numerical terrorists and understand them as best we can, because, I assure you, they will not change and we need them to do our jobs.

Let's take a look at a hypothetical report an animal facility manager has prepared, based on figures received from the accounting office. This document could

Table 4.3   Budget Control Report for a Laboratory Animal Facility

Budget Control Report			
Jan. 1, 2016–Jan. 31, 2016			
Item	*Budgeted This Month*	*Actual This Month*	*Variance This Month*
Animal food	$400	$475	(19%)
Bedding	$75	$30	60%
Salaries	$4000	$5000	(25%)
Telephone	$85	$87	(2%)

Table 4.4   Budget Control Report with Reasons for Variances Added

Budget Control Report				
Jan. 1, 2016–Jan 31, 2016				
Item	*Budgeted This Month*	*Actual This Month*	*Variance This Month*	*Explanation*
Animal food	$400	$475	(19%)	Unanticipated price increase
Bedding	$75	$30	60%	Received double order last month
Salaries	$4000	$5000	(25%)	Temporary help
Telephone	$85	$87	2%	

be called a budget control report. Table 4.3 is the report as received from the accounting office.

In Table 4.4, I added another column (Explanation) to give you an idea of what you might include in the budget control report to account for the variances. In this particular report, I have included only expenditures. Most budget control reports would list income and year-to-date activity as well.

You should be able to account for *significant* underspending as well as overspending. Within the big picture, a few dollars either way just isn't worth worrying about. They always occur, and it is rare to have income and expenses exactly match each other. Too much income, however, may indicate more research activity than expected, that you have set the *per diem* rate too high, or perhaps you can give more or better service.

You may have a boss who does not want a written report each month. However, if you ever have to explain or justify your budget, a control report can be very helpful. It is a tool for you to use for whatever reason you see as appropriate. Even if you're the only one in your entire organization who tries to justify variances, if it suits your style of management, do it.

You may have to do more than write down the reason for a variance if you want to avoid having the same problem in the future. For instance, you might not have been able to anticipate the increased price of animal food, but what about hiring temporary help for technicians on vacation? Was that in the budget? If it was, it should be noted as such. If not, why didn't you budget for it? Become introspective and think about whether you could have altered some work assignments so that people could have covered for each other. Perhaps you didn't even have to hire temporary help. Planning and organizing are part of every manager's job.

If you are the person who has to account for variances in the budget, you know you have no control over certain environmental factors. The unexpected rise in the price of animal food is a good example. It may be that a change in federal price supports is what led to the increase. Not only can you not do anything about price supports, but also there is just so much you can reasonably be expected to know about the external environment of your animal facility. Under these circumstances, your performance should not be judged as inadequate. On the other hand, you can (and should) be held responsible for not budgeting for temporary help if you knew (or expected) you were going to need it. However, if a large and unanticipated contract was awarded to your organization, it would be ridiculous to hold you accountable for the cost of hiring temporary help, the cost of additional food, and so forth.

I suggest that you gather only *necessary* information as part of your budget controlling function. Don't take up room in reports with irrelevant items. For example, reporting income by species is usually valuable. Reporting income from different strains or stocks is irrelevant in most organizations, but may be important to animal breeders. You have to decide what information you want to receive and what information you want others to have. Meaningless information rarely impresses upper management and just wastes your time.

Although you may not be able to change certain aspects of the internal or external environment, you should keep alert for environmental changes that can directly affect your budget—for example, one major environmental factor (the so-called *key variable*). Consider what could happen if a large percentage of your budget relies on providing animal care services to a single investigator. Obviously, the fate of that investigator's finances will directly affect your animal facility's finances. If the investigator has a large grant that is not renewed as anticipated, undoubtedly your budget will eventually suffer. If you are an astute manager, you will try to monitor this investigator's grant to try to detect a possible problem for your animal facility before it actually occurs, then discuss the situation with your superior, and subsequently attempt to make adjustments to your operations. And, by all means, don't forget to let the rest of your staff know what will probably happen if a major grant isn't renewed. Keep communications open and ask people for their thoughts.

A *key variable* is an internal or external business factor that can have a major impact on your operations. Good managers carefully monitor key variables.

Inventory and Inventory Control

Taking an inventory of supplies and equipment is a part of the total financial management control picture. I am discussing it separately to emphasize its importance in laboratory animal facilities.

Most of us are familiar with an inventory of the drugs that are used in a laboratory animal facility. It may require recording the disposition of every dose dispensed, or it may require nothing more than recording when only one bottle remains so that a new one can be ordered. Another example is keeping track of feed and bedding. Counting the number of remaining bags of food or bedding is an example of *supply inventory*, which is continuous, because it is concerned with supplies that are replaced quite frequently. Rodent cages may be considered supplies or equipment; it's not critical to be designated one or the other.

Another type of inventory is an *equipment inventory*. This is often performed at least once a year. It usually includes more or less long-lived items such as metabolism cages, animal racks, and gas anesthesia machines. I have intentionally used the phrase "more or less" because all of these items need periodic replacement, but they are not consumable in the sense that pencils, paper, bedding, and drugs are.

The purpose of an inventory is to be able to estimate for future budgets the typical replacement requirements for your animal facility. It can also help you determine if theft or abnormal usage has occurred. Over time, you will probably get to know the expected life span of most of your supplies and equipment. Once you know how often to replace certain items, you have to estimate their periodic replacement cost in your budget. Therefore, if you know that a new rat cage lasts four years, you might consider replacing 25% of your rat cages each year, so that every four years you will have replaced all of your cages. Having read this, I hope you don't throw out 25% of your new cages next year. As you may have guessed, the cost for cage replacement is calculated into your *per diem* charge.

When taking inventory, there are no shortcuts to counting, even if everything is bar-coded or otherwise identified for automated counting. It is advisable to have a second person randomly spot-check your count. In many retail stores, a recount is needed if there is more than a 1% difference between the two people counting. You might wish to set certain percentage limits for yourself.

The good news is that numerous computer-based programs can aid your inventory. At the least, they should be able to compare one inventory period with another, show numerically or graphically the differences between the inventory periods, and be able to show purchases, sales, or discards that have occurred during the period. Even if the computer automatically subtracts supplies you have used, you will still have to perform a periodic manual count to ensure that the computer-generated numbers match the actual numbers counted.

Animal Census and Its Use in Billing

Another type of inventory is an *animal census*. The number of animals (or cages) that are being used for a particular study is periodically recorded. Sometimes, this recording is done by an actual animal count, and sometimes by subtracting (or adding) animals that have been removed (or added) since an earlier head count. In the latter method, an actual "nose count" is still periodically needed (perhaps every two weeks) to account for any addition or subtraction discrepancies. Some institutions substitute a cage count for an individual animal count. Doing an inventory of cages (particularly with rodents), rather than of individual animals, makes for a quicker inventory and simplifies most of your record keeping. In the long run, it probably pays for itself.

An example of a cage inventory is shown in Table 4.5 (only the first 16 days are shown, but a real inventory would cover all 31 days in January). In this animal facility, the animal technicians do the cage count at the middle and end of the month. On a daily basis, the investigator or animal care technicians record the number of cages that have been permanently added or subtracted (in the latter instance, it is usually because the animals in the cage were euthanized). The animal facility's business office (not the investigator or the animal care technicians) fills in the balance column at the end of the month and uses it for billing purposes (that's why it's shown with shading). With computerized billing programs, the balance column can be automatically calculated. Different animal facilities have different ways of adjusting billing if there is a discrepancy between the actual cage count and the number shown in the balance column. Typically, *the investigator is charged (or credited) for the correct cage count (as determined on the day cages were actually counted) back to the date of the last addition or subtraction of cages.* In the example shown, the investigator would pay a *per diem* charge for five additional cages on January 14 and 15, because the last subtraction of cages was on January 14. Whoever filled in the census form made a mistake somewhere, since the actual count was 65 cages, not 60. The correct balance was entered on January 16. The investigator was lucky that it was only a one-day difference. Some animal facilities aren't that kind and charge the investigator the difference back to the very first day animals were added or subtracted since the previous census (in this example, it would be January 1).

Table 4.5   Sample Animal Census Report for a Laboratory Animal Facility

Great Eastern University—Daily Animal Cage Census					
Investigator: Smithwick, J.					
Month: Jan. 2016					
Protocol no.: 01-001					
Beginning census: 100					
Species: Rat					
Date	*Add*	*Subtract*	*Balance (Office Use Only)*	*Cage Count*	*Recheck*
1		10	90	XXXXX	
2			90	XXXXX	
3		5	85	XXXXX	
4		20	65	XXXXX	
5			65	XXXXX	
6		5	60	XXXXX	
7		5	55	XXXXX	
8		10	45	XXXXX	
9	20	5	65/60	XXXXX	
10			60	XXXXX	
11		5	55	XXXXX	
12	20		75	XXXXX	
13			75	XXXXX	
14		10	65	XXXXX	
15		5	60	65	65
16	10		75	XXXXX	

From the above example, you can see the problem of not performing a daily animal census. In this case, the animal facility didn't count cages from January 1 until January 15, and on the 15th, the person doing the census actually counted five cages more than the number recorded in the balance column (i.e., a total of 65, not 60). In the interim, people adding or subtracting cages just entered what they believed to be the correct numbers. To straighten out this problem the animal facility charged the investigator for five additional cages beginning on January 14. It could have been worse. But how can you do a daily inventory to eliminate (or at least try to eliminate) these problems? Many animal facilities have turned to bar coding of cages. Others are considering radio frequency identification (RFID) chips imbedded into cage cards. Either way, a scanner has to be able to record the presence of each cage. This is usually accomplished by scanning the cage card of every cage. The remainder of the process is largely computerized, and at the end of the month, a bill is printed. Although scanning can be labor-intensive, some animal facility managers feel that it helps oblige the animal care staff to look in each cage. In the future, I am reasonably confident that scanners will be available that can reliably scan all the cage cards in an entire room in a matter of seconds without interference from the amount of metal that is typically in an animal room and without unintentionally counting the cages in an adjacent room. Progress has been slow in developing and implementing this technology.

In many organizations, the animal census, irrespective of how it is compiled, is used to determine the monthly animal care bill for a particular investigator. If an investigator has kept 200 rat cages for 30 days, at a *per diem* rate of $0.90 per cage per day, then the animal facility will charge the investigator $5400:

$$200 \text{ cages @ } \$0.90 \text{ per day} \times 30 \text{ days} = \$5400$$

If an investigator used 200 rat cages for 30 days, 100 for 10 days, and 50 for 20 days, the total *per diem* charge would be $7200:

$$200 \text{ cages} \times \$0.90 \text{ per day} \times 30 \text{ day} = \$5400$$
$$100 \text{ cages} \times \$0.90 \text{ per day} \times 10 \text{ days} = \$900$$
$$50 \text{ cages} \times \$0.90 \text{ per day} \times 20 \text{ days} = \$900$$
$$\text{Total} = \$7200$$

I noted earlier that the facility either bills the investigator directly or informs a central financial office of the charges. In the latter instance, the financial office will transfer funds from the investigator's account into the animal facility's account.

Although not-for-profit organizations most often use an animal census to calculate *per diem* bills, profit-making organizations occasionally use them for the same purpose. In some profit-making organizations, their central administration provides an animal care budget for the entire year. Under this arrangement, the animal census may be used to provide investigators with information for their own future budget planning. The investigators then present their animal care budget to central administration.

An additional reason for keeping an accurate animal census is to comply with certain laws and regulations, such as the Animal Welfare Act and the Good Laboratory Practices Act. These laws can be applicable to both profit-making and not-for-profit organizations. Federal officials, as part of their responsibilities, may want to see your animal census records and compare that information with an investigator's experimental records. The Association for Assessment and Accreditation of Laboratory Animal Care International (AAALAC) also requests information about the animal census as part of its accreditation program.

To summarize, an animal census can be used in planning your budget, in helping plan a budget for certain investigators, in calculating animal care bills, and for complying with the requirements of federal and voluntary agencies.

Other Financial Control Systems

So far, we've seen that budgets and inventories can be used as parts of an overall management control system. You will recall that developing and utilizing management control systems are important roles of all managers. I then went off on a little tangent and showed how the animal census, which is a type of inventory, is used as part of the process of establishing a *per diem* rate. Now, let's look at other financial activities in animal facilities that require the kind of oversight that we call financial control systems.

Sometimes, when developing financial control systems, you have to think like an auditor. An *auditor*, for those of you not familiar with the word, is a person who, among other duties, examines financial records (such as payments for products received) and compares them against the supplier's bills for the same items. Auditors also review financial balance sheets and many other financial reports to help ensure that a business is operating lawfully and in a financially responsible manner. Whether you do your own auditing, have special in-house people who do it, or have an outside firm do your audit, the end purpose is always to help protect your assets, limit or control your financial risks, and assist in achieving your animal facility's goals. How important is this? I'll give you one quick example. Soon after I took a new position in a laboratory animal facility, the departmental administrator (who was also new) and I recognized that we were doing a reasonably good job of sending out bills every month, but there was no control system in place to see if the bills were actually being paid. If a payment was received, it was properly recorded, but if a payment wasn't received, nobody knew the difference. The monthly bill that was sent was only for the work we performed during the past month, never the accumulating charges. That lack of oversight was inexcusable. In fact, it was so bad that we had to threaten legal action against a company that was contracting for our animal care services in order to collect well over $100,000 that was owed to the animal facility over a period of years. As you can guess, we quickly remedied the problem.

Animal facilities can develop and monitor financial control systems for policies or procedures that are set by the institution (e.g., your school or corporation) or division (e.g., the animal facility). Let's take a look at a limited sample of how such controls might be implemented.

■ *Control system for the receipt of purchased goods.* Personnel must match a completed order requisition to the packing slip (or actual items) that came with the newly arrived order to confirm that the shipment actually contains what was ordered. The person doing the comparison then signs the packing slip to confirm the accuracy of the contents or notes any discrepancies. If problems are found, there is a formal procedure for resolving the issue.

■ *Control system for accounts receivable.* On a monthly basis, a person reconciles income received from an investigator against the amount billed to that investigator. This can be for *per diem* charges, veterinary services, animal orders, or any other charges. Procedures are developed to handle problems arising from a lack of payment or incorrect payments. To avoid any conflict of interest and to strengthen the control system, it is preferable to have one person do the billing and another the accounts receivable.

■ *Control system for invoicing of animal users.* At the end of each month, a person ensures that a bill for *per diem* charges has actually been sent to all investigators who used animals during the month. This is typically reconciled through the IACUC protocol number, not individual investigators, as one investigator may have more than one active protocol. This assurance can be as basic as counting the number of bills sent against the number of active protocols during the month.

■ *Control system to ensure that the entire cost of purchased animals is properly recharged to an investigator.* As an example, the animal facility will not charge an investigator for the purchase of animals based on a telephone estimate of costs from the vendor. Rather, the animal facility waits until the vendor's bill arrives with charges for animals, shipping, crating, and anything else.

■ *Control system for spending limits.* In this financial control system, there is a standard operating procedure (SOP) for the amount of money that can be authorized for a purchase by a person with a specific job title. For example, an animal care supervisor may have the authority to sign a purchase order for amounts up to $100. An associate director may have authority to authorize a purchase of up to $1000 without a second signature, but can authorize an expenditure of up to $5000 if the director cosigns the purchase order.

■ *Control system for authorization of overtime.* No employee can decide that he or she will work overtime without specific written authorization from that person's supervisor or a higher-level supervisor. Standardized forms can be used for this purpose. This control system prevents unanticipated financial expenditures or compensatory time, while allowing for overtime when it is truly needed.

■ *Control systems to ensure competitive pricing for large-dollar items or multiple small-dollar items.* This is simply a way of saying that every animal facility wants to get the best price for comparable items, whether it's a cage washer or shoe covers. One way to do this is to execute written specifications for a type of product (or even for a specific brand name product) and send it out to bid. You may be surprised how much money you can save. Is this really a control system? It most certainly is. It's another way of overseeing that one of your resources (money) is being utilized efficiently and effectively.

Each animal facility will have a different set of control systems because the facility can be in a corporate environment, academic environment, federal or state agency, or other venue. As with all control systems, it goes without saying that somebody has to have the responsibility for making sure that they are being used. That is the human element; someplace along the line, a manager, thinking like an auditor, has to make sure that everything is actually being done as it is supposed to be done. It may not be the most exciting part of the job, but it's one of our most important roles.

What Do I Do If I'm Told to Cut My Budget?

As this edition is being written, the United States is emerging from years of economic downturns, and for research universities during those lean times, it resulted in substantially less income arising from federal research grants. Since grants usually include funds to pay part of a researcher's salary, fewer grant dollars means the institution has to help support researchers until they can reestablish their grant funding. Similarly, less grant dollars usually means that there will be less income for the animal facility, but animal care expenses (for food, supplies, etc.) continue to rise. Also, animal facility personnel have to be paid and provided with benefits, all of which costs money. In response to these problems, university presidents or deans try to develop methods of saving money. Should they simply lay off people and save on salaries and benefits, which are the largest part of most institutions' budgets? Sometimes, that is the knee-jerk reaction we see, but we can do better.

There are at least two ways of approaching this problem, and often they must be used at the same time. The first involves prevention. Developing honest and transparent budgets year after year and promoting that transparency to your own boss and others in upper management will help engender confidence that the animal facility is not padding its budget and can be trusted to need what the facility is requesting. Speaking from personal experience, that helps (and often helps a lot) but certainly isn't infallible.

The second method is more pragmatic. You have to be a team player, and if an employee leaves, don't fill that position unless it is critical to do so. You can probably get by with one less person, especially in larger animal facilities. Look at your planned equipment purchases and see if used but refurbished equipment is available. There are companies that specialize in refurbishing lab animal equipment, and my facility has saved many tens of thousands of dollars purchasing highly serviceable used equipment. When you read Appendix 1, you will find that Lean management, and for that matter any management system, should rely on all of its employees to make suggestions on how to become more efficient. Having decisions made only by managers is not the way to go; coworkers who use the equipment and supplies are familiar with daily operations and often can provide excellent input to managers.

But should you lay off people or delay strategic plans? Whenever possible, don't do that. Layoffs demoralize and cause stress in the remaining employees, and that's the last thing you want to do. Also, laying off a higher-paid, more senior employee can result in a loss of animal facility knowledge that may be hard or impossible to replace. It also is better to try to find ways of reducing certain ongoing operational costs than to delay a strategic plan, such as gaining AAALAC accreditation for your facility [2].

I'll give you a final example of poor financial and personnel management. I am familiar with an institution that, like many others, was facing hard financial times. One of its first decisions, made without any staff input, was to demand higher productivity by increasing the workload but without any additional compensation. You can guess how that was received. Then, it cut back on continuing education, such as travel to professional meetings, and then it withheld even minimal salary increases for all employees except top management. There were no layoffs, but as normal attrition occurred, replacements were not hired, further burdening the existing staff. Need I continue, or do you see the problem that thoughtless financial management was causing? The outcome was that the institution went bankrupt.

References

1. National Center for Research Resources. 2000. *Cost Analysis and Rate Setting Manual for Animal Resource Facilities.* NIH Publication No. 00-2006. Bethesda, MD: National Institutes of Health.
2. Kaplan, R.S., and Norton, D.P. 2008. Protect strategic expenditures. *Harvard Bus. Rev.* 86(12): 28.

Chapter 5

Management of Information Resources

Information availability is not the issue ... we have too much of it.

Ethan Rasiel and Paul Friga

Information surrounds us. It permeates our lives. Meetings, memos, telephone calls, gossip, journals, books, web browsing, email, text messages, instant messages, spell checkers, and on and on. It was reported that the average worker receives 11,680 emails per year [1]. Unfortunately, I get much more than that. Sometimes, the goal of information management is to keep it away from us before we become overwhelmed. Nevertheless, information is a valuable resource, and we have to use it efficiently and effectively. In terms of the roles of a manager (planning, organizing, directing, making decisions, and controlling), managers use information as one of their planning tools: they organize information into usable items, they make important decisions based on the information they have, and they direct projects and obtain feedback based on the continuous flow of information.

General Aspects of Information Management

Information is always available to us, and the manager is often the person who needs the most (and has the most) available information. Interestingly, there are some managers who don't let needed information reach them, or people working with them, for a variety of reasons, such as not wanting to hear bad news, fearing having to give negative feedback, and a generally poor organizational culture. If you sequester information that's important to others, you have created an information gap and time waster for others, and you have demonstrated poor management skills. However, the astute manager always has his information antenna raised and functional. The astute manager also tries to develop a culture where people are willing to share information with each other. In an earlier discussion (Chapter 3), I used the analogy of national security concerns to describe the extent that some people don't want to share information. They think that sharing diminishes some of their power or makes them less valuable to their organization. Nevertheless, managers have to share information if they want people to give them information and minimize possible distrust. Unfortunately, more often than not, managers use the information they gather intuitively rather than in an organized systematic manner. If you ask me what time it is, I'll look at my watch and tell you. It's a simple, almost thoughtless process. In many instances, that's all you need. At other times, you truly have to systematically manage information, particularly when it comes to the information you need to run a laboratory animal facility (and remember that organizing is one of the roles of a manager). Preparing a budget, for example, is not a quick, automatic process. Information has to be gathered, evaluated, and used. The purpose of this chapter is to describe some of the characteristics of information and how you can manage its use.

You have to be willing to share information if you want others to provide you with information.

Almost everything that has been discussed thus far in this book requires the use of information. We use information to learn about internal and external problems (i.e., from the internal and external environment) that might affect our work, to learn about possible opportunities, and of course, for planning. For example, Chapter 4 described budgets and budgetary controls. Looking at any variances between what was budgeted and what was actually spent in a given month was an example of how you could use information. It's interesting, but studies have shown that there is a poor correlation between many managers' perceptions of how their companies are faring financially and how they are really doing [2]. Perhaps better information, or paying closer attention to the available information, would rectify some of these misperceptions. The gathering, transmission, and proper use of financial and other information are all parts of information management.

The modern laboratory animal facility has a large amount of information at its disposal. In fact, there is so much information that companies exist to provide needed information in a manner that is easily usable. For example, if you are a veterinarian, you might be able to review the current veterinary literature on a compact disc or MP3 player while driving to work. Even if you work for a small organization, you may have a need for specialized information-providing companies. Research facilities may subscribe to a web-based information source such as Current Contents Connect. Among other features, it lists the tables of content of many other journals. By reviewing this information source, you can request or download reprints of articles on animal diseases, biochemistry, pathology, or many other areas that might be of value to you. You can use the power of your computer to establish specific search strategies and thereby save yourself substantial time. Google, the largest of the search engines, has more than 70% of the search engine market share and performs more than 100 billion searches per month. Yet, it has been reported that only 5% of people who use Google go past the first search page [3]. The information you need may be on the second or third or other search page, so don't be afraid to dig a little deeper when you use Google, Yahoo, Bing, or any other search engine.

One of the best, but often overlooked, sources is simply asking people for information. In fact, managers place heavy reliance on people as a source of information [4], and maintaining a network of contacts, whether through social media such as Facebook and Twitter or through emails, phone calls, and personal contacts, is an important part of a manager's information base. However, as important as it is to obtain information from colleagues, the wise manager will learn to ask a simple question that is pertinent in some situations, which is "How did you find that out?" All too often, people present idle gossip to us in a manner that seems to be a definitive and true statement. Consequently, when you think that you will have to incorporate information given to you by a colleague into an action you will take (such as a presentation to your boss or a significant decision), don't be afraid to politely ask, "How did you find that out?" If the answer is "I heard it in the hallway," you should think twice before accepting the information as being accurate.

Managers learn a lot by asking specific questions. At the very worst, you will receive nothing of value, but, at the best, you may get all the information you need. And don't forget that idle gossip is another great source of information, but only if it turns out to be true [5]. If you can synthesize the information you get from formal reports, gossip, asking questions, and other sources (such as books, journals, and meetings), you will be doing exactly what most successful managers do.

Managers synthesize information from formal reports, gossip, asking questions, and many other sources.

Sometimes, the manner in which information is presented can be a clue to how much trust people have in you. When my usual information sources start to dry up, I know I'm on the outs with somebody. Conversely, the manner in which we give information to other people can be revealing. If, for example, you ask me what I think of a person and my best response is "He's a nice guy," then it should be apparent that I'm beating around the bush and not giving you a complete answer.

Obviously, some information does not come to you; it originates with you. The animal census, animal health reports, clinical observations, *per diem* bills, and descriptions of new techniques are just some examples of information that can originate with you or your staff and be passed on to others who need it. These recipients might include your own staff, the staff of other divisions, your own supervisor, and the Institutional Animal Care and Use Committee (IACUC).

There also is an information phenomenon that we all know about but usually don't recognize, which is following the lead of others when we aren't quite sure of a right choice. For example, if we don't know which vendor we should use to purchase zebrafish equipment, perhaps a large crowd at a zebrafish equipment booth at a laboratory animal science meeting will give you a starting point. This may not be a bad way of making a decision if the people at the booth independently decided to visit the booth because they had some knowledge of the field and they liked the products that they saw. This is the wisdom of crowds that was described by James Surowiecki [6]. However, if the crowd started to form and then get larger because of a giveaway of free candy bars, then the crowd may not know anything more than you know and it can be wrong [7]. We do have to be careful at times about following the leaders.

Six Characteristics of Managed Information

When you have a lot of information coming across your desk, you want to be able to understand and use it (or discard it) without having to wade through piles of words before you get to the important part. Sometimes, you never do get to the heart of the matter. Still, there are certain characteristics of information that can make it more user-friendly and help you develop good habits for providing useful information. Here are six characteristics that I think will be useful. There is an additional discussion on information usage in Chapter 7, although that discussion is primarily focused on what can happen when you get advice from others. For our present purposes, we will assume that the information you receive is accurate. However, that's not always true. In fact, managers and leaders often test the validity of information they receive before running with it. For now, assume most of the information is correct.

1. Information Should Be Relevant

The information you receive or give to others should not be superfluous. It should help you resolve a present or potential problem. In a laboratory animal facility, you may want to know about upcoming legislation that will affect your work, but it is unlikely that you will be interested in the fluctuations of the dollar's value overseas.

You must be able to tell those people who provide you with information, such as a librarian or a research administration office, just what kind of information is relevant to you. There is simply too much information available on almost any topic for you to be able to digest all of it. The web-based information source Current Contents Connect, mentioned earlier, is a good example. It would be a waste of your time to read the titles of every article in every journal it indexes. You must be selective as to which journals you scan and what information you receive.

You also must be selective as to what information you retain. Even though managers listen to gossip as much as anybody else, they must be able to sort out what's important and what isn't, because sometimes gossip turns out to be true. At a meeting such as the American Association for Laboratory Animal Science (AALAS) annual conference, most of us keep our eyes and ears open to find out what is happening at other institutions. In essence, we're exploring current information in our external environment to help us in our jobs. The chances of picking up some highly relevant information as part of a general conversation may be minimal, but if you sit down with a colleague over a cup of coffee and get specific about what you want to discuss, it may surprise you how much you can learn.

2. Information Should Not Be Excessive or Insufficient

With too much information, even if it is relevant, you can still have a problem separating the wheat from the chaff. In fact, information overload has become a serious problem in knowledge-based organizations such as research institutions [4,8]. Although there is no law saying that I have to read all of my emails, I do have to look at the headings of each to see if it is a message of importance or something I can trash. Sometimes, a summary of the relevant information will suffice, while at other times, a full reading or discussion is required. Consider a disciplinary meeting with an employee. Do you want a summary, all the facts, or something in between? Even though some people argue that too little information is worse than too much, I suggest that the minimum amount of information required to do the job often is the right amount of information.

In most cases, the minimum amount of information needed to do the job is the right amount of information.

It has been reported that the most frequent causes of information overload are an excessive amount of information, difficulty in managing information, irrelevant information, lack of time to understand the information, and getting the same information from multiple sources [8]. Therefore, you should request the appropriate amount of information needed to complete your task—not too little, not too much. As a general statement, you do not need all the information reasonably available to successfully complete most tasks. Usually, having 20% or 30% of the information is more than adequate to start (and often finish) a project. If you wait until all the available information is in your hands before starting something, that project may never get off the ground. However, if you don't know what you need, err on the side of asking for too much information rather than too little. This was the method used by Henry Kissinger in the anecdote I gave in Chapter 3.

Budget reports are another area where you have to be very careful about how much information you request. Depending on how your particular animal facility and organization function, you may need anything from summary data to a detailed listing of all income and expenditures.

3. Information Should Be Timely

Not only do you want accurate information, but also you want it in a understandable format, it has to be up to date when fresh data are relevant, and it should be available when you need it. It may be old news, but if you need to see a copy of your budget from two years ago, you should be able to retrieve it rapidly. Perhaps one of the reasons we like gossip (and often use it) is because it's usually timely.

At one time or another we have all experienced a problem providing a timely communication to large numbers of investigators. Nowadays, we often overcome this problem by sending group emails. Similarly, some animal breeders send emails to their clients when rapid communication is necessary. That method is appropriate since email has become a common method of communicating information to many recipients via just one email message. You may already know that there is the IACUC Forum email discussion group sponsored by AALAS, which also hosts the Compmed and TechLink discussion groups. If your organization has the proper equipment, voice mail can be sent to many people at once. Additionally, numerous other electronic resources for veterinarians and laboratory animal facility managers are available.

When I was doing clinical veterinary work, I was particularly attuned to the timely receipt of vendor health reports. Most major vendors can provide them quarterly, and they provide much important information. Some vendors have their reports on a website. I received a report from one vendor that was very impressive except for one problem: the report wasn't dated. I called and found that the vendor checked its animals once every six months for viral pathogens and once a year for bacterial pathogens. Sorry, but I wanted more timely information than that vendor provided. I purchased my animals elsewhere.

4. Information Should Reach the People Who Need It

Information has to reach you or other people who need that information. This is why it is becoming much more common to see an open office arrangement where there are no private offices with closed doors, or if there are private offices, they are clustered near each other so people can readily interact and share information. It's a variation of the water cooler analogy, where we often get or provide more information standing around the water cooler or coffee pot than we do from any other source. Very often, the people on the front lines, such as technicians, are not kept in the information loop. However, it has been argued that these are the people who should be the best-informed individuals in the organization, rather than the least informed [9]. It makes good business sense to keep these people informed because they are often the ones to provide solutions to the day-to-day problems they encounter.

Sharing information among different groups can also be of great value in managing a laboratory animal facility. It's not unusual for an IACUC administrator to need concurrent information from the animal facility, the Institutional Biosafety Committee, or the Institutional Review Board (the human studies equivalent of an IACUC). Alternately, the animal facility may need to share or obtain computerized information from the IACUC administrator. One way of approaching this need is to have a local network of computers that can provide pertinent information to multiple users. Another way is to have computerized links to relevant information. For example, if an investigator is filling out an IACUC protocol form that includes a question about the volume of blood to be withdrawn, wouldn't it be nice to have a link to information about the volume of blood that can safely be sampled on a daily or weekly basis?

I wrote earlier that you have to be willing to share information if you expect people to give you information. You have to engender a culture of openness to encourage people to provide you with certain types of information. People have to know that you need certain types of information. You also may need appropriate equipment (e.g., a computer) to get certain types of information. Nevertheless, even with everything in place, there are some people who attempt to gain or keep power by withholding information from others. For example, the supervisor who doesn't tell employees about pending changes in a work schedule until the last minute has certainly exercised his power to withhold information—whether it was the smart thing to do is another story. There also are people who may withhold information from you because they like to see you squirm when a minor problem elevates into a major problem as a consequence of the withheld information. Others may withhold information in an attempt to diminish the power of others [10], whereas still others may interject their own biased opinions into information they eventually pass on. If you withhold information simply to prove that you can, you have a significant managerial personality flaw and you may have to pay the piper later on. As you may have guessed, your management style (authoritarian, micromanager, etc.)

can affect the dissemination of information. A recent study concluded that abusive managers or supervisors are a barrier to information sharing [11]. That finding is somewhat intuitive, but interestingly, the same study reported that a person's perception of high-level organizational support can ameliorate the negative effect of abusive supervision. This is yet another reason to try to develop a strong and caring organizational structure.

The truth is that in most instances, all information should be disseminated to the people who will be affected by it. There are times when this cannot and should not be done. If you work for a pharmaceutical firm, you may be advised about health effects and safety precautions involved with a drug under development, but it is very unlikely that you will be told the chemical structure and other details, even if you will be involved in animal studies that use it.

At one time or another, all of us have said, "Why wasn't I told?" Sometimes, as I just wrote, you weren't told because the information was confidential at the time. In other instances, you weren't told because the information meant for you was given to the wrong person. If that person had no idea why she received the information, it could have ended up in the trash. On many occasions, I have been sent copies of memos that, for the life of me, I couldn't understand why they were sent to me. Sometimes, they were filed, but more often, they were discarded. A good way of providing information to animal-using investigators is via a monthly or bimonthly newsletter addressed to that person. If the "Why wasn't I told?" question comes up, you have a black-and-white answer—or perhaps an email copy—in front of you.

There are two sides to every coin, and so far I have only discussed the problem of not sharing information. The other side of the coin is that we can waste a person's time by giving them information they don't need. In Appendix 1, there is a discussion of Lean management, and I wrote that one of the pillars of Lean management is the reduction of waste. Providing a person with unnecessary information can easily waste the time of the person receiving the information, along with the wasted time of the person who sent the information. For me, this happens most often when I am copied on emails that have little or no importance to me. We have a department of more than 75 people. Why do I have to be copied on an email from a supervisor to her manager that explains how the supervisor is going to rearrange the afternoon break because one person has to leave early? Is it really necessary for my own boss to have me sit in on a meeting about how she plans to refurnish her office? Sometimes I don't know if I'm being punished in some way, if people are anxious to keep me in the loop, or if I project an image of micromanaging everything so I have to know the minutia of what is happening. The message to you is to think about what you are doing when you provide information to a person. Does that person really need the information, or are you just covering your bases "in case" they might want to know?

Here's another facet of information sharing. Many readers of this book work in a biomedical research environment, and it is interesting to note that scientists are very careful about the people and sources from which they obtain information, and they are equally or even more careful about with whom they are willing to give scientific information. Many of their choices are based on personal relationships, the perceived scientific credibility of the person with whom they are considering getting or sharing information, and the perceived "approachability" of other scientific colleagues [12]. None of this should surprise us because we know that science is as much of a competitive business as is banking or higher education. And as much as most scientists love to discuss their work, they are very careful not to disclose information that directly or indirectly may give another scientist the ability to outshine them.

One final thought on disseminating information: Although I said it previously, I want to reemphasize that you must keep your immediate supervisor informed of significant events in your area of responsibility, whether the event has already happened or it is still in the planning stage. This is another one of those circumstances where too much information may be better than too little, although you have to know what your supervisor wants. In Chapter 3, I commented on some of the ways to integrate your own management style with your supervisor's style, and this may be a good time to review those comments. One way or the other, you have to provide needed information to your supervisor in a timely manner. Equally important, when you provide that information, you should try to know more about the situation than the most basic facts. For example, if you learn that a nearby new private laboratory is doing research that is very compatible with research being performed in your own institution, part of your job becomes trying to find out some additional details about that research and the appropriate persons to contact. All of this should be done well before you meet with your boss. The additional information certainly may help your boss and your institution, and it also speaks positively of your own initiative and commitment.

5. Information Should Be Cost-Effective

The management of your information resources may be costly. Business computers and their associated software, printers, Internet access, email, voice mail, professional meetings, and the like can cost many thousands of dollars. We have to make managerial decisions as to whether the value of the information is worth its cost. Sometimes, the answer is obvious. For example, professional journals are important in animal facilities. Because it may be too expensive to receive individual copies for everyone, many people can share one journal subscription. A less obvious example is the cost of installing a computer link to a library database, such as the National Library of Medicine. Would it be cost-effective for your organization? It has been said that in the not too distant future, biomedical journals may all be digitalized. No more reading them on the train or before going to bed—unless, of course, you read them on your laptop, tablet, e-book reader, or smartphone. Perhaps it's the end of an era that shouldn't end, but the ability to do rapid searches for articles and passages certainly offers a compelling case for having electronic journals in addition to print journals.

In some instances, it's almost impossible to evaluate the worth of information. Consider a professional meeting in another part of the country. Although you almost always learn something new, there is more to meetings than the formal sessions. The contacts you make with other professionals may, in the long run, be worth far more than the cost of the meeting. If a professional meeting is important for you, is it also important for your technicians? Have you considered budgeting money for them?

6. Information Should Be Used

Assuming that the information reaching you is timely, relevant, and adequate in quantity, you now must do something with it. It does very little good to have it pile up on your desk. It's like going through the mail. Sometimes you do nothing with the mail, and sometimes action is required. As an example of information that's often not used (but should be used), think about our laboratory animal science journals. Chances are that you read some articles in at least one journal. If resources are available, do you or other managers implement information from the journals that might help your department? Does the animal facility have a culture that openly shares and discusses potentially relevant information? Even if you are current with relevant literature, if the managers of the vivarium do not utilize journal information that is important to the mission of their vivarium when resources are available, they should sit down and rethink their roles as managers and leaders.

Some information simply confirms what you already know, and if everything is under control, or if it is "for your information" only, no managerial action may be needed. One way or the other, you will have to decide what to do with the information. It's not always necessary for you to wait until confirmatory information comes across your desk before taking action. Most managers develop the skill of piecing together snippets of information from multiple sources in order to help visualize the bigger picture of what is actually happening. If you think that immediate action is required, take it, and worry about the confirmation later. An excellent example of this point occurred when a filovirus was found in imported cynomolgus monkeys. Because of the potential zoonotic hazard of certain filoviruses, the Centers for Disease Control and Prevention (CDC) imposed stringent requirements for the importation and quarantine of those animals. The CDC continued its investigation into the source, dissemination, and potential hazard of the virus, but it acted rapidly and decisively.

Perhaps it's just my personality, but I always like to use the information I have to perform in my mind "what if" scenarios, imagining the worst outcome and the best outcome. If the worst outcome is really horrendous, I have to plan damage control ahead of time. If the best scenario is favorable to everybody concerned, I mentally sketch out the next step to take once that outcome arrives. Most often, I'll use whatever resources I have available to identify the most likely outcome, and focus my energies in that direction. For example, in broad terms, I know the major challenges that are confronting the University of Massachusetts Medical School. I'm also initiating the annual budgeting process and, after meeting with various people, I'm mentally making decisions on financial allocations. These decisions are based on what I know about my department and the university as a whole, about changes in animal numbers, about researchers coming and going, and so on. My budget decisions will be made on our best guess as to what is likely to happen, but I keep in the back of my mind some contingency plans in case things are better or worse than I'm predicting.

As a manager, there will be times when you want to be decisive, but the corporate culture says "slow down." You have to know your own organization. No book can tell you that. But you should not use your corporate culture as an excuse for either doing nothing or taking reckless risks. Use some common sense.

Desirable Characteristics of Information

- Information should be relevant.
- Information should be neither excessive nor insufficient.
- Information should be timely.
- Information should reach the people who need it.
- Information should be cost-effective.
- Information should be used.

Information Commonly Needed by Animal Facility Managers

I mentioned earlier that you probably are not concerned about the value of the dollar overseas. Currency fluctuations may be a concern of your chief financial officer, but, for the most part, you have other things to worry about in your job. Yet, don't think for one second that because you are a manager of the animal facility, you should be concerned only with that little piece of real estate. This idea is absolutely not true. If you think it is, this is the time for you to skim through Chapter 2 to refresh your memory about environmental variables and how they can affect your operations.

Let's take a look at some information requirements that are common to most laboratory animal facilities and are used to help keep the facility running smoothly and to establish goals. Some of these, such as financial statements and personnel policies, have already been discussed.

For starters, as noted in Chapter 3, you and all the other employees should know the chain of command and how your institution's administration is organized. Strange as it may seem, many people in any given organization have very little idea of who anybody is, other than their immediate boss. You certainly don't have to know everybody, but you and your coworkers should at least know the names and titles of those people who can immediately affect your operations.

A common information need is knowledge about your employees' working hours, as well as their efficiency and effectiveness in performing their jobs. The assessment of efficiency and effectiveness was discussed in Chapter 1 and is expanded on in Appendix 1. This information helps us assess and discuss performance with our staff and gives us objective criteria, if needed, to increase or decrease personnel based on the projected workload.

Standard operating procedures are quite important, and in some cases, they may be required by law or regulation (e.g., the Good Laboratory Practices Act). It might seem obvious, but it's worth stating that you should be aware of and understand other pertinent laws, regulations, and recommendations (e.g., the *Guide for the Care and Use of Laboratory Animals* [13] and the Public Health Service policy on the use of animals in research). The information contained in those laws, regulations, policies, and guidelines may affect your organization's research effort if the animal facility is not in compliance. It's unfortunate that more than one laboratory animal facility "manager" believes that all he has to do is make sure the required work is done efficiently and effectively. Everything else is somebody else's problem. Some of these managers hardly know what is in the guide. This attitude is really unfortunate for a manager to have. These people are more like construction bosses than true managers.

Budgets are of major importance in most animal facilities. If you are *not* the one who prepares it, you must know who does, where to get the needed information, how to help that other person prepare a budget, how to use it, and if appropriate, how to develop *per diem* rates. Some managers compare their expenses and *per diem* rates with those of other institutions, even though this usually provides nothing more than very rough information. Different institutions do not always include the same items in their *per diem* rates, so at the least, you should know what is included or not in your own rates.

Cage inventories, cage washer temperature charts, cage cleaning schedules, and equipment maintenance charts are also among the information the facility manager must have available. Periodic animal health reports may be required, and therefore, you should keep health records (either on a group or an individual animal basis). For some species, such as dogs, you must have records that are specified by law. You may be asked to trace the location and treatments on a single animal (e.g., a dog) from the time it was ordered to the time it left the animal facility. If you have good records, fulfilling this request should not present a problem. A typical record for individually housed animals is shown opposite.

Great Eastern University
Individual animal health record

Species/strain _____ Investigator _____

IACUC number _____ Arrival date _____

Vendor _____ Birth date _____

Animal ID _____

Appetite: √ = normal   F = fair   P = poor

Bowels: √ = normal   D = diarrhea   S = semisolid   X = other problem

Urine: √ = normal   B = with blood   X = other problem

Date	Appetite	Bowels	Urine	Comments	Initials
1					
2					
3					
4					
Etc.					

Not all animal health records originate with you. Most vendors of common laboratory animals are able to provide you with health reports on their animals. Even if you have an in-house quarantine program for arriving animals and perform your own periodic general health and serologic monitoring of animals, receiving vendor health reports can only help you. These reports must be timely and reach the person (usually the veterinarian) who is most able to interpret and act upon the information they contain.

Here's another information-related consideration you should be aware of. Many animal facilities are unionized. If you are employed by such a facility, you should know the appropriate rules for working with the union. For example, you may need to change some employee's days off for a few weeks to accommodate a woman out on maternity leave. Does your union contract allow you to do that? If a union employee wants to make the change to help a fellow employee, is it permissible? Do you know the contract's provisions on regular working hours, overtime, promotions, attendance at union functions, arbitration of grievances, and so on? If you don't, you should.

What about the research or teaching programs of your institution? You certainly must know how to work with the investigators and instructors who use your facility. You must know what types of animals are usually needed, the quality of vendors, and the availability of animals. You will probably have to know some specifics about different animal models and alternatives to the use of live animals. It's all part of the job.

It should be obvious that I have not and cannot possibly discuss all the information managers need in laboratory animal facilities. It's your responsibility to determine what information is needed for your day-to-day operations. One thing is clear though: information is as much of a resource as money and people. As such, it must be managed by you so that it helps you to reach your short-range and long-range goals.

Organizing and Retrieving Information with Computers

One of the desired characteristics of information management, as described above, is that information must be timely. That is, the information you have available must be managed so that it can be rapidly located when you need it, and it must be in a usable form. Every office and every manager has some type of an information organization and retrieval system. My personal system is to put important information in the top tier of a three-shelf basket on my desk, and put progressively less important information in the baskets below it. Then, of course, there is the junk pile on the table next to my desk. I am the only one who knows where things are, which sometimes works to my detriment.

Filing research protocols by using a code number and alphabetical filing of personnel records are typical examples of basic office information retrieval systems. Of course, computers are a major means of storing and retrieving information. They can sort information and present it to you in a multitude of forms. Thus, you may be able to track a study with a database that sorts it by investigator, protocol number, species, type of procedure performed, and many other parameters.

Remember, computers can only do what you can do manually. They simply do it faster, and usually with greater accuracy. As a rule of thumb, if you do not understand the basic concepts of how to do something manually, you won't understand how to do it with a computer. There are exceptions, of course. You may not understand the mathematics behind a particular statistical test, but you should know what information to give the computer to get an answer. Let the machine worry about the formulas and calculations. You must, however, know what statistical test to use before you enter the values into the computer.

Every manager should have a basic understanding of computers and how they can be used. This is usually termed being "computer literate," and nowadays it is almost impossible to manage an animal facility without computer literacy. Computers can also give you access to laboratory animal science electronic bulletin boards, many other types of laboratory animal and scientific information, and other managers throughout the nation and, indeed, the world. In today's world, you simply cannot operate an efficient animal facility, even a small one, without computers.

Still, computers should not become a substitute for the basic skills I have previously described for managing your information resources. But, once those resources are in place, you should be leading the troops toward computerization. As I said, you simply can't be particularly efficient or effective without them. If one or more people you work with are afraid of computers and don't want to invite change, you're the person who has to lead by example, by offering training opportunities (even one-on-one, if necessary) and doing whatever else it takes to bring your facility into the modern world.

References

1. Gill, B. 2013. E-mail: Not dead, evolving. *Harvard Bus. Rev.* 91(6): 32.
2. Mezias, J., and Starbuck, W. 2003. What do managers know, anyway? *Harvard Bus. Rev.* 81(5): 16.
3. Bradley, S.V. 2015. *Win the Game of Googleopoly: Unlocking the Secret Strategy of Search Engines.* Hoboken, NJ: John Wiley & Sons.
4. De Alwis, G., Majid, S., and Chaudhry, A.S. 2006. Transformation in managers' information seeking behavior: A review of the literature. *J. Inform. Sci.* 32: 362.
5. Grosser, T.J., Lopez-Kidwell, V., and Labianca, G. 2010. A social network analysis of positive and negative gossip in organizational life. *Group Organ. Manage.* 20: 1.
6. Surowiecki, J. 2004. *The Wisdom of Crowds: Why the Many Are Smarter than the Few and How Collective Wisdom Shapes Business, Economics, Societies and Nations.* New York: Doubleday.
7. Easley, D., and Kleinberg, J. 2010. Information cascades. In *Networks, Crowds, and Markets: Reasoning about a Highly Connect World.* Cambridge: Cambridge University Press. 425–444.
8. Farhoomand, A.F., and Drury, D.H. 2002. Managerial information overload. *Commun. ACM.* 45(10): 127.
9. Hamel, G. 2012. *What Matters Now: How to Win in a World of Relentless Change, Ferocious Competition and Unstoppable Innovation.* Hoboken, NJ: John Wiley & Sons.

10. Pettigrew, A.M. 1972. Information control as a power resource. *Sociology* 6: 187.
11. Kim, S.F., Kim, M., and Yun, S. 2015. Knowledge sharing, abusive supervision, and support: A social exchange perspective. *Group Organ. Manage.* 40: 599.
12. Andrews, K.M., and Delahaye, B. 2000. Influences on knowledge processes in organizational learning: The psychosocial filter. *J. Manage. Stud.* 37: 797.
13. National Research Council. 2011. *Guide for the Care and Use of Laboratory Animals*. 8th ed. Washington, DC: National Academies Press.

Chapter 6

Time Management

> There are far too many people that waste their time telling themselves
> that they don't have enough time.
>
> **Daniel Willey**

It certainly is a rare day when a laboratory animal facility manager can sit back,
look at the ceiling, and feel guilty about having nothing to do. Most of the time,
facility managers are juggling three or four major projects, along with their routine
responsibilities, while trying to maintain the productivity of the animal facility.
That's just the beginning of the time crunch, particularly for a seasoned manager
who may get more and more work delegated to him. People are working longer
days than ever before, taking their work with them on their vacations, taking their
work home, and neglecting their family life (although the actual divorce rate has
not been increasing). Then, of course, the inevitable emergency arises at the worst
possible moment. What can we do to remain sane? How do we find enough time
to do our work? This chapter will discuss day-to-day time management. Time and
time limits are also part of the vision and strategic plans discussed in Chapter 1.

General Considerations in Time Management

It's very easy to forget that time is a resource that must be managed to help us reach
our goals. Time is a limiting resource in the sense that there are just so many hours
in a day. The word *efficient* is easily applied to time management, but let's not forget
that we must balance efficiency with effectiveness. I've always been fascinated by
budget-cutting managers who expect people to be as efficient and effective as they
were before their budget was slashed by 30%. "How you do it is your problem; just
do it," they say (or imply). This is no small trick, and in some cases, it's impossible.

Often, you have to increase your efficiency or decrease your service. If your staff is already working at reasonable efficiency and effectiveness, a budget cut will usually lead to decreases in service or effectiveness. Unfortunately, almost everybody who has worked in a laboratory animal facility has experienced budget cuts, and I will do my best to try to help you manage one part of that problem—increasing your efficient management of time to help you effectively accomplish your goals.

New managers, up to their necks in more work than they can handle, may fall back on the old saw that when you are being chased by alligators, you can't worry about your swimming form. In other words, they feel they are too busy to worry about any fancy theories on how to manage time. That, they say, is for people who have nothing better to do. Yet, this is precisely the reason that time management is so important. It allows you to get out of the swamp and think more clearly. More time for planning may actually lead to more productivity, not less. If you are one of those people who claim to work 60- to 70-hour weeks, you should read this chapter carefully. Your timekeeping is wrong, your ability to manage your time needs to be improved, or you have to consider another job. For the vast majority of us, there is simply no need to regularly put in ridiculously long work hours. Although most people who put in long hours do so because they like their job and find it challenging, their home life often suffers [1], and I am willing to speculate that their day-to-day work quality will suffer as the number of hours worked increases beyond some as of yet unspecified breaking point. I am not suggesting that putting in more than a 40-hour week and working hard throughout the regular workday is inappropriate; I do mean that working ridiculously long hours and loving your work to the near exclusion of everything else, including your family, is not a healthy state of affairs.

I'm sure that some of you reading this book feel that your life is just getting out of control with people making demands on you. There is only so much multitasking any person can do, and although we can delegate some tasks, there are certain ones we must keep for ourselves. But people, including our bosses, continue to make demands on our time. How do we handle this? One simple (perhaps simplistic) answer is to politely say no, explaining that you are absolutely swamped with work. But can you do that if you're the low person on the totem pole? It's hard to know because each situation is different, but it's fair to say that if you don't try to push back a little once in a while, you'll never know if you can. In fact, one way of raising yourself a couple of notches on that totem pole is to exert a little authority by not being afraid to decline a new task when you know you won't be able to do it justice. As I will emphasize a little later in this chapter, it's better to do a few things well than many things poorly.

Time must be managed like any other resource. It is more than working faster or putting in more hours to complete your work. Working smarter can save time.

Most managers of laboratory animal facilities have worked their way up through the ranks. Even for those with a college degree in animal health technology or laboratory animal science, almost invariably there has been an apprenticeship period where hands-on work was the rule rather than the exception. Now, as a manager, you're probably still spending a certain percentage of your time as a "doer," rather than as a manager. This is not at all unusual, particularly if you are a first-line or new supervisor. A lot depends on the size of your facility. It's hard to envision one person sitting behind a desk all day and managing two others (although this is not unknown). On the other hand, if you manage a very large facility, you may rarely have time for hands-on animal work. It's been estimated that supervisors in most professions spend 39% of their time "doing things," rather than managing [2]. Another report indicated that top executives averaged less than 40 minutes a day of actual accomplishment [3]. That may not be so bad. In fact, Tom DeMarco points out that those managers need some extra time for thinking or handling unique problems. Without what he calls "slack," you can keep things efficient, but not necessarily better [4]. In other words, you need some slack time to see how things can be improved, although achieving only 40 minutes a day of accomplishments seems markedly inefficient to me.

One of the keys to successful time management for the new manager is to learn relevant new skills, such as management skills. You should not be completely involved in performing and refining skills that don't have a significant impact on your daily activities. To rephrase this, you have to stop functioning as an expert in laboratory animal technology and start functioning as a manager. Don't steadfastly refuse to give an injection because it's no longer in your job description, but, at the same time, don't do it just because you want to prove that you're still one of the guys. Do it because there is a particular need at a particular time. We all have to chip in when it's necessary to do so. I've changed cages, washed down runs, cleaned filters in the cage washer, and fed animals when we were short-handed. My wife, who is a veterinarian, will insert catheters or do other technical tasks when her technicians have trouble doing so. Some managers make a point of periodically working on the production line or doing other jobs within their company to keep themselves in touch with what's actually happening outside of their office. They don't do these other jobs day in and day out, but the occasional stint also helps keep employee morale high. If you are first and foremost a manager, spend most of your time managing, but don't be afraid to pitch in when the situation calls for it.

If you are first and foremost a manager, then spend most of your time being a manager.

There are numerous ways that time gets wasted. For example, it has been stated that recurring work problems and hassles cost employees 40% of their work time [5]. Not having enough cages available when you need them is a recurring problem in animal facilities. Ordering animals by telephone rather than doing the same electronically and taking an animal census manually are time wasters. Waiting for a cage washer to finish its cycle is another example of a time waster. "The goal is not time 'saving' per se, but rather reallocating hours for more important tasks. This involves understanding what is important and what is not" [6]. It's entirely reasonable for you to have more than one goal. In fact, it would be unusual if you did not. You don't have to completely stop working toward one goal if another is somewhat more important, but you do have to allocate more time to the more important goal. That is, you must establish what your priorities are. This is part of your role as a decision maker, and it is an important facet of your job. Similarly, your goal and the steps needed to reach it should be crystal clear to you so you don't waste time trying to work out the details of your strategic plan after you begin working to reach your goal [7]. Some of the needs you must consider when establishing decision-making priorities include the following items [8]:

- Is there a need to make the decision by a certain time, and why? Perhaps waiting a little longer and getting a little more information would help you make the best decision. Alternatively, if you wait too long, you might not be able to get the deal you can get today. As a manager, you will have to make decisions, but you want to make the best decision in the most appropriate time frame.
- How long will it take for you to get enough feedback to determine whether a decision you made is the right decision? Will it be possible to correct the wrong decision? If the wrong decision cannot be corrected, you may have to wait until you receive more information, but if the wrong decision is readily correctable, you might find it in your best interest to make an early decision.

Here's a simple example of decision making by an animal facility director having multiple ongoing projects.

> *It was early March, and Dr. Randy Smith, the director of the laboratory animal facility at Great Eastern University, was working on three projects. There was a monthly newsletter to be sent to investigators, a grant application for capital improvements in the animal facility, and his annual budget.*
>
> *The budget was due in three weeks, and although he wasn't short on time yet, he knew he had to pay more attention to the budget than to the other projects because his boss, the vice provost for research, needed to integrate the animal facility budget into the larger university research budget. Dr. Smith continued working on the two other projects, but spent the bulk of his*

*time working on the budget. When the budget was complete,
he was able to spend more time on the other projects.*

Smith appears to have been an experienced manager. He had a ton of work, but he didn't get flustered. Whereas an inexperienced manager might have wanted to be helpful to others and take on more and more work, Smith limited himself to three projects. He set priorities and got the most important project, the budget, completed first. Maybe he even shut his door to keep people away from him for part of the day. Like any good manager, Smith knew where he was, knew where he wanted to be, and developed a short-range plan to get his budget completed within three weeks. Smith understood that time management does not mean managing for one day at a time. It implies that you must set your short-term and long-term goals and properly allocate the time needed to reach those goals. As with all aspects of management, you are constantly reevaluating where you are, where you want to be, and how you are going to get there.

Establishing priorities is an important aspect of time management.

There are some activities that do take a limited amount of time, but often have a positive overall impact on another managerial resource, information management. One author believes that talking around the water cooler or gossiping with coworkers wastes time [9]. However, as I wrote in Chapter 5, managers often gather valuable business information from office gossip and chats around the water cooler. Clearly, too much gossiping can be a time waster and otherwise detract people from their work, but a little informal talking between employees is not a time waster and should not be discouraged.

Thinking about Work Processes

If, as I noted above, the goal is not saving time per se but spending time on the most important task facing you, does that mean that some waste of time is acceptable? No, it does not. Wasted time is just waste, nothing more, and one of your goals should be to eliminate waste from the workplace. Spending time on the most important task assumes that the time will be used efficiently and effectively; it's not a license to waste time. One common time waster in animal facilities is not taking into consideration the individual parts of a process that make it work. The goal is to have as few steps as possible, and for those steps to be accomplished in a continuous and smooth progression whenever possible. Let's look at ordering animals. Do investigators order animals any day of the week by calling your front office and telling a person what they need? If they do, they interrupt a person's ongoing work in order for that person to take the call, write down the order, and then return to the work they were doing and refocus on that work. This wastes time. Next, consider how the front office person actually orders the animals. Does that person call the vendor repeatedly during the week or save up the orders and place all of them at one time, the latter being a process that will save time? Perhaps the most efficient way of ordering animals would be to use a computerized system that allows the investigator to fill out an online animal order form. This form, when sent electronically, reaches a special computer "mailbox" that keeps the orders filed until the person who orders animals is ready to place the order, after viewing it to make sure that there are no readily observable problems. In the best of all worlds, the order would be placed electronically with the vendor, who would return an electronic confirmation or indicate its inability to fulfill the order. Each step I have just described is part of the process of ordering animals. We try to look at each step in the process to eliminate any wasted time and to provide value to the investigator ordering the animals. Other examples of stepwise processes include the handling of cages before, during, and after they are washed; the process used to get health reports to a veterinarian or veterinary technician, and how that person responds to those reports; and many other common processes performed every day in an animal facility. Whenever possible, the process should be examined, step-by-step, to eliminate wasted time by being as simple and efficient as possible. Here's an example of how one manager saved time by keeping his cage washers working for an extended day, rather than having cages backed up every morning, waiting to be washed.

> At the end of the day, Angel Gomez, a supervisor at Great Eastern University, always found there were cages still to be washed. It seemed that only working overtime would solve the problem. After reviewing various options with the cage wash staff, he solved the problem by having some people work from 7:00 a.m. to 3:00 p.m., while others worked from 9:00 a.m. to 5:00 p.m. All Angel had to know was that the employees could begin at 7:00 a.m. without a supervisor being there, and that his own supervisor would approve the plan. From that point, Angel delegated all the details to a subordinate. The decision did not require any knowledge of which person would be involved, who was making how much money, or any secondary facts.

This was a simple solution to a common problem.

Recording Time Usage

Some years ago, I tried to become more efficient in my daily time usage by keeping a time log. Every hour, for two weeks, I recorded how I spent my time. I followed the same advice I'm going to give you. I did it because I believed it was standard doctrine in time management. It helped me for a few months, and then I began to slip back to my old habits. It was like going on a diet; it works at first, and then the slightest little slip and you're back to square one. Nevertheless, it was not a total waste of time. I picked up and kept enough good habits to make it worth describing to you. I do this with a little hesitation because, after speaking informally with other managers, my impression is that recording time usage works for some people, but not others. With that caveat, let's move forward.

To properly manage your time, you should begin by knowing how it is currently used. How much of your time is spent as a manager? A technician? A teacher? How much is spent in meetings, on the telephone, looking at your email? An easy way of making such a determination is to keep a time log. For one workweek, record what you are doing every 30 minutes or every hour. Record this as you go along, not at the end of the day, because nobody can remember the events of an entire day. Some people have forms that divide their time into telephone, meetings, planning, and so forth. Others simply list their activities and summarize them later. But one way or another, it is necessary to define what you are doing and when you are doing it before you can consider making changes.

Here's an example of a time log for a manager with minimal animal care responsibilities.

Great Eastern University time log							
Date: March 23, 2016							
			Enter all times as minutes spent				
	Species	Animal Care	Adminis-tration	Records	Phone	Meeting	Personal
9–10 a.m.	None	-----	40		20		
10–11	Cat	30		30			
11–12	None	-----	30		15	15	
12–1 p.m.	Lunch	-----					60
1–2	None	-----	15		25	20	
2–3	Rat	15	30		15		
3–4	None	-----	60				
4–5	None	-----				60	
5–6	None	-----	45		15		

In addition to keeping records of your own time, all members of your animal facility may have to keep time records. This process can serve a multitude of purposes. The most obvious is to help your employees manage their own time. In addition, having employees record their time is immeasurably helpful in setting *per diem* rates. The biggest part of the *per diem* rate is labor, and it is imperative that everybody be accurate about recording the amount of time that is allocated to specific tasks. Although I'm somewhat ambivalent about the use of personal time logs, I am a strong proponent of knowing, in general, how my staff spends their time. This is discussed in greater detail in Appendix 2, where a special time log for people working with animals is included.

Asking your staff to record their daily time usage can lead to some raised eyebrows unless you clearly communicate what your goals are. If someone handed me a sheet of paper and told me to record my time usage (which has actually happened to me), my first impression would be that someone is out to get me. If the economy were bad, I would be paranoid. So be sure to let your staff know why they are recording their time usage. Be sure to give them periodic feedback about your findings.

Some years ago, when working at the American Health Foundation, I did a small informal study of time usage by our research technicians. We found that approximately 30% of their workday was occupied by meetings with investigators or with their own supervisors, work breaks, personal telephone calls, going to the restroom, and the multitude of items that make up a typical day in a laboratory animal facility.* This came as a great shock to the heads of the various research divisions, who were convinced that their own staffs worked 70 minutes every hour and had no bodily functions. Yet, none of them would do a similar study. Perhaps they didn't want "bad news." Perhaps they didn't realize that talking to each other, going to meetings, and doing other normal daily activities helps build the camaraderie and skills that are so important to an effective and efficient operation.

You should evaluate staff time usage records for other reasons. Record keeping can help you justify the need for additional staff or to reassign staff to other areas. In the facility where you are currently working, can you justify how many people are required for general animal care? In those facilities where the staff salaries may come from more than one source, time records help allow for the correct allocation of charges. Under these circumstances, it may be necessary to have, on an ongoing basis, hourly (or other intervals) recordings of how time is spent during the day.

Many laboratory animal facilities charge for administering medications or for the special care of an animal. When this occurs, labor is often charged to the nearest quarter hour. To do this, a record must be kept, particularly if you find yourself having to document a charge. A final advantage to keeping time records is that it provides information to compare your time usage, and the time usage of your staff, with that of other institutions (assuming, of course, they have time records).

Utilizing Time Records

I just gave you some reasons for keeping time records. Now that you have this information, what are you going to do with it? Begin by realizing that time management

---

* A few years after the first study, I performed similar ones at two different animal facilities. The results of the two later studies were quite similar to the first one. These findings are similar to those noted for managers by Marvin [2].

is an important part of your planning function. Think about what your goals are and how you have planned to reach them. Then review how your time, or your animal facility's time, has been used. Try to determine how much time has been used for repetitive functions. For example, did two different people check to see if the surgical packs had been autoclaved? How much of your daily work could, and should, be delegated to others? How much work could actually be deleted without harming the animal facility or the organization? Even though you are trying to find ways to save time, it may turn out that you and your staff are being efficient, but not very effective. In laboratory animal facilities, given a choice, we would have to choose effectiveness over efficiency. It may turn out that you need *more* people or equipment.

As a starting point, make a list of what wastes your time relative to what you want to accomplish. In most animal facilities, major time wasters include

- Telephone calls
- Unexpected visitors
- Unanticipated work problems
- Special requests from superiors
- Excessive memos, including reminding people about what you already told them
- Emails that are irrelevant to you
- Meetings
- Delays while waiting for information from others
- Your own procrastination

I'll return to this list in a moment.

Next, set up a list of priorities for each day, month, year, or whatever other time period you feel is appropriate. For example, on a daily basis, you might define three goals that are essential to both your animal facility and your organization. These goals become your most essential ones for the day. Deciding on which applicant to hire might fit into this category. These are the items you should put a star next to on your priority list.

A second grouping can be those objectives that are important to your facility, but not the organization as a whole. This group has a somewhat lower priority. For example, setting new work schedules might fit into this group. A third group, with still lower priorities, might include objectives that are of importance, but do not need immediate attention. For example, evaluating the need for additional animal caging might be appropriate for this group. As a rule of thumb, if by *not* doing something there are no severe consequences, then that item can receive a lower priority.

Needless to say, low-priority items cannot be delayed indefinitely. Time management, like everything else, is in a constant state of review. An item that has low priority one day may have a higher priority on another. Remember, your priorities should be related to your organization's goals. Buying a new television for your home does not belong in any priority grouping.

Maximizing Time Usage

Let's return to the list you made of time wasters and discuss some methods by which they can be avoided. Obviously, you should eliminate those items that are totally unimportant to the functioning of your facility. Buying the television I just mentioned is an example. Now, begin to schedule your own time or, if appropriate, the time of your coworkers. You have to be very careful when you schedule somebody else's time, as it may be taken as exhibiting a lack of trust on your part. Once again, good communication about telling people why you are doing what you are doing, and asking for opinions, is of inestimable value. Your staff may perceive problems that you have not thought of. For example, you may have considered the surgical prep and actual surgery time required for a given procedure, but not the time needed to take the animal from the cage into the prep area, and the time to bring it back to the cage area.

Scheduling Your Day

Don't try to schedule 100% of your day. It's a fool's errand because the many unexpected interruptions that occur every day or every month will disrupt your plan. Try scheduling no more than 75% of your time, and indeed, 50% may be a safer figure. As I mentioned previously, it's better to do fewer things well than to do many things poorly, so if time is really tight, try to plan on doing fewer things well. Don't take on some many tasks that you simply can't get them done.

Do not schedule more than 50%–75% of your time. Unanticipated items will take up the rest of it.

Determine for yourself the time of the day when you are at your best. For many people, this time is in the morning. For you morning people, this is a good time for you to handle your roughest chores, particularly those on your highest-priority list. Don't procrastinate in tackling problems, but make sure that you have enough information to accomplish your task. Procrastination can have a snowball effect for you and other people. It's all too easy to put things off. Personally, I like to get one or two problem items out of the way as soon as possible. I'll tackle something I don't like to do before handling things I do like to do. It's akin to a having sweet dessert after a main course you really don't like. Some people disagree with this, trying to get some easy items out of the way first. I think that's appropriate for a meeting, but for my day-to-day schedule, I'd rather tackle the hard stuff first. See if this works for you.

If necessary, set a deadline to complete the task, as Dr. Smith did in the earlier example. Many people work better and with more intensity if they know there is a deadline to meet. You may remember that many students begin their term papers only when the deadline is upon them. My wife is one of those people, and she does some of her best work when she starts at the eleventh hour. I'm just the opposite. I have to get everything done as soon as possible so I have time for revisions. Whether you do things my wife's way or my way, stick with the job. Shut the door, if necessary, to avoid interruptions. Turn off or forward your telephone if you can (and there's no law that says you must answer your telephone if it rings). You don't always have to be writing or reading while you work. It's perfectly reasonable to stop and think. Look out the window if you want to. If you feel that a task is particularly difficult or objectionable, schedule it to be accomplished in more than one work period, *but you must work during the scheduled time.* The preparation of a budget is a common high-priority task that might be scheduled over more than one time period.

Work on a project during the time you allocated for it. Do not procrastinate.

How else can you maximize your scheduled time? I've already discussed setting aside time to tackle your priority list, which may include innovative items or repetitive chores (such as analyzing budget variances), but they all have to get done. As a practical matter, a good deal of your scheduled time is often taken up by these chores. You might also consider setting aside certain times of the day for specific managerial functions, such as planning or controlling activities. This can include your time for meetings or answering emails.

Let me say a few additional words about email because it is the bane of my existence, yet I can't live without it. I know I shouldn't read emails as soon as they come in, but I do a lot of work at my desk and, for practical purposes, I cannot avoid them unless I shut down the application or shuttle them into an email digest, and that would make me more frustrated than reading all my new emails. It may sound trivial, but in terms of time management, every time I open an email, or even delete an email, my attention gets diverted from what I'm doing. That's probably not overly important if what I am doing does not require my full attention, but it does becomes an issue when I'm busy and every diversion requires me to reset my attention on the more important work. Yet, even if I read my emails as they come in, it doesn't mean that I have to answer them (especially since about 95% is junk mail that is a total time waster). In general, I try to hold off answering my email until the latter part of the day, unless a more timely response is required. I like to be brief but concise with my email responses, so when things have quieted down at work, usually beginning at about 4:00 p.m., that's when I can give a little extra thought to what I am writing. In fact, if the email indicates that there's a significant problem requiring my attention, whenever possible, I will put off a response or other action until the next day—just to give myself some thinking time. Email, like other information sources, is for us to control, not the other way around.

My final comment about email concerns "reply all" messages and "me too" messages. I infrequently employ the "reply all" icon unless I feel that it's important for everybody copied to read my message. I do that because I strongly dislike being copied on emails that have no good reason to be sent to me other than a person wanting half the world to know her thoughts. Likewise, I don't want to fill up my inbox with messages that have no value other than to agree with someone else's posting (e.g., "Yes, we do that too"). I certainly understand that there are times when a consensus of opinion is helpful, and when that is appropriate, I suggest that you don't include the entire string of messages on your email. My mailbox is always nearly full, and I waste a lot of time cleaning it up. Thankfully, I learned how to divert certain messages directly into the trash bin.

Do You Need Time for Planning?

For the longest time, I was disappointed with myself because I couldn't get around to sitting down in my office to "plan." I felt much better when I learned that many managers are spur-of-the-moment planners who do some of their best preparing in brief spurts and in response to specific situations [10]. Now, I save my scheduled time primarily for items that I know have to be accomplished while sitting at my desk. (I actually do my best thinking in the shower.) This does *not* mean that you should jump from one project to another. You still must set your priorities toward your most important ongoing projects.

As part of your scheduling, consider setting specific times for receiving and making telephone calls or answering emails. Many physicians, private practice veterinarians, and purchasing departments have learned the value of such a policy. Politely inform colleagues, salespeople, and others of this policy. An office assistant can be of immeasurable help here. They can screen calls and let you know what information to have ready when you return a phone call. It's also important for you to know just when to call certain people, even when it doesn't fit into your usual schedule. For example, some laboratory animal facility managers order their animals on Friday mornings. This requires substantial time in getting through to some suppliers, and may require callbacks from suppliers to confirm the availability of animals. If it doesn't present a problem, try placing your calls on a Thursday, or have an understanding with a particular supplier that they will call *you* at the same time every week.

I know that some people, when they're away on a trip, arrange ahead of time when they will call their office. I don't do that, and in fact, I rarely call my office when I'm on vacation or at a meeting. I like to think that my animal facility can survive without my being there every day and without my having to constantly call in. Maybe that's just my ego, but I really believe that a good manager should have things running smoothly enough to be able to take some time off without the entire operation falling on its face. If it's an emergency, I can usually be reached on my cell phone. If it's not an emergency, wait until I get back or leave me an email (yes, I usually do check my email while I'm vacation or at a meeting).

If you normally have an open-door policy for your office, close it while you are busy. This simple procedure will inhibit many people from walking in on you. If your office door must be open, or you work in a relatively open cubicle, position your desk so you don't make eye contact with everybody who passes by (that's why I keep my office door only half open, with my desk in the opposite corner). If you have an administrative assistant, try to have that person sit between you and anyone who might be coming in to see you. An "admin" can be most helpful in stopping unnecessary interruptions.

While discussing the management of information resources (Chapter 5), I stated that you must have adequate information to complete your task. You probably do not need 100% of all possible information. In many cases, you can complete 80% of your work with only 20% of the total available information (this is sometimes called the 80/20 rule). That is, to do most of your job, you may need only 20% of the total information you would like to have. You don't have to be absolutely perfect. More often than not, the attempt for perfection becomes nothing but a time waster. You become too anxious and nervous about your work. If you wait until you have 100% of the information, you may be old and gray before that happens. Just think about all the progress we have made in biomedical research, and we still don't completely understand how a cell functions.

There is another time management topic that is a spin-off of the 80/20 rule for information needs. It is fairly well known in management circles that about 80% of the time you spend resolving personnel concerns involve 20% (or less) of your staff. In terms of time management, you cannot let 20% of your staff consume most of your time because you will be shortchanging the management of the 80% of your staff that are not problematic.

In Chapter 3, I wrote about the need for communicating with people, establishing trust, providing motivation, and so forth. I also provided you with some thoughts on how to approach difficult conversations with people. There is absolutely nothing wrong with using that information to help prevent or resolve work-related problems with employees; in fact, that's part of a manager's job. But when you have tried your best yet the problems remain, it may be best for an unhappy coworker to leave your vivarium and seek a more suitable position. Under most circumstances, you simply cannot spend an inordinate amount of time on a small number of people.

Meetings

Let's turn our attention to another common time waster: meetings. I say "time waster" only because most meetings either accomplish very little or accomplish their goals in twice the needed time. Meetings can be useful for many reasons, but they have to be run efficiently and effectively. There are some meetings, such as general staff meetings, that you might want to have even though there's nothing critical to discuss. That kind of a meeting keeps communication lines open, builds solidarity, and is something I believe to be necessary. I have a weekly managerial meeting with those people who report directly to me. Sometimes, we have important things to discuss, sometimes not, but we often go off on fruitful tangents. Also, during the week I often accumulate information that might be valuable to some managers. Once I "download" that information to them and we discuss it as much as necessary, I empty it out of my memory bank. But, for most other meetings, the following advice should prove helpful:

1. *Before you call a meeting, consider whether the meeting is even needed.* If the probable result of not having the meeting is inconsequential, then why have it? If it is needed, carefully consider who should attend. Don't leave out any key persons who have a stake in its outcome. It is often not necessary for every person on your list to be present for the entire meeting. Let people know about the parts of the meeting for which they are needed, so they can make optimum use of their own time. Before anyone leaves the meeting, ask if there are any final comments they would like to make. This technique frequently brings out important points. At other times, it starts a "postmeeting meeting" that goes nowhere.

2. *Meet in a physically comfortable area.* People should be focused on the topic being discussed, not that the room is too warm or too cold or the chairs are unbearably uncomfortable after 10 minutes.

3. *Set specific times for the meeting to begin and end.* For some meetings, it's a good idea to circulate an agenda before the meeting and stick to it. The agenda should say something more than "budget discussion." Try enlarging it to "consequences of board of trustees proposed 3% budget cut to animal facility operations." The curt "budget discussion" notation does not give people a chance to formulate their thoughts ahead of time, thereby wasting everybody's time with nonproductive side discussions. If you have background information that will aid the discussions, attach it to the agenda. It will save time at the meeting and make it more productive. I like to put some easily agreed-to items at the top of the agenda. The usual agreement on these items gives the group a sense of accomplishment and paves the path for tackling the harder issues.

4. *Consider brainstorming before the meeting.* There are some meetings that require ideas emanating from brainstorming (getting all ideas on the table,

no matter how important or unimportant they may seem) before a detailed discussion ensues. Based on my own experience, it appears that most brainstorming goes on during the meeting. This takes time, but it also helps break the ice and open up a general discussion after the brainstorming is complete. Nevertheless, if you deem the situation to be appropriate, consider having people submit their ideas ahead of the meeting. They can then be neatly summarized and discussed at the meeting itself. In many instances, this will save a tremendous amount of time and allow people more time to focus on the ideas the group deems most appropriate.

5. *Come prepared.* There are few things that frustrate me as much as attending a meeting where people routinely say, "I'll check on that and get back to you." Of course, sometimes that's the best possible answer, but when it happens too often, it's usually an indication that someone didn't prepare for the meeting. If there is information that will be helpful at a meeting, learn about it or bring it with you. How would you feel about going to an Institutional Animal Care and Use Committee (IACUC) meeting where there's a planned discussion about a particular federal regulation and nobody brings a copy of that regulation?

6. *If possible, set a time limit for each topic to be discussed.* Write that time limit on the agenda. In general, I like to hold meetings in the afternoon, and I try to keep them to an hour or less, but sometimes that's not possible. Meetings that drag on much longer start to lose focus and people. If you want to be sure a meeting will end on time, schedule it about an hour before most people leave to go home. It works wonders. Try eating a light lunch, without alcohol, so you stay alert during the meeting.

7. *Begin your meetings on time.* Unless there is an overriding reason to wait for a particular person, go ahead and start. I've been known to start meetings with nobody present. I wouldn't do that if it would be political suicide, but if I'm in control, I'll do it. I've done this with classes that I've taught. You would be surprised how quickly students learn to be present on time. If you do not routinely start on time, you may find yourself always starting late because people expect you to start late.

8. *If you called the meeting, it is part of your responsibility to ensure that superfluous topics and prolonged discussions do not interrupt its flow.* Some meetings aren't really meetings; they're gatherings designed to rubber-stamp already-made decisions. Nobody talks except the most senior people in the room because the others are afraid of rocking the boat. Not only is this a waste of time, but it's a sick organizational culture, and if you are one of the meeting's leaders, you should try to get things back on track. I'm not suggesting that you avoid discussion and dissent for the sake of keeping on time or pushing through your own agenda. Under those circumstances, meetings can become an exercise in frustration. Open communication is as important during a meeting as at any other time. Even if people disagree with you, even if they know you

will disagree with them, they want their opinions to be heard. Team spirit is built, not dictated.

9. *It might be important to keep formal minutes of your meeting,* especially if policy decisions have been made. Minutes help ensure that there are no misunderstandings about what was discussed. One hoped-for outcome of any meeting is a consensus of all those present. Formal records (and subsequent minutes of the meeting) are often vital to document this consensus. If you are going to keep meeting minutes, I suggest making sure that somebody is given the responsibility of being the recorder. I've been to more than one meeting where it was assumed that the minutes would magically appear at the end of the meeting. Just keep in mind that not all meetings require minutes. Often, the formality of taking minutes can inhibit open discussion and stifle camaraderie.

Keep your policy decisions in an easily retrievable form, such as a separate book, so that you don't spend hours going through old meeting minutes to find out when and if a policy decision was made. I've sat through many IACUC meetings where far too much time was spent trying to remember what a particular policy said, or even if there was a policy.

10. *Not all meetings need to be formal.* Try having a brief meeting with somebody standing up. It tends to shorten the meeting and makes it easier to leave. Some managers find it convenient to go to somebody else's office to have a meeting, as that can make it easier for them to get up and leave. You might consider having a conference telephone call or video conference, rather than getting people around a table for a meeting. It can save both time and money.

Here's a summary of the points I just made about having a successful meeting.

Hints for Successful Meetings

- Make sure the meeting is needed and decide carefully who should be invited.
- Meet in a comfortable area.
- Set specific times to begin and end the meeting.
- Consider if brainstorming ideas will be needed and if it might be completed before the meeting, rather than during the meeting.
- Come prepared with any information that is likely to be useful during the course of the meeting.
- Circulate an agenda before the meeting. Include information that might facilitate discussions.

- **Set time limits for each topic to be discussed, but do not arbitrarily halt productive discussion.**
- **Solicit all viewpoints.**
- **Consider whether formal meeting minutes are needed.**
- **Consider whether a video or telephone conference call can substitute for a face-to-face meeting.**

Not everybody will agree with my philosophy on meetings, particularly if you reside outside of the United States. In a study of attitudes toward business meetings around the world, it was found that, in the United States, the five most important elements of a business meeting are [11]

1. Clearly set the desired meeting outcomes.
2. Have an effective moderator.
3. Be adequately prepared.
4. Have agreement on follow-up action.
5. Start on time.

As you can see, these five elements are part of the hints for a successful meeting.

Spontaneous Meeting Dilemma

Before leaving the topic of meetings, let me address one more type of a meeting: the one that pops up spontaneously when a person says, "I have a quick question for you" or "Can I have a minute of your time?" If you have the time, say yes. But if you don't have the time, how do say no without the other person being annoyed at you, especially if the other person is your boss or another power broker in your organization? Well, the truth is that there are times when you will have to bite the bullet and give that person some of your time even if it's not what you really want to do. However, you often can remove yourself from that situation by saying something as simple as "Sure. I'm working on the weekend schedule right now. Can we talk a little later when I can give you my full attention?" That takes the burden off of you and places it on the other person. You have said yes, but it's a delayed yes. The other person may say that it's real important and she needs to talk to you now, or she may be willing to let it go until later. This is one of those situations where it helps quite a lot to know the person with whom you are talking. There are some people who can fool you once about the importance of having a spontaneous meeting, there are some people who you know that you have to honor their request, and then there are many others with whom you can delay a meeting until a mutually convenient time is found.

Paperwork and Reading Material

I will say only a few words about how to efficiently use the time that is spent with your general paperwork and reading material. The simplest solution, and one that frequently is best, is to handle important work as it comes in. If it can't be handled on the spot, establish a priority filing system for handling it, as I previously described. That means you determine what is most important, what is next in importance, and so on. You might also put it in a file to be handled on a specific day, and stick to that schedule.

I'm not much for writing hard-copy memos unless I believe that formally putting something in writing is necessary. I do most of my in-house correspondence by email and, to a lesser extent, by telephone. For key items, it's usually the telephone. It's faster, more personal, and usually gets a better response. It's easy to ignore an email or a letter, but it's harder to ignore someone with whom you are talking. You can therefore understand my prejudice when I suggest that you should not add to the paperwork clutter by writing memos when a telephone call or email will do.

Let me return for a moment to our understanding of an organization's culture (Chapter 2). In some organizations (or with some bosses), an email is the only way to get things done. If you believe that an email is needed, but waiting for a return email will delay your work, try sending one that reads "Unless I am informed otherwise, I will proceed with this plan on such and such date." That gives you the opportunity to move ahead while keeping others informed of your intent. If they choose to respond, that's fine; you can act on their comments. This type of tactic can turn on you if you are planning to do something that you know is against the grain of either the organization or your superior. As I have said many times in this book, use a little common sense.

Many managers become bogged down by excessive amounts of reading material, such as journals, junk mail, junk email (spam), important mail, and memos. A few years ago, someone told me about a report he heard on the radio. In essence, it said that people who leave messages on email or voice mail thought that what they had to say was very important. But these same people were annoyed at the length of messages that other people left for them. Obviously, it's all in the eye of the beholder. At least with email, you often have the option of using a "digest," which is often just the message header and one or two sentences from the email. It makes it easier for you to determine what you do and don't want to read. It also tells you that when you send an email, get your key idea into the header and the first sentence because the recipient may be using an email digest.

Suggestions that might help you cope with the mounds of reading material that will come across your desk include learning not to read all of your junk mail and reading only that material that is pertinent. It drives some people crazy when they see me discard about 75% of my mail without even opening it. The reason is simple; I just don't have the time to read everything I'm sent. I have to be selective. In journals or books, scan through the table of contents or read the abstract. If it is not important to you, move on. If it is important, you might want to underline or highlight specific passages so you don't have to reread the entire article in the future. You can do the same thing with some email chat groups; set the program to just give you a list of the email subjects (and sometimes a first line) so you quickly scan it and pick and choose what you need to read (this is the digest just noted above). Some managers have found it useful to have a filing system on cards, or on a computer, that lists key words from an article, where it originated (e.g., the journal and its date), or where it is filed. This is to save time in trying to relocate the article. As an example, you might have one 3 × 5 card entitled "Facility Design" that tells you the key words in the article that interested you (e.g., *mice* and *cancer research*) and in which journal it appeared (e.g., *Lab Rodents* 1: 101, 2016). Nowadays, with the ability to locate most anything online, it's less likely that you would need such a system, although some journals are quite restrictive with what you can access for free. Other managers, particularly those who also are active scientists, make use of electronic journals that "automatically" set up a key word retrieval and article citation system. To help you save time in reading, you might even consider a speed reading course, if you think you require that kind of help.

Delegation

After following all or most of the suggestions that were made above, you still might not have enough time in your day. There are only so many things you can do, and if you try to solve too many problems or become involved with too many projects, there is a good chance that you will succeed at nothing. Part of the problem could be that you are trying to do those jobs that could or should be done by others. Because most animal facility managers have worked their way up through the ranks, some have a tendency to try to bite off more than they can chew. They worry that nobody except themselves can do the job right, so they attend to their new managerial responsibilities while clinging to many of their old nonmanagerial ones. The parts become greater than the whole. In other instances, there is such a large amount of work that no one person could handle it alone, even if projects are prioritized. If you find yourself in this situation, you may have to delegate some of your responsibilities. It's fair to say that all good managers have learned to delegate certain tasks.

If you decide to delegate, don't give up those activities you do best. For example, if you are good at writing proposals and that's part of your job, find someone else to do your photocopying or go to the library for you [12]. Please don't interpret this to mean that you should stick only to those areas of your competency, because that would be foolish and can lead to a managerial disaster in the long run. In fact, a good manager should always try to develop at least some skills in his or her areas of weakness. The point I am making is that you should delegate support activities that have to be done, not your central role as a manager.

Don't delegate the activities that you do best.

Mackenzie [13] has provided a fine description of the value of delegation. He notes that delegation extends results from what a person can do to what a person can control. It releases a manager to do more important work. It develops in subordinates a greater amount of initiative, skill, knowledge, and competency. Finally, it maintains the decision level. That is, it allows operating decisions to be made at the operating level and major policy decisions to be made at upper management levels. I'll add one additional thought to Mackenzie's: delegating responsibility and authority demonstrates trust in a person. Trust, as you know by now, can go a long way to increasing motivation, morale, and productivity.

Many tasks can be delegated, and they need not all be yours. You can work with supervisors who report to you to see which of their tasks could be delegated. Some simple examples of what might be delegated include monitoring food supplies, ordering bedding, and checking study records. This takes the burden off of you or other managers and allows other people to show off their competencies. No matter who does the delegating, it should be axiomatic that nobody likes being saddled with the dirty work while their boss sits back and takes all the soft jobs. I remember how the dean of a medical school took his turn on emergency call like everyone else. Emergency call was not among the faculty's favorite pastimes. The dean did this not because he had to (he certainly could have delegated this responsibility), but because he wanted to give himself credibility when he found it necessary to delegate unpleasant tasks.

At times, even your best employee doesn't want to take on more responsibility. Perhaps there is a problem that needs to be solved, or perhaps you are just giving too much work to a small number of people. In the latter instance, the "good old boy" network comes into play. We tend to give jobs to our friends, and sometimes we just overload them. There is also the possibility of demoralizing the quiet person who wants to be part of the team, gain a little authority, and work his or her way up the ladder. Don't overlook these people when work is delegated. They may blossom later on.

Some quieter employees may want to take on delegated work, but they are too shy to ask. Ask them if they would like to contribute.

Delegated projects should be completed in a timely manner, not when a person "gets around to it," because if that happens, there's a good chance the project will never get done. If you delegate a responsibility, set a mutually agreed-upon schedule and stick to it.

It seems logical to delegate a task to the person who can do it best. Sometimes, logic loses out to friendship or other considerations. The results may not be what you might have wished. In the following example, Erin Rourke got more that she bargained for.

> *Erin Rourke managed the laboratory animal facility at Ethical Pharmaceuticals, Inc. Although she worked hard, there came a time when she realized she would have to delegate some of her responsibilities to one of her group leaders. She had come up through the ranks with George Novak, a good friend who shared many of her views on facility management. George was still a group leader on the first floor of the facility. He was able to work with people and he could follow directions, but he was not very detail oriented.*

> *Erin asked George to help her out by doing some of the supervisor's weekly inspections and reports. He would do the first and second floors of the facility, and she would do the third and fourth floors. She explained that his job would include evaluating the daily operations to make sure they complied with the* Guide for the Care and Use of Laboratory Animals *and the* Animal Welfare Act *regulations. This help would save her a substantial amount of time and effort. George was a little apprehensive, but he agreed.*

> *George did his best. He made some superficial comments about peeling paint in a rat room and that the hose in the cage wash room needed fixing. Erin was tolerant and asked him to be more thorough the following week, but the next report was not much better. Erin had to redo the inspection herself, and now she found herself in the position of telling George that she would do the work herself or assign it to someone else. Her decision was a double-edged sword. George was somewhat relieved to get rid of the task, but he had a deep-seated feeling of failure.*

In this example, it appears that George simply didn't have the motivation (and perhaps also didn't have the needed skills) to do the work, and Erin should have used more forethought about using George for the job. She didn't do him any favor by giving him more responsibility, and in fact, the deep-seated feeling of failure that resulted was likely harmful to his position and feelings about his work.

While discussing morale, I pointed out the need to give the right job to the right person. We have to make sure the person wants the added responsibility. Was

George Novak the right person for the job? Erin chose him because of friendship, shared values, and the fact that he could get along with people. But were these the characteristics needed for the delegated job? Did George Novak really want the responsibility, or did he accept it out of loyalty to Erin? The bottom line is that Erin Rourke should have thought about the qualifications needed before she thought about who she wanted for the job.

At times, you will have done your best to define the responsibilities of a job and who would be best for the delegated work, but you may still have lingering doubts about your choice. If you are not sure about a person's ability, you can set limits on the delegated responsibility and authority. For example, an animal care technician may be given authority to reorder cleaning solutions for your cage washer, but you are not obliged to give that technician the authority to choose a different supplier or a different brand. That limitation of authority should be clearly communicated to the technician; otherwise, the authority and actions taken may not be in the best interests of the facility.

When you delegate a task and the authority that goes with it, take the time to let other people know what has occurred. Don't send the new person on a journey without your staff, and possibly your own supervisor, knowing that somebody has been handed part of your authority. I'm sure that you would appreciate the same courtesy. Likewise, it's the delegating manager's responsibility to ensure that the person with the delegated task has the resources (people, money, time, etc.) to get the job done. I wouldn't want to guess the number of times I've seen a manager, with a figurative wave of his or her hand, delegate a responsibility where failure was the likely outcome because the needed resources simply were not there. Then, of course, the manager would blame the failure on the other person.

Let your staff and other people know when you have delegated authority to another person.

Here are some additional tips about delegation. First, make sure you give the person with the delegated work enough information to get the job done. Obviously, you don't want to micromanage what you are delegating, but at the same time, it's your responsibility to get periodic feedback about the assigned task. The frequency of feedback will depend on the specific task and may require you to go to the job site and see for yourself how things are progressing. And lastly, give some feedback once the task is completed. It doesn't have to be a formal sit-down session; rather, some informal comments may be all that is needed.

One great source of delegated assistance managers often overlook is an office assistant. A good assistant can do more than screen your calls. They can schedule your meetings, draft letters for you, keep or review financial records, interact with investigators, and perform a myriad of other functions. Good assistants are very much a part of your team. Delegate responsibility to them the same way you would to anybody else.

Whether you delegate full or limited responsibility and authority, you cannot blame all failures on the person who did the delegated work. It may work once or even twice, but pretty soon it will be obvious that you, the manager, must be accountable for the actions of your staff. If you are not willing to be accountable, you should reconsider being a manager.

In most cases, a manager will be held accountable for the actions of his or her staff, even if the work was delegated.

Returning Delegated Problems Back to the Manager

Some employees may ask for your opinion or advice about projects you have delegated to them. Assume you have passed both the responsibility and authority to an employee to purchase the best cleaning compounds available, at the best negotiated price. That person now comes to you and asks, "Who do you think I should buy descaler from?" At first, you may be flattered, but be careful. In many cases, that person is inadvertently (or intentionally) giving the responsibility back to you. I recommend that you resist direct answers to these questions. Suggest that they look into the problem for themselves, come up with alternatives, and act on the best choice. This also helps that person learn how to make decisions and emphasizes your trust in the person.

This problem of returning the responsibility for making a decision has an interesting spin-off that involves the IACUC. As most of you know, the IACUC represents its institution's animal care and use program to the federal government, and most research institutions in the United States that use animals have an IACUC. For many years, the laboratory animal science community has been advocating the implementation of performance standards (i.e., doing what is best for the animal given the circumstances of the research) and not engineered standards (formal, one standard for everybody, irrespective of specific needs). Little by little, the federal government has been allowing certain performance standards to be implemented. But what have I heard from many IACUCs and investigators? It is "Just tell us what to do and we'll do it." Well, we can't have it both ways. The federal government delegated to the IACUC the authority for making certain decisions. The IACUC can't turn around and say, "Tell us what to do." That's returning the delegated authority to the manager. It's asking for an engineered standard because of presumed problems (or perhaps laziness or fear) in developing a performance standard.

**Be on the alert for innocent questions or a casual discussion
that may actually be returning delegated tasks back to you.**

Don't make somebody else's problem your problem unless it is absolutely necessary to do so. This doesn't mean that you should be harsh or inconsiderate by avoiding a direct answer to a person's questions. There are times when people have done their very best and have hit a roadblock; this is when you should become involved. Becoming involved doesn't always mean taking direct responsibility for the problem. You can help stimulate a person's thinking about the problem by asking key questions, in addition to "What have you already done?" For instance, you might ask about the positive and negative consequences of two possible solutions to a problem. Or, even a simple statement such as "Give me more details" gets some people thinking in the right direction. On the whole, though, avoiding returned questions establishes where the responsibility is and obviously saves you time since you aren't trying to resolve someone else's problems. And, as I noted, it demonstrates that you have confidence in the person. If you wanted the burden, you wouldn't have delegated it in the first place.

This may be a good time to review the section of Chapter 3 that discusses the need for positive reinforcement in human resource management. When tasks are delegated, there is a possibility that something may not work out exactly as you wanted it to. This is rarely a tragedy, but it does require that you express your concerns about the problem while being generally supportive of the individual. In the same light, praise should be given when all goes well.

I hope you are convinced that time management is not working faster or longer hours, but a skill that every manager must learn. Time is a resource that is to be used with care. It is part of the process of asking where we are now, where we want to be, and how we can get there. As a resource, time must be planned, controlled, organized, and directed, and decisions made about its use.

References

1. Hewlett, S., and Luce, C.B. 2006. Extreme jobs: The dangerous allure of the 70-hour workweek. *Harvard Bus. Rev.* 84(112): 49.
2. Marvin, P. 1980. *Executive Time Management.* New York: AMACOM.
3. Kennedy, D. 2014. *No B.S. Ruthless Management of People & Profits: No Holds Barred, Kick Butt, Take-No-Prisoners Guide to Really Getting Rich.* 2nd ed. Irvine, CA: Entrepreneur Press.
4. DeMarco, T. 2001. *Slack: Getting Past Burnout, Busywork, and the Myth of Total Efficiency.* New York: Broadway Books.

5. Harnish, V. 2010. *Mastering the Rockefeller Habits: What You Must Do to Increase the Value of Your Growing Firm.* Ashburn, VA: Gazelles.

6. Ginsburg, S. 1982. *Management: An Executive Perspective.* Richmond, VA: Robert F. Dane.

7. Bak-Maier, M. 2012. *Get Productive! Boosting Your Productivity and Getting Things Done.* Hoboken, NJ: John Wiley & Sons.

8. Hamm, J. 2012. *Unusually Excellent: The Necessary Nine Skills Required for the Practice of Great Leadership.* Hoboken, NJ: John Wiley & Sons.

9. Darden, D.C. 2015. Time wasting activities within the workplace: Don't be a part of them. *Open J. Bus. Manage.* 3: 345. http://dcdarden.com/author/dcdarden/ (accessed March 10, 2016).

10. Mintzberg, H. 1975. The manager's job: Folklore and fact. *Harvard Bus. Rev.* 53(4): 49.

11. Elsayed-Elkhouly, S., and Lazarus, H. 1995. Business meetings in North America, Asia, Europe, and the Middle East. *Am. J. Manage. Dev.* 1(4): 15.

12. Griessman, B. 1994. *Time Tactics of Very Successful People.* New York: McGraw-Hill.

13. Mackenzie, R. 1972. *The Time Trap: How to Get More Done in Less Time.* New York: McGraw-Hill.

Chapter 7

Leadership

The task of the leader is to get his people from where they are to where they have not been.

Henry Kissinger

It has been said that 51% of employees believe that business initiatives succeed in spite of their leaders, not because of them [1]. Why? Because there are leaders who don't clearly communicate what they are trying to accomplish, who establish useless policies, and who don't consider the capacity of their organization to accomplish goals when initiating new programs. Therefore, it seems reasonable to assume that there are some pretty good self-motivated employees who work hard to make poorly planned new programs work, even if they don't get any of the credit for their successes. But how can it be that some leaders don't lead, yet they can claim success? It can happen, at least in part, when managers move up the ranks into more and more visible leadership positions, and eventually reach a position where they think that they don't have to pay attention to the details because that's somebody else's job. Basically, these so-called leaders are taking credit for the work of other people. Unfortunately, there is a fairly obvious downside to this: eventually, these self-motivated employees leave and the leader is left with a weaker team, if there is any team left at all [1]. The take-home message is that leaders are expected to be positive role models and lead their organizations to new and important goals. They must solicit honest and critical feedback from others and must have a solid understanding of their organization's day-to-day operations. Leadership, like management in general, is a daily job; it's not an opportunity to relax and let others do the work while you take the credit for successes and shift the blame for your failures. In this chapter, we'll look at what leaders do, how they do it, and some of the differences between managers and leaders.

Are Managers and Leaders the Same Person?

Managers and leaders are not necessarily the same person. Whereas *manager* is a job title that a company gives to you, good leaders are good leaders, with or without a title. There are good managers who are weak leaders, good leaders who are weak managers, and in the best of all worlds, good leaders who are also good managers. Can you imagine a person who is supposedly a good leader but is a complete failure at planning, organizing, or making decisions? Not likely. Similarly, in a business organization, it's highly unlikely that a good manager will be a total failure at all aspects of leadership. Some leaders may be better than others in one or more areas of either leadership or management, but widespread weaknesses will not result in true leadership. Therefore, I think it is fair to say that most leaders (although not all) can apply basic management skills when needed, and most managers (although not all) can show a degree of leadership when it's needed. It's also fair to say that good leaders know that they are not omnipotent, and they will gather a support group that will help compensate for any of their own weak areas. This support group can come from people within or outside of the organization, such as employees, vendors, and other leaders.

Before we go too far, let's see if we can come up with a working definition of a leader and leadership. I've heard as many definitions of these words as I've heard of managers and management, but most definitions of leadership tend to revolve around the process of pulling people, rather than pushing them, toward new goals. In fact, one of my favorites is that "management controls people by pushing them in the right direction; leadership motivates them by satisfying basic human needs" [2]. As much as I like that definition, I understand that it's somewhat of a catch-phrase because, as I wrote in Chapter 3, any organization has to meet the basic needs of its employees to prevent poor motivation and morale. As a manager, you can only push people so far before they rebel. I would prefer to define a leader as a person who can motivate others to accomplish a goal. As a corollary, leadership is the methods a person uses to motivate others to accomplish goals.

- A leader is a person who can motivate others to accomplish a goal.
- Leadership is the methods a person uses to motivate others to accomplish goals.

The above definitions clarify that leaders do not reach their goals by themselves. Though a manager, by the definition provided in Chapter 1, may work alone, this is not so for leaders. Leaders lead people; they are leading and motivating the people who are needed to get the work done. If you go back to Chapter 3, under the discussion of motivation, I wrote, "If an employee *wants* to perform efficiently and effectively … you have helped give that person the inner drive to reach a personal goal." And that is what leaders do; they motivate people to do things by instilling within them the desire to do whatever it is that the leader wants them to do. Leaders are goal-oriented influencers of people. Managers may have to motivate one person, but leaders usually have to motivate a group of people. People don't have to follow a leader; rather, they want to follow a leader. Here's what D.R. Clark wrote:

> Although your position as a manager, supervisor, lead, etc. gives you the authority to accomplish certain tasks and objectives in the organization, this *power* does not make you a leader … it simply makes you the *boss*. Leadership differs in that it makes the followers *want* to achieve high goals, rather than simply *bossing people around*. [3]

People who follow a leader feel that the leader's influence over them is reasonable at that point in time. There is no need to coerce a person to follow a leader [4]. The leader is satisfying the needs of those who follow him. Therefore, although managers (and usually leaders) have a certain amount of authority that is vested in their position, it is the *quality* of leadership, not the *authority* of leadership, that will determine whether (and when) all or part of those goals will be reached. Leaders, like managers, are judged by results (i.e., the outcome of their decisions), and it is what the followers do, not what the leader does, that defines a leader's success or failure.

For leaders, success or failure is ultimately determined by the actions of their followers.

Leaders are always leaders. Even when they're doing basic managerial functions, they don't turn their leadership skills on and off like a light bulb. Sometimes, they have to go into a higher gear, but on a daily basis, it's their obvious motivation and drive, how they talk and listen to people, their mannerisms, their intuition, and their overall personality that helps set them apart from others. I'll discuss some of these traits later on in this chapter. Nevertheless, business leaders, including those in laboratory animal facilities, don't walk into their office every morning and start to lead things. They have many of the same basic management responsibilities that we all have, and they often get involved in mundane work issues, but at the same time, they have that something extra I just described that distinguishes them as leaders. Still, sometimes it's a little hard for us to see just where basic, good management differs from leadership. It's important for you to understand the basics of management, corporate culture, and the use of resources before launching into those characteristics that define a leader and leadership. In this chapter, you will be constantly reminded about the basics of management and how they relate to leadership. That's why this chapter on leadership is toward the end of the book, rather than in the beginning. As I have preached for many years, leadership relies a good deal on the ability to motivate people, but it also encompasses all of the basic roles of a manager (planning, decision making, directing, organizing, and controlling).

So a leader is a motivator, but leadership does not start and end with motivation. I might be able to motivate you to jump out of a window if your house were on fire, but that would hardly qualify me as a leader. The fire did most of the motivating; I just urged you on a little bit. In an organizational setting, the leader is focused on developing and accomplishing goals. The leader can communicate a vision statement in a way that makes people want to accomplish its goals. The leader is able to convince the entire organization that he symbolizes trust, open communication, and the importance of the organization's goals. He projects a sense of urgency. The leader projects an image of "I will make you happy to follow me. My beliefs are your beliefs. My values are your values. You can trust me. Follow me!" The talented leader is a highly self-motivated and inspiring person, and that enthusiasm to achieve rubs off on others. There is almost a tinge of religious fervor; after all, following an inspirational leader requires a dose of faith because very often, we don't know where we're being led. Typically, we trust that our leader is leading us down the right path. As Kouzes and Posner [5] put it, "If you don't believe in the messenger, you won't believe the message."

Another difference between leaders and managers is that leaders are always in the spotlight, whereas managers are in the spotlight less often. Managers try to minimize risk, while leaders open the organization to risk in order to move it in a new or enhanced direction [5,6]. From the moment a new leader comes on board or is promoted into a leadership position, she is being scrutinized. People are wondering what she is like and how her actions will affect them. Here's a true happening with only a name change.

> When Dr. Laura Phelps was named dean at the Great Eastern University School of Veterinary Medicine, tongues quickly were flapping about what might happen when she assumed her new position, even though very few people had actually met her. Of course, those people who already knew her were on a pedestal of importance for a short while because they had a modicum of information about the dean. When Dr. Phelps finally arrived and settled in a little bit, she clearly articulated some of her short-range goals, told the faculty how she planned to reach those goals, and asked for their help and feedback. She was astute enough not to try to make too many changes in too little time. To do so would have alienated the faculty and department chairpersons, who are the basis of support for a dean.

There will be more about Dr. Phelps later in this chapter. Veterinary schools, like medical schools, animal facilities, pharmaceutical companies, and aircraft carriers, have a certain inertia that keeps them on their current course. It takes time and effort to change direction, and if the change is too sudden, everything will fall off of the shelf. There are exceptions to this, such as when a company is nearly dead and drastic action is needed and needed quickly, but generally speaking, a new leader has to tread carefully at first, build trust with honesty and open communication, and then motivate the entire organization toward higher goals.

Do We Need Managers or Leaders for Animal Facilities?

It's a fact that more often than not, an animal facility is a monopoly within its own organization. Therefore, why does an animal facility need a leader? Can't we just get by with talented managers? I can think of a number of reasons, but let me give you the first three that came to my mind.

1. *You may have a monopoly today, but there will be competition tomorrow.* There are private companies that would be more than happy to take over part or all of the operations of your animal facility. A company making disposable animal caging and an animal facility management company joined forces to entice animal-using institutions to use their services. This was done because both organizations believed they could gain more of a share of their markets by working together and offering an institution more bang for its buck than was currently possible. You may like the status quo and think that you're doing a fine job as a manager, but those companies are looking at innovation in equipment, supplies, personnel methods, and overall delivery of service. Animal facility leadership has to do the same. In Chapter 2, I suggested that every organization and every animal facility has its own culture, and that culture is the animal facility's "brand name." Ask yourself, would your customers (i.e., the animal users of your institution and the people you lead), if they had the opportunity, abandon you for a different brand [7]? If you think the answer is yes, then you should start making changes that are of value to your customers. Strong leadership and an associated well-managed animal facility are potent reasons to keep outside contractors at bay, if that is your intention.

2. *There is a constantly changing business environment.* Managers and leaders both look for changes in the political, socioeconomic, information, and physical environments that can affect the animal facility. It's not unusual for the direction of research to change, and leadership is needed to prepare the animal facility for that changing environment. We can't remain in a horse and buggy age when there are rockets flying to the moon and planets. Just think about the burgeoning use of interference RNA (RNAi). If your animal facility was already set up for genetically modified mouse studies, there may not be a need for significant leadership in order to perform RNAi studies. However,

if much of your work has been with larger animals, and your organization's researchers recognize the importance of RNAi in biomedical research, true leadership will be needed for the significant changes that will likely have to occur in the animal facility.

3. *There is often a need to initiate change.* Whereas the business environment itself might be changing, and corporate competitors are out there, it is not unusual for an animal facility to just plod along year after year, getting its work done, but not doing much more. If RNAi studies are being done, that's fine. We'll take care of those animals. If behavioral testing or gene therapy is being done, that's fine. We'll take care of those animals too. But are your business processes in the twenty-first century? Is the equipment being used up to date? Are people being properly trained? Are services being offered that should be offered, and so forth? If the answer to any of these questions is no, then good leadership is needed.

Are You a New Leader?

In Chapter 1, I gave you some hints on how to succeed as a new manager, and I added that with practice, you most likely will succeed. You may want to review the new manager section from Chapter 1, as much of that discussion also applies to a new leader. Like general management, leadership also is a skill that can be learned. When leading a laboratory animal facility, whether or not you are a veterinarian, there is a good chance that you have already demonstrated to others that you are a competent manager and that you have leadership skills or potential. Still, when taking on a new role, particularly in a new location, people will want to know who you are, why you were chosen, and what you will be able to do for them. Concurrently, it is very likely that you will have some anxieties of your own about how people will view you and what you will have to do to gain their confidence.

If you guessed that effective and open communications are necessary to help establish a new leader, you guessed right. Earlier in this chapter, I told you about Laura Phelps, a new dean in a veterinary school, and how she began to break down walls by communicating with her faculty. The same holds for any new leader, even one who has risen through the ranks of the same organization he or she is now leading. People may know the type of person you are when you are at work, but if you want to gain additional confidence, I suggest you open up a little and tell your coworkers a bit about your life outside of work. Just a little bit is fine. You can let people know if you enjoy sports, if you're married, if you have children, if you have a hobby, and other information that's not deeply personal. You're just trying to establish a link with many of the people who are listening to you. On an interpersonal level, your communication style should reflect support of your coworkers, self-assuredness, nonaggression, and precision in what you say. Being supportive of coworkers is particularly important [8]. Your interpersonal communication style as a leader is independent of the somewhat more affirmative communication style you may use as a manager when setting goals, planning, directing, and so forth.

I also suggest you remind people why you are the new leader, even if they fully expected that you would be in that role. What experience do you have? Is it obvious how that experience links with your new position? If it's not obvious, tell people how it is linked.

Let people know if you have a vision of where and how you would like to lead the vivarium. If you currently have no such plan, let people know that you will be developing a plan after gathering information and receiving feedback over the next few months. Your coworkers want you to provide them with job security, someone who will support them if and when support is needed, and someone who appreciates them and will make known their role in supporting research, teaching, or product safety testing. Lastly, even if they don't say so, they want to know your management style, so why not tell them? Will you have an open-door policy? Will you be on a first-name basis with people, or something more formal? Whatever your style, plans, or aspirations, you must communicate honestly and frequently.

What about your knowledge and technical skills in laboratory animal science? Are they important when you become a leader? In Chapter 1, I reminded you that if you were a technician before becoming a manager, you should concentrate primarily on your work as a manager. The same can be said about your responsibilities as a leader. You should focus on your new responsibilities as a leader more than the responsibilities you had with your previous job. But your knowledge of laboratory animal science will still be important, and coworkers will appreciate your skills and knowledge about laboratory animal science when there is a need to interact with accreditation or federal regulatory agencies. To look at this through another window, it can be said that if you work in a university or a pharmaceutical company, you probably would not expect the institution's president to know much about laboratory animal science; however, the leader of the animal facility is at a very different management level, and your coworkers expect you to have a technical skill set that they can relate to and is important to them.

Quick Summary

Let's summarize where we are so far. The primary responsibility of a leader is to lead significant organizational change, not to break a tie vote on which coffeepot to buy. An example of a significant change in an animal facility might be a three-year vision of the facility's leader (such as the attending veterinarian) to save money and streamline animal care activities by moving away from manual work to the use of robotics and other advanced technologies. This can include the entire cage wash operation, zebrafish feeding, taking the animal census, billing for *per diem* charges, and cage-side observations. Perhaps fewer people will be needed, and perhaps more technically skilled people will be required. Will there be layoffs? What will happen to me? Can the veterinarian actually get this done? There may be significant repercussions unless the leader sets the stage for the entire department *and* other senior members of the department accept the vision of the leader. For such a bold and expensive vision, it is unquestionable that upper management of the institution has been part of the decision-making process and also will support the vision. Nevertheless, the person leading the effort will be the attending veterinarian, and he or she will continually communicate the vision, take the lead in establishing the plans and strategies necessary to fulfill the vision, and make sure that all employees are kept abreast of progress as the vivarium moves toward implementation of the vision. Most of all, the attending veterinarian, or whomever the leader is, must be decisive and convince everybody that the vision can be accomplished and it will benefit the entire department and the institution as a whole. To quote the biographer Walter Isaacson, "Vision without doing is hallucination." That's another way of saying that talk is cheap; leadership entails putting the talk into action.

Framing a Situation

We know that good management has many facets. There is a need for managers to know the personalities and needs of the coworkers who report to them, to teach and coach employees, to help ensure a fair wage is being paid, to reward uniquely superior performance, to network with others in the field of laboratory animal science, and perhaps to have an annual holiday party or monthly guest speakers who describe their research to the entire staff. In other words, there are many aspects to management, and it's not unusual for a manager to be involved in most of them. Likewise, there are many characteristics to leadership, and a good leader may have to use one or more of those characteristics, depending on time and circumstances. One of the simplest yet most logical ways of envisioning the need to use different features of leadership is based on the thinking of Bolman and Deal [9]. These authors point out that a good leader has the ability to consider a business situation from more than one point of view. The four major points of view (which they call *frames*) are formal structures, human resources, politics, and symbolism. A good, experienced leader can quickly determine which frame or frames will work best for a particular situation. Of course, making this decision takes time and practice, but the concept of determining which frame or frames might work best for a particular leadership requirement really isn't that different from what a manager is used to doing on a daily basis. Therefore, let's consider Bolman and Deal's four frames of leadership, and then I'll comment on how they can enhance your skills as a leader.

1. The *structural frame* is what a lot of us are used to. There is a hierarchal personnel structure with one "big boss" at the top. There are well-defined work roles for everybody, and everybody knows their own role; there are specific

rules and regulations to be followed; and so forth. In this structural frame, the leader sees a need to make sure that everything is running smoothly and the work being done is in line with the leader's or the institution's vision.

2. The *human resources frame* comes into play when a leader sees a need to better align the work of the organization with the needs of its employees. When a leader puts emphasis on his role as a coach and a tutor, when he acts as a motivator to lead people to a goal, he is working within the human resources frame.

3. The *political frame* is pretty much what it sounds like. In every organization, there are political considerations (we discussed this in Chapter 2 as being the politico-economic environment) and the leader must be a strong advocate for her vision. Sometimes, she must be a negotiator to get things done or to gain resources, such as money or space. This may require her to make alliances with certain people or groups or solicit letters of support from outside groups. If the leader can negotiate a deal where everybody wins (the so-called win–win negotiation), that's the best outcome. In Chapter 3, there is a discussion about the importance of negotiation for managers. Negotiation is equally important when you are in a leadership position.

4. Finally, there is the *symbolic frame*. A good leader can use the traditions and symbols of the organization to aid his leadership efforts. For example, the University of California at Berkeley has a statue of its Golden Bear mascot in the middle of the campus. It is an ideal place for a football coach to give a pep talk to students before a game.

A strong leader considers all four points of view (all four frames) while evaluating a situation. Essentially, he is asking, "What will it take to get my vision accepted? Do I need to be more specific about the details of the vision? Do I need to get the department heads on my side? Do I need to do both?" The leader (or manager) who is stuck with only one way of approaching a problem (e.g., using only the structural frame) is not as versatile of a leader as one who approaches the same problem considering all four points of view. There may be a need for a strategy that requires a structured approach with details of every step along the way spelled out. Sometimes, this works in an animal facility, such as when developing a plan to resurface all of the floors. But a structured approach is not likely to work if the senior veterinarian tries to develop standard operating procedures (SOPs) that dictate to the other veterinarians how to treat a prolapsed uterus in a pig. Perhaps a human resources frame that supports the skills of the other veterinarians and gives them the authority to treat animals as they see fit would be a more appropriate leadership method for that particular need. Even though every leader tends to have a preferred style of leadership, the best leaders automatically look at significant decisions using structural, human resources, political, and symbolic frames.

A leader considers using one or more points of view (frames) that might be best for different situations. These frames are as follows:

- **Structural: Emphasizing that everybody knows their role and they are moving ahead to reach the goal**
- **Human resources: Putting an emphasis on coaching and encouraging people**
- **Political: Negotiating, influencing, and networking to ensure resources are available and obstructions are removed**
- **Symbolic: Using the organization's symbols, traditions, and culture to help reach a goal**

A perceptive leader or manager will develop narratives that fit one or more of the frames of reference just discussed in order to develop the best means of promoting his vision to those he wishes to lead. Now the question is, how does a person even begin to develop such a narrative? Early in this book, in Chapter 1, I wrote that managers are thinking about three questions when considering goals that may be pertinent to the animal facility. Those questions are, Where are we now? Where do we want to be? How are we going to get there? Your coworkers will have similar questions about your vision, wondering how it will affect them. In essence, as they are sitting and listening to you roll out your vision, they will be asking themselves:

- Where will you take us (i.e., what is the goal)?
- Why are we going there?
- How will we get there?
- What's in it for us?
- What's in it for me?
- What do you want from me?
- Why should I trust you?

The narrative that you develop to promote your vision should address all of these questions [10]. You may have to vary the story a little, depending on to whom you are speaking (i.e., the framing may have to change a little), but the basic elements can and should remain the same. In fact, weaving your vision into an interesting story is an excellent way of promoting your vision. And don't forget—every animal facility has influential people (managers and nonmanagers) who can sway others to accept your vision. Therefore, working with these people ahead of time to have them agree to support the vision may save you a lot of extra work in the long run.

Managers versus Leaders

As you can see by the frequent mention of the importance of having a vision, leaders must have a focus on the future, whereas managers give heavy consideration to the present. Nevertheless, leaders and managers have to base their actions on trust and open communications. They must work together to develop long-range plans to serve our animals and those who use them, but it is the leader who has to be able to secure the needed resources to help accomplish those goals. The leader must be attuned to the internal and external environment to determine which opportunities to seize and which threats to avoid while seeking those resources. And, as was noted in the previous section, the leader must have the political savvy and sufficient authority to be a strong advocate for the animal facility.

A talented manager may now and then have the opportunity to show her leadership skills, but as I noted at the beginning of the chapter, leaders and managers aren't always the same person. Your vivarium needs a good leader, along with good managers, in order to stay competitive.

For those of you who want some concrete examples of differences between leaders and managers, Table 7.1 provides a side-by-side comparison. As I mentioned previously, it is a matter of emphasis, because, depending on the situation, at one time or another almost all managers will have to exhibit leadership and all leaders will have to be able to manage.

As expected, there are many traits shared by good managers and good leaders. Here are some of them:

■ They communicate clearly.
■ They establish a culture of openness and trust.
■ They clearly state goals.
■ They define and disseminate what is meant by success.
■ They foster collaborations whenever and wherever they are needed.
■ They work to remove hindrances to achieving goals.
■ They remain honest, even though it has been said that increasing status and power are associated with decreasing honesty [12].
■ They know what they don't know and use others to fill in knowledge and skill gaps.
■ They actively solicit feedback before making significant decisions.
■ They understand that the ideas of others may be as good as or better than their own.
■ They are willing to learn.
■ They are self-motivated.
■ They are a stable force in the face of uncertainty.
■ They hire the best people possible.
■ They project an aura of confidence.
■ They are friendly.

Table 7.1   Some Side-by-Side Comparisons of Managers and Leaders

Managers	Leaders
Can motivate a person	Can motivate a group
Is a key player in establishing or influencing the culture within his or her area of responsibility	Is a key player in establishing or influencing the overall organizational culture
Creates a clear purpose and direction for the area of his or her responsibility	Creates a clear purpose and direction for the entire organization (i.e., a vision)
Looks for potential problems, but is more involved in obtaining results for what is already decided	Looks for potential problems and is more involved in initiating significant organizational changes
Focused on limited area of responsibility (e.g., animal husbandry)	Focused on the company or department as a whole
May, by definition, work alone and not oversee the work of others	Is, by definition, leading one or more people
Has a primary concern that existing business processes are working efficiently and effectively	Has a primary concern to look to the future and develop new business processes
Primarily interacts with subordinates to accomplish one or more goals	Interacts with subordinates, peers, and outside contacts who can contribute to accomplishing (or thwarting) one or more goals
Knows the skills and personality of each direct report and works to maximize that person's potential	Understands the basic needs of all employees and unites them toward achieving a goal [11]
Occasionally in the spotlight	Almost always in the spotlight

- They are aware of their own strengths and weaknesses and do what is necessary to minimize weaknesses.
- They are not afraid to embrace constructive conflict as a means of improvement.
- They recognize that people are assets, not objects.
- They give credit to others.
- They serve as mentors.
- They are consistent in their management and leadership style, but can change when necessary.

Authority of Leaders

It's relatively easy to envision great leaders coming from politics, armies, and business. However, not all leaders have to come from these fields, and not all leaders even have to know that they are leaders. Just look around the organization where you work, and I'm fairly sure you will be able to identify at least one person whom you consider to be a leader, even if that person has no formal authority. For instance, there may be "ring leaders," who always seem to be planning something you aren't happy about, but people do follow them—and since they do make and execute plans (even when they have a malevolent intent), they can be considered leaders. It's obvious then that not all leaders have to be at the top of their organizational structure. There are excellent leaders who are the number two people in an organization. Other leaders not only help lead people above them in the corporate hierarchy, but also prove their value by influencing their peers, their peers' subordinates, their boss's peers, and of course, their own subordinates [13].

There are all kinds of leaders; some are good and some are bad (and sometimes good or bad depends on your point of view). I think many people would agree that George Washington, Adolf Hitler, Golda Maier, Winston Churchill, Steve Jobs, Jack Welch, and Bill Gates qualify as important leaders, but their impact on history will be vastly different. It should be apparent from the above group of people that not all leaders are good people, even if they are good leaders. I often think about Lord Acton's observation that "power tends to corrupt, absolute power corrupts absolutely." Our nation's Founding Fathers certainly recognized and were concerned about this possibility well before Acton's time and devised a system of checks and balances in government that were designed to prevent any one person from assuming absolute power. And yet, there were those (such as Senator Huey Long) whose corruption was ostensibly channeled to the welfare of the man on the street, but if left unchecked might have tried the foundations of our democracy.

Part of the leader's ability to lead is due to actual or implied authority, while another part is due to leadership skills (what I called the quality of leadership). The two are intertwined even in an authority-laden environment such as the armed forces. A leader with authority will not get very far if that's all he has. Indeed, as I indicated in Chapter 1, gaining the trust of your coworkers is a critical ingredient for leadership. Nevertheless, we can all agree that authority can catch your attention. For example, France's King Louis XIV allegedly proclaimed that he was the state. He would not have done that unless he believed that he had the power to do so. Let's take a look at the types of authority leaders have.

■ *Positional authority* is the authority that goes along with your job title. Elizabeth Alexandra Mary has the title of "Queen of England" and has the positional authority that goes with her title. A supervisor has a certain degree of authority over nonsupervisors. An associate director has authority over supervisors, and the director probably has positional authority over the entire animal facility. The importance of positional authority is that people believe you have the right to make certain requests of them, and that they have a responsibility to comply with those requests. When positional authority is abused, we often see low morale and high employee turnover. It is usually better for a leader to work toward accomplishing a departmental or institution-wide goal by motivating people, rather than by using positional authority.

■ *Knowledge authority* is the ability of a leader to gain the respect of followers as a consequence of her special knowledge or skills. In academia, for example, it's typical for department chairpersons to have distinguished careers in the same general field as the other members of the department. It's also the reason why we see advertisements for board-certified veterinarians, as opposed to nonboarded veterinarians, to lead animal facilities. The board-certified veterinarian is expected to have advanced knowledge and skills. It's also why scientists like to be led by other scientists, veterinarians by other veterinarians, technicians by other technicians, and so forth.

With knowledge authority, people follow you because they believe that you can teach them something, that you have already experienced their unique work concerns, and that you will provide needed resources for their work because you understand their needs. When Jimmy Carter first was campaigning for the presidency of the United States, my boss was urging all members of our research institute to vote for Carter because he was a scientist and would understand our needs. I voted for Carter, but whether he understood our needs turned out to be immaterial because federal research grants were harder to get under Carter's presidency than in previous years.

■ *Coercive authority* is a type of authority that emanates from coercion; if we move out of polite society, this is sometimes reflected in raw brutality. Al Capone's special skill was killing those who didn't do things his way. It certainly gave him authority among those who understood the consequences of not following the leader. A supervisor who threatens to suspend you for three days has coercive authority. An Institutional Animal Care and Use Committee (IACUC) chairperson who tells you to use a particular anesthetic or your protocol will not get approved has coercive authority. Coercive authority occurs when people require a reward from a leader, or they do not want to experience a punishment from the leader. A good leader never makes demands that are not consistent with the overall vision [6]. A coercive leader is not really a leader; he is a tyrant.

Leaders use all three types of authority, as needed, in order to reach their goals. As I will discuss a little further in this chapter, a leader's leadership style requires flexibility because not every style of leadership will meet every situation. The same holds true for framing a situation (discussed earlier) and authority. Sometimes, the leader who normally depends on knowledge authority will have to use coercion, and there are times when simple positional authority isn't sufficient and the leader has to rely in part on knowledge authority. Nevertheless, the ultimate source of power for a true leader comes from the people who follow his leadership [14]. This was drilled into my head by a former faculty colleague who emphasized that the administration of a veterinary school is there to serve the faculty, not the other way around. Likewise, a leader has no real power without willing followers. The best leader may very well be a person who is in a powerful position and is a skilled politician [15].

A Leader's Personality Traits

Leaders lead by a combination of authority and leadership skills. Sometimes, authority is more important, and sometimes skills are the key, but usually both are used at the same time. Some (perhaps most) leadership skills are learned. In fact, the good news is that some traits that may seem to be innate (e.g., motivation) can also be learned.

Leaders lead by combining authority with leadership skills.

Earlier in this chapter, I listed many traits that are common to both managers and leaders. Both are good communicators, trustworthy, and self-motivated; state clear goals; want to learn new things; and so forth. Now, let's dig a little deeper and take a look at motivation and other personality traits that are common to good leaders [16]:

1. *Motivation.* I've already noted that being self-motivated is a hallmark of a good leader, just as it's a hallmark of a good manager. Good leaders want to get the job done. They set goals and are motivated to reach that goal. If you're not happy being in a leadership role or you feel you are unmotivated in your work, then chances are you're not going to be a good leader, and perhaps you should reconsider what you are doing. Those who are motivated often become good leaders.

   I'll tell you about one person in my department who routinely did superior work. She was intelligent, self-motivated, technically competent, and everything else that any manager would want from a coworker. At her annual review, I gave her a glowing report, which was largely little more than a summary of what I had told her during the past year. However, there was one small aspect of her job over which she had no authority and limited dealings, so I gave her a "meets expectations" rather than an "outstanding" rating for that item. It wasn't meant to be an indication that improvement could be made, but rather, I meant that it had little impact on her performance. Nevertheless, she asked what it was she could do to improve her work in that area. If anything, I was the cause of her query because I was not clear enough in stating that she did not have any true weakness in that area. Yet, that constant drive to achieve was what spurred her on and is one of the hallmarks of leadership potential.

Good leaders simply don't want to lose, whether it's at playing a game or leading a company. They are motivated to win. They push themselves to succeed, and they are totally committed to fulfilling their vision or specific goals. Equally important is that their motivation can become contagious and spread throughout the vivarium.

2. *Self-awareness.* Good leaders are typically self-confident, assertive people. They are willing to accept criticism as being constructive, not destructive. This goes along with asking for feedback, whether positive or negative. Leaders continually gather, accept, and respond to feedback [17]. Talented leaders can look in the mirror and see themselves for what they really are, accept who they are, and try to better themselves whenever possible. They are not trying to make themselves into a textbook description of the perfect leader. They know how certain problems might affect them, so they work to find ways to constructively work around these issues. For example, in Chapter 3 I discussed the need to gain the respect of people by calmly and politely expressing your true feelings when having a difficult conversation. A self-aware person can do this and draw out the feelings of the person with whom he or she is talking. In a leadership role, being honest with yourself is as important as being honest with others. This personality trait (being able to understand other people's needs and willing to accept all forms of feedback) is of particular importance to laboratory animal facility leadership, where we are relatively rigid in how we care for animals and typically have to clearly define everybody's roles and responsibilities. The leader has to explain the need for this rigidity and provide direction on how to achieve consistency in animal care and other technical procedures. From this need, we can also see that in an animal facility, the authority that comes with a knowledge of facility operations is often quite important in establishing a person's leadership.

A self-aware leader not only accepts feedback, but also actively solicits it. Ed Koch, the former mayor of New York City, was famous for asking people on the street, "How am I doing?" It was a brilliant way of giving people the opportunity to provide him with feedback on most anything, even if they chose not to take advantage of the opportunity. Now consider how feedback can be used in the animal facility. Because leaders are judged on results, and results emanate from the decisions leaders make, it is almost self-evident that asking for employee input about decisions that will affect them can build morale, help open communications, and give employees a degree of ownership of those decisions. It's reasonable to assume that most employees know that they often do not have the final say in implementing a decision, but in my experience, that has not been a deterrent to their providing useful feedback. Still, it's a good practice for the leader to make it clear that he is asking for input, not a decision. If a leader tries to incorporate everybody's opinion into a final decision, it may be very difficult, or impossible, to make progress. This is because many people simply don't like changes; however, a leader has to make decisions that often lead to changes.

When asking for feedback, leaders should make it clear that they are not asking for decisions.

You may find that you have to set up a formal meeting with one or a few people to get the feedback you want before making a final decision. There is no requirement as to who you might meet with; it can be your senior staff or other people you know will give you honest feedback (not just the feedback they think you want to hear). Your listening and other communication skills may be put to the test, but the result will be worth the effort. Here's an illustration of how feedback can be used.

During the national American Association for Laboratory Animal Science (AALAS) meeting, a few attendees got together to discuss problems that were common to larger laboratory animal facilities. One of the topics discussed was how to provide water for animals in the event of a significant disaster. After we returned to work, I put this same topic on the agenda for our weekly management meeting. I was a little (but not overly) concerned that we could have a water supply problem for animals in the event of a major human disease outbreak (such as an influenza pandemic) or even from domestic terrorism. During such an event, water company workers might not be able to work if the water supply was cut off or contaminated. I presented one possible solution—stockpiling 55-gallon barrels of clean water—to start the discussion. I asked for everybody's thoughts about the need to even consider this issue, and if it was important, what could we do to ameliorate the problem if it occurred?

The senior managerial staff first asked themselves if water alone was the potential concern, or if we should also be concerned about food, bedding, or anything else. After some open discussion, it was agreed that an interrupted water supply was a legitimate concern and was more of a problem than food, bedding, or other items. Some quick solutions (brainstorming) were put out for consideration. The associate director for animal care was given the responsibility for meeting with his supervisors and getting their feedback on both the problem itself and the suggestions that were already made. In turn, the supervisors decided to discuss the issue with the animal care technicians and get their opinions. After all was said and done, a contingency plan was developed and there were even two backup plans for the contingency plan. Everybody who wanted to have a say had a say, and although it was the senior management team as a whole that made the final decision, it was really a departmental effort. Now, of course, it was my job to make sure that the plan was actually carried out. If there were to be a catastrophe requiring alternative sources of water for animals and our plan didn't work, then I alone will probably be taking the heat. I will be judged by the results of my decisions, not the steps I took to get there.

Here's the addendum to the above story. In 2013, there was a major water main break and the water supply to all the animal facilities was cut off for nearly 48 hours. The emergency plan worked nearly flawlessly.

As we have seen, leaders not only need people to lead, but also need to get honest feedback and honest advice. Therefore, managers who are leaders need to engender a feeling of freedom (trust) among staff members to speak up without getting dirty looks, cold indifference, a quick thank you and you're out the door, and so on. People aren't dumb; they know if you are sincere about asking for feedback or if it's just a ruse that looks good on paper. Herb Kelleher, former executive chairman of the board of Southwest Airlines, would go the extra mile and get right in with his employees, periodically work beside them, and obviously talk to them. All in all, it created a level of comfort for employees to able to provide feedback to one of the bosses. Although Southwest is now led by its president (Gary Kelly), one can only hope that Kelleher's example and constant admonitions about teamwork have filtered down throughout the company. Kelleher did what Warren Bennis advocates, which is "to lead people, you have to enter their world" [18].

3. *Extroversion.* Although most leaders of for-profit corporate business tend to be extroverts, this is not invariably the situation. In laboratory animal facilities, personal experience suggests that there are as many introverts or just "neutral" personality people as there are extroverts who are leaders.

4. *Self-regulation.* The best leaders I've seen are typically calm people who look for solutions to problems, rather than looking for people to blame. They're not afraid to take risks (in fact, a good leader must take some risks), but they're also not afraid to say no to out-and-out bad ideas. Either they have learned from experience or they know intuitively that it's wiser to sit back and think about a problem, rather than make a snap judgment that they will regret later. My personal rule of thumb is not to make an important decision until I'm confident that I have enough information to make that decision *and* that I've mulled it over for a day or so. Of course, there are times a quick decision is needed, but these are the exceptions. My other rule of thumb is that I'll typically delay or reject a possible solution to a problem if it simply doesn't feel right to me. A good example of the last concept is my tendency not to send an email in haste. If it's an important email, I'll write it, save it as a draft, and then look at it again the next morning. When leaders are capable of self-regulation, their demeanor sets a corporate climate for people to work together toward a common goal, not to try to save their own skin when something goes wrong. If the boss is willing to take his or her share of responsibility for an error, then the remainder of the staff will be that much more willing to do so as well. And, as I wrote in Chapter 3, the smart leader knows not to take all the credit when something goes right. When you are the leader and you give most of the credit to others, you will get your share of the pie, even if you don't ask for it.

There are exceptions to self-regulation. Some leaders, such as the late Steve Jobs, are successful even with a domineering "in-your-face" personality. I'm not talking about the out-and-out tyrant who uses nothing but coercive authority (as I described earlier for Al Capone), but rather, an individual with an abrasive personality toward those within and outside of the organization, yet one who understands politics and can get things done. President Lyndon Johnson was another classic example of such a person. These people, who use power, intellect, taunts, temper tantrums, slurs, and demeaning looks to make their points and get their way, are those Kramer [19] calls the "Great Intimidators." There is no doubt that sometimes this type of a personality can result in effective leadership, particularly when a rapid change is needed to bail out a sinking organization.

5. *Empathy.* Whereas sympathy is feeling sorry for another person's predicament, empathy is the ability to see a problem from the other person's point of view. If you wish to reword this, we can say that the leader can relate to the man on the street. Perhaps one of the reasons for Shakespeare's greatness as a writer was that he could think like the characters he wrote about. And he could also think like the person in the audience watching his plays. Together, it was a winning skill. As a leader, you also must be empathetic. If a group of employees was late for work because of a power failure on a train line, you have to put yourself in their shoes and consider

what choices they realistically had. If you are working for a pharmaceutical company that was planning on giving everybody a bonus when its new wonder drug came on the market, what would you do if your company had to pull it off the market due to adverse reactions that were found during clinical trials? The good leader would stand up and tell everybody that she's as disappointed as everybody else because she could also have used the extra money, but it's more important that the drug was taken off of the market before more people were affected. Pfizer, Inc. found itself in that situation in 2006 when its new cholesterol-lowering drug had to be withdrawn from further consideration due to unanticipated problems found during clinical trials.

6. *Social skill.* When I was writing my management thesis, I quickly learned a critical fact: people in any given field, such as laboratory animal science, want to be led by others in the same field. But, I wondered, why was it that on occasion some intelligent, pleasant, hardworking people were left in the dust, while others who seemed less motivated and less talented became department chairpersons? At the time. I was convinced that there were two answers. The first was that those who quickly rose through the ranks were more articulate than most others, and second, that they knew how to "schmooze" with the best of them. I still think that communications are crucial and being articulate is important, but being articulate is certainly not as critical as I used to think it was. However, I do believe that social skill, including being part of the "in-group," is an important but unsung attribute of leadership. You simply have to have friends and acquaintances, be at ease with other people, know the customs of where you live or where you are, and as shown above, have empathy, self-regulation, and self-awareness. You may not want to bow or curtsy when you are introduced to the queen of England, but the astute leader knows that this is something one simply does. The astute leader goes to an AALAS meeting and learns more from socializing than she does from sitting in on scientific sessions. The astute leader knows how to build personal and political relationships that can only help in the future. The astute leader knows that she needs friends and alliances both outside and inside the animal facility. A leader networks to be sure that today's course of action is correct and to get insights into the course of action that should be taken tomorrow.

Chapter 2 provided information about the environments that surround an organization. I indicated that there was a social environment, a politico-economic environment, a physical environment, and an informational-technical environment. They all impact our ability to reach the goals we set for our animal facilities. Our general social skills—and in particular our ability to network within and outside of the animal facility and our own organization—can provide us with information about these various environments that we can never get from books, newspapers, or journals.

Networking helps leaders make decisions about future courses of action.

As important as networking is, you can overdo it and spread yourself too thin. It has been suggested that effective "core" networks have about 12–18 people. Ideally, these should be people who can provide you with needed information, who may be able to provide political impact for you, and who are willing to challenge your decisions [20].

7. *Prudent paranoia.* This personality trait is in addition to those presented by Goleman [16]. I don't know it's extent in management and leadership, but until I read two articles that, in part, described the business paranoia of Andy Grove (former Intel chairman and CEO) [21,22], I was always too embarrassed to admit to my own paranoia because it seemed too much like micromanagement. To be sure, I am not concerned that everything that goes on behind my back is a threat to me, or that people who don't say good morning to me know something that I don't know. Rather, I continually fret about the details. I worry that all cages are being changed correctly, that every animal gets the best possible medical care, that investigators aren't complaining, that we provide enough services, and so on. Of course, all leaders continually sniff the political and social environments where they work to determine if there are business or personal problems to unravel or opportunities to grasp, so maybe a little "prudent paranoia," as Kramer [9] calls it, is a healthy personality trait for leaders.

If we put most of the information from the previous discussion together, we can say that good leaders know themselves. They know their strengths and weaknesses and don't try to hide them. They don't try to become somebody they are not. They actively work on making themselves better leaders by doing the right thing, not by forcing themselves on people by the sheer power of their position. All of this leads to trust, which is one of the keys to good management and good leadership. "People will tolerate honest mistakes.... However, they won't forgive lapses in character" [23].

Nonpersonality Characteristics of a Leader

1. *Leaders can convince people of the importance of their vision.* Leaders tend to have the personality traits that were just described, but that is not the end of the story. Leadership is about setting new directions, and leaders have to be able to convince people of the importance of their organization's mission and their vision for change. Therefore, leaders—like managers—must

communicate well. That doesn't mean that they have to be eloquent with every word and phrase that they use (not that it would hurt), but they do have to be convincing. When presenting a vision or most anything else, they must be decisive and not hedge their words, lest they weaken their position as a leader. For instance, it's not recommended for a leader to speak about a vision by saying, "We hope that by next year we will have the funds to start this renovation of our animal facility." That's not very encouraging to the listener who wants to know that the institution is fiscally sound and would like to start the renovations tomorrow. If there is no fiscal problem, then the leader should be saying, "I have directed John Mulcahy to immediately initiate the planning for the new vivarium. He will work with Laboratory Animal Resources to meet the current and future needs of our researchers by providing an outstanding animal facility. It will facilitate the important medical research that we do, and it is where you will be proud to work." The leader has raised her own stature by using the word *I*, is speaking with confidence, and has mentioned the importance of the research being done, and she has also described why the animal facility staff will want to be part of the planning and use of the new vivarium. For many leaders, that kind of upbeat and positive rhetoric seems to flow easily from their mouths, but I can assure you that there was a lot of practice that went into learning how to be so smooth.

Every day in every way, leaders have to repeat their vision and emphasize its importance. They cannot sit back and hope things happen. They have to get out there and walk their own talk (i.e., do what they are asking others to do). If they cannot convince those who will be carrying out the vision about the importance of the vision, then it will be near impossible to have success.

Leaders must be able to convince their followers of the importance of their vision. If they cannot, the vision and its goals have little chance of success.

As you know by now, leaders have to make their vision sparkling clear to everybody who hears it. If it takes more than a minute or so to clearly state a vision, then it's too long or not clear enough. Leaders have to describe their vision as if it was a story, and not just any story. The story has to go to the heart and mind of the listener because the leader wants that listener to become a follower. It has to be relevant to the listener, in the same way that politicians target their speeches to their audience. If the political candidate is speaking to an audience that strongly shares her own beliefs about a controversial topic such as abortion, then the message the audience will hear is what they want to hear. But if the audience and the speaker differ in their views on abortion, the astute politician will find another topic, almost always one where she and the audience can find common ground. An online article [24] made the same argument for scientists. It stated,

> Scientists must learn to focus on presenting, or 'framing,' their messages in ways that connect with diverse audiences. This means remaining true to the underlying science, but drawing on research to tailor messages in ways that make them personally relevant and meaningful to different publics. For example, when scientists are speaking to a group of people who think about the world primarily in economic terms, they should emphasize the economic relevance of science.

Good leaders and good scientists are often good politicians. Here's another example, once again using the dean of the school of veterinary medicine.

Dr. Laura Phelps was the new dean of the Great Eastern University School of Veterinary Medicine. The school was already well known for its excellence in clinical medicine, but it was just another school as far as basic, translational, and clinical research was concerned. During the time she was interviewing for her new position, she told many people, including the search committee, about her belief that the veterinary school should be as active in research as was the university's medical school. There was general support for this, but now that she was the dean, many of the faculty wanted to know how her vision would be accomplished.

Phelps began by meeting with the faculty as a whole and talking about how their academic careers would be enhanced by their becoming more involved in research. She emphasized how, as an academician herself, she understood the intellectual needs and desires of the faculty. In other words, she showed empathy with the faculty. She appealed to their excellence as clinicians and reminded them of the prestige that would come to them by being "dual threats" and being on the faculty of a school whose reputation would rapidly grow. Here, she was motivating them. For those who were already researchers, she thanked them for their existing work and told them of the positive changes for the better that were coming, including more laboratory space and hopefully more grant dollars.

Phelps also knew that she had to continue to solidify the support of a skeptical faculty, so she had low-key, informative individual meetings with every single faculty member and found out from them what they needed to make the research vision a reality. She was demonstrating social skills, self-regulation, and empathy. From these meetings, she learned that the biggest concern was that the school's clinical caseload was so demanding that the clinical faculty had no time to even think about their research interests. Armed with this knowledge, Phelps continued to articulate her vision, and each time was able to provide a little more detail on how it could and would be accomplished. She also worked behind the scenes, making sure the department chairpersons and the school's board of trustees shared her enthusiasm and were kept up to date. Little by little, the faculty began to feel like academicians rather than private practitioners in a large not-for-profit setting. More faculty were hired, and eventually things began to fall into place.

Toward the beginning of this chapter was a true story about Dr. Phelps. The story above is hypothetical, but typical of the activities of a new leader. Phelps could not transform the veterinary school's research endeavor overnight. The purpose of the story was to highlight Phelps's activity as a leader who understood the desires and needs of the faculty and was able to motivate the faculty to support her vision by appealing to those needs. The story also highlights that in most instances, a leader is setting a new direction, not improving on what already is in place (managers do much of that work). It's true that there was already some research at the veterinary school, but it certainly was not a major facet of its work. Enhancing research was a major new direction that required leadership, as well as good management.

Can an animal facility make major directional changes requiring leadership rather than managerial skills? Admittedly, the type of work we do does not have as many opportunities for initiating major changes as do many other enterprises. As a profession, we often need good managers more than good leaders. Nevertheless, although much of our work is routine, the same was true at the veterinary school until Phelps came in and saw a need and an opportunity that previous deans had not acted on. An analogy would be having a laboratory animal facility begin a division of comparative medicine that would have research but no animal care responsibilities. Another example would be an animal facility that expands its in-house training program into a national training program. Also consider the leadership that might be needed to motivate your scientists to want a mouse phenotyping core that might raise their *per diem* rates, or an *in vitro* center for making monoclonal antibodies. I'm sure you can think of more examples where leadership is more important than management, and I gave two additional ones earlier in this chapter. Even relatively small changes, such as trying to have your entire animal care staff obtain some level of AALAS certification, take leadership. It is a forward-looking change that is focused on the animal facility as a whole, it requires people to be motivated, and it requires a true leader to make it happen.

2. *Leaders can adapt to the needs of the organization.* One of the most crucial assets any leader can have is to be adaptable to the situation at hand and not assume that his or her preferred leadership and personality style is suitable to every situation. As with managers, there are times when a leader has to be a chameleon and change his or her usual leadership style to meet the immediate needs of their organization. Farkas and Wetlaufer [25] studied many leadership approaches, but the two that we see most often in animal facilities are those they termed *box leadership* and *expertise leadership.* Box leaders establish and communicate a set of controls that ensure predictable results for customers and employees. It's similar to the structural framing discussed earlier in this chapter. Those of us in laboratory animal science can relate to this because most of us rely on SOPs as control documents for helping to get predictable results. Box leaders in animal facilities are aware of the regulatory pressures in laboratory animal science and help ensure that their customers (animals and researchers) do not have to worry about compliance issues. Thus, the box approach leader expands on the managerial control role that all managers have. This person is constantly looking over reports, supply requisitions, and the like. The box leader gets involved in almost all of the day-to-day activities of the organization to help ensure consistency. There is a chance that a box leader can get so wrapped up in controlling variables that he or she will develop a micromanagement personality. On the positive side, box leadership results in a good deal of predictability in how the animal facility will react to any external situation, such as a new direction for research or a new regulatory requirement. The users of the animal facility can be reasonably assured that sufficient controls are in place, so their own research will remain on track even if the organization as a whole moves in a different direction. Therefore, the box leadership approach may be best when an animal facility has had a heavy turnover of its upper management and research consistency is suffering.

On the other hand, leaders who primarily implement the expertise leadership approach spend most of their efforts improving expertise, such as developing new training programs for managers and animal care personnel, and looking at technology that can yield a competitive advantage. This is similar to human resources framing that was discussed earlier. These leaders design programs to track their progress toward meeting their mission and vision. As with box leaders, they may have quality assurance teams in place to review processes and suggest more efficient and effective methodologies. This approach to animal facility leadership may work best when there is a rapid expansion of research and new people and new technology have to be brought into the facility.

These two approaches are just examples. Both of these methods (and others) may be appropriate, and choosing the best one will depend on the needs of the animal facility at a particular time. Sometimes, you need multiple approaches, just like leaders who sometimes have to mix positional authority with knowledge authority and, once in a while, a little coercive authority. However, once you decide on the leadership approach with which you are comfortable and which meets the current needs of your animal facility, then just like a manager, you have to be consistent in your leadership style. Continuous vacillations in managerial or leadership styles lead to confusion and frustration among employees. It then becomes easier for employees to look out for their own interests and relegate the animal facility's, school's, or company's interests to the rear burner.

Leaders, like managers, should be consistent in their leadership style.

---

3. *Leaders understand the needs of those who follow them.* Just as leaders have to adapt to the changing needs of their organization, they also have to recognize the needs of those who follow them. We discussed a somewhat related topic in Chapter 3, pointing out that different people seek different types of rewards, based on their own needs. The earlier discussion of empathy, in this chapter, also touches on the fact that leaders have to relate to the needs of their followers. We know that different followers need different types of support, some of which are tangible and others nontangible. For example, in terms of nontangible support, the best support my boss (the associate provost for research) can give me is to meet with me periodically so we can discuss emerging issues, make sure I have rapid access to her when I need it, and support my decisions if they deserve support, but otherwise just leave me alone and let me do my job. Fortunately, she does just that. This type of a relationship meets my needs, but not necessarily those of everybody else. Some of her other direct reports may need a shoulder to cry on for personal issues, while others may want to be mentored to improve their skills. In fact, you may recall that I said earlier in this chapter that leaders often get a certain amount of their authority by having special knowledge or skills. They often have to show their support for a person or group by sharing this knowledge with them to help the group accomplish its own goals, giving it information that fills in some blanks or puts their minds at rest. Sometimes, the leader has to cater to an entire group, sometimes to direct reports only, and sometimes to individuals who aren't direct reports. But a leader has to be attuned to people's needs and wants, and be willing to meet those needs whenever possible.

Leaders usually have to lead groups, but sometimes they have to gain the individual support of key group members. If we can agree that an important part of a leader's job is to implement change through a clear vision, then the leader must be able to have a core of supporters who will be leading the charge. It was interesting for me to learn from Jim Collins's research that the best leaders attended to people first and strategy second [26]. To gain the support of people, you, as a leader, must understand and be adaptable to the needs of many different individuals at the same time. Wherever possible, you want your core coalition to be composed of other senior managers, but also those key individuals (managers and nonmanagers) who have connections with, and can influence, the opinions of other employees. There's a lot of inertia in some animal facilities, and it can be hard for any one person to implement significant changes without an initial broad base of support.

Supporting the specific and general needs of people is an important characteristic of good leaders and good managers. Whatever your style, support is particularly important when leading an animal facility where the work is often repetitive, requires following specific rules, and can be stressful [4]. This support might be manifested by showing respect for an employee's opinion, by talking to everybody as peers rather than as underlings, or by pitching in and cleaning cages when needed. To return to a theme in Chapter 3, meeting the needs of a person or a group satisfies two important basic needs of people: the desire to have respect and to feel that their work is important.

Leaders, like managers, also have to understand that there are times when tangible rewards might be best for an individual or a group. At the group level, think for a moment about AALAS certification. Do your animal facility and the person who leads it provide any rewards for attaining a particular level of certification, such as a bonus or a salary increase? If the answer is no, ask yourself if that type of reward would be meaningful to people passing their certification exam. In many animal facilities, the answer is yes, as few of us are getting rich in our jobs. But there may be behind-the-scene issues that prohibit a bonus or salary increase, such as union contracts. A union may oppose a bonus or increase for one department that provides the opportunity to take a certification exam, whereas another unionized department may not have a certifying organization and therefore cannot provide such an opportunity. When this type of a scenario occurs, the leader must have a body of facts ready and explain to the animal facility employees what he did to try to secure a raise or bonus, and provide a logical explanation of why the plan could not be implemented. People are often willing to accept an honest effort that led to disappointment, as opposed to a limited effort with no explanation and no success.

I am of the opinion that for a leader to understand the needs of those that follow him, he has to get out there and find out for himself what's happening, and "press the flesh" with his coworkers. This is similar to what I described earlier for Herb Kelleher of Southwest Airlines and his propensity for periodically working alongside employees. This is partly management control and partly socialization. Speaking for myself, I believe it's important for me to periodically drop by and informally chat with my coworkers, at all salary levels. I actually enjoy doing this. I'll also sit next to different employees during our group meetings and not just next to my trusted lieutenants, to help build morale and learn about their concerns. Equally important, I'll periodically drop in on researchers just to chat for a few moments and see if there's anything I can do to make their life a little easier. Every little piece adds up.

4. *Leaders get things done.* This may seem obvious, but in reality, it's not as obvious as you may think. There are people who are in leadership positions who can get some things done but fall down in other significant areas. As an example, let me use strategic planning retreats, which some of you may be familiar with. These retreats are designed to help set the direction for an organization or part of an organization, whether it is for profit or not for profit. I've participated in about six such retreats that have been led by professionals in strategic planning and who help provosts and deans develop strong strategic plans. They run excellent meetings in which they knit together the outline of some exciting institutional goals. Everybody would leave those gatherings singing "Kumbaya," and with big smiles and renewed motivation. But for almost all of those I've attended, nothing was subsequently implemented. The retreats led to excellent vision statements and fine strategic plans associated with the vision, but there were no specific strategies that were developed to implement the plan. There were no deadlines, no accountability, no reports of progress, and in general, no follow-through. Soon afterwards, it was business as usual. Skillful leaders do not let this happen. They take charge and personally advocate for their vision, as they know that they, not someone who reports to them, must be the prime advocates of the organizational vision. When they do delegate responsibility, they make sure there is follow-through. Great leaders lead; lesser leaders falter.

5. *Leaders are mentors.* Leaders (and managers) have to think about the future of their organization, such as replacing themselves before they move on or retire. They also have to work with others who simply need the skill and experience that a mentor can provide to help them advance to higher levels of management and leadership. Mentoring benefits the mentor, the student, and the organization. The mentor gains satisfaction from the feeling of being needed and respected. He also is forced to evaluate what he says and what he does more than ever before. It's almost like a self-imposed refresher course in management and leadership. There is also a tremendous sense of accomplishment when one sees his or her protégé becoming successful. When I was teaching many more graduate students than I do now, it was a true emotional high for me to place their master's degree hood on them at graduation. Their success was my success. Their happiness was my happiness. I'm not ashamed to say that it invariably would bring tears to my eyes. Of course, the person being mentored benefits from the knowledge and experience gained and the close

networking with the mentor and the mentor's colleagues. The organization gains because now there is increased camaraderie and a more knowledgeable, integrated, and skilled employee.

How do leaders decide who to mentor? It's usually a two-way process. Sometimes, leaders actively seek out people who seem to have promise and who might benefit from mentoring, and sometimes an astute novice seeks out a mentor. That is, the learner tries to find a mentor in the organization that has the skills, knowledge, and position that she believes can help her if the two of them are compatible. It's not often that one person directly asks another to mentor her (although this does happen); more often, a relationship develops as a result of the proximity and access between people. In some organizations, a mentor is routinely assigned to a new manager, which is often a valuable practice. Mentoring doesn't have to be anything formal, although it can be. It often entails shadowing the mentor, talking informally to the mentor, taking on projects for the mentor, and so forth. The mentor has to share her knowledge and skills, guide the apprentice away from danger, introduce the apprentice to the "right" people, and even let the apprentice fail when failure has a positive effect on learning. The person who has the desire to be mentored and the person who has the desire to help a coworker together set the stage for an important aspect of leadership success. There is more valuable information on mentoring in Appendix 4.

How Leaders Use Advice

Advice is a form of feedback. We all need advice at one time or another, and a good adviser can help a leader refrain from making a poor decision. Even though one of the roles of a manager and leader is that of a decision maker, it is not unusual for leaders and those they mentor to listen and take advice as needed. You should not hesitate to seek advice when help is needed to make a decision, but you should not get caught up in the trap of asking for advice that will simply validate your own preconceived opinions.

Do not seek out advice simply to validate a preconceived opinion.

There are basically two types of information that are used when advising a person: factual and nonfactual. The best advice is always based on facts, even if those facts are based on little more than experience. The worst advice is based on wild guesses or providing knowingly wrong information. Because leaders are so often in the spotlight, and because many people enjoy the status that often comes with a leadership position, it becomes all too easy for some leaders to become surrounded by yes men (those people who tell the boss what she wants to hear in order to promote their own careers or beliefs). But if the leader cannot listen to trusted colleagues, who can she depend on for advice and direction? To begin with, the basic premise has to be challenged. Most people surrounding leaders are *not* yes men, and most leaders are astute enough to know who is providing honest feedback, as opposed to a person who is providing self-serving or strictly ideological feedback. A good and experienced leader is an intuitive person. He can "read the crowd" and read a person. He seems to understand what people are thinking without having to be told what they are thinking, by being attuned to hidden meanings during a conversation, by reading body language, and of course, through listening and gathering information. Although a leader cannot always follow the crowd's wishes (because to do so often leads to maintaining the status quo), there are times when the astute leader (like a manager) finds that a group opinion may be the best opinion. A few of you may recall the television show *Who Wants to Be a Millionaire*? If a question stumped a contestant, the contestant could poll the audience members to see what they thought was the right answer, and more often than not, the audience was correct. So, a good leader is often intuitive, but she also knows when to back off and ask for advice.

There is a caveat here. Decisions (or advice) emanating from the consensus of a group can be helpful, but not every group can give good advice. For group advice to be useful to a leader, the individuals in the group have to be able to act independently (i.e., there is no "groupthink"), the group has to consist of people from many different backgrounds and areas of expertise, and of course, you need to devise a means to gather the input from the group [27]. These requirements can be difficult to fulfill because our colleagues in laboratory animal science often have similar backgrounds and expertise, as do the investigators that often comprise "user advisory groups." Being able to gather the group's input to try to detect a consensus may also be difficult, but not insurmountable. You can always consider using an online survey program such as SurveyMonkey to solicit feedback.

What kind of advice does a leader need? The quick answer is good advice, but good advice has many parameters to it. For example, is good advice that is given at the wrong time any better than no advice at all? In the remaining parts of this section, I'll discuss three aspects of advice. First will be comments on what constitutes good advice; next, what a leader (or a manager) can do to avoid bad advice; and last, some warning signs that suggest you may be getting bad advice.

Getting Good Advice

As stated in Chapter 5, one characteristic of managed information is that the information should be relevant. That means it should help you solve your problem because the information provides you with the key facts you need. I also wrote that information should be timely. That is, it should be available when *you* need it, not when someone else decides you need it. A good advisor will give you both relevant and timely information. Additional characteristics of good advice are that it's grounded in the solid expertise of the advisor and you can use the advice without additional extraordinary help [28]. I'll focus on solid expertise for a moment because there are far too many people who act as consultants, gurus, and coaches, who have book knowledge but haven't really walked their talk. Try to avoid these people (I hope I'm not one of them). If you've attended college, you can relate to this because you probably suffered through graduate assistants teaching a lab or a discussion section and they didn't know much more than what came out of a textbook or the lab in which they were working. Wouldn't it have been nice to have had the professor who could have said, "Yes, that's right, but let me tell you what *really* happens when you do that." It's the difference between real experience and limited experience. I had a similar experience with a very fine gentleman who came highly recommended for his ability to teach basic management and leadership skills in laboratory animal science. His name was fairly well known in the field. I should have done my homework a little better and checked into his credentials, but foolishly, I didn't. After one month, I had to terminate his contract because it was all too evident that he had carved out a little niche in the lab animal world, but he had no practical experience in the field. He couldn't respond to questions in a manner that gave me any confidence in his ability. However, had I not known anything about management in laboratory animal science, I could have been influenced by many of the things he said, because on the surface, they seemed reasonable. I suspect that's how he got as many jobs as he did. In case you are wondering, I've always been adverse to teaching management to my own staff.

Just as we need different types of advice at different times, we may need different advisors at different times. Using the same advisor for every situation can be dangerous, for as I just noted, not everybody has experience in every field, although most of us (me in particular) are willing to voice an opinion on almost anything. There are times when I'll ask for my own boss's advice on how to handle a matter, whereas I'll ask a school architect for advice on how to reconfigure our limited office space. Advisors have limitations, and we have limitations. We can't assume that because we are managers and leaders that we know it all. We have to listen carefully to people who know more than we do about a particular subject.

Listen carefully to the advice of people who know more than you do about a particular subject.

Have you ever been reluctant to take someone's advice because he's much younger than you, she's your wife's cousin, or you just don't like that guy? It happens to all of us, but the budding leader has to overcome these biases. It's not a matter of swallowing your pride or dislikes; it's really a matter of leadership. A good manager—and of course a good leader—recognizes that information comes in many different colors, and he does not let personal prejudices override his business judgment.

Preventing Bad Advice

Sometimes, a leader has to know when to take advice even if it isn't directly offered. Consider the dilemma faced by former president George W. Bush. The people he trusted the most were advising him to push forward with the Iraq portion of the War on Terror, while opinion polls, the 2006 midterm national election results, and the Iraq Study Group Report indicated that the majority of Americans were advising him to do otherwise. It's easy for us to be Monday morning quarterbacks and make decisions for the president, but in terms of reading the climate of the nation versus accepting the opinion of his trusted advisors, the president had a problem. Would "staying the course" in Iraq, which was what the president said he was going to do, prove him to be a wise leader or a foolish one? Was the information he was receiving filtered to give him what he wanted to hear? Was he being misled or properly led by his advisors? Was he continually relying on the same group of people for advice? "Dangerous confidants come in all shapes and sizes.... They habitually lie and cheat to achieve their aims without any apparent constraints of conscience" [29]. Can this happen in an animal facility?

We all know that leadership in a laboratory animal facility can be nerve racking. There are investigators pushing us in one direction, administrators pushing us in another, an IACUC going its own way, and our own coworkers requiring that their needs be met. There have been times when I just wanted to crawl into a hole and let someone else make my decisions, take the heat for me if things went wrong, and give me the credit if they went right. Since that's a pipe dream, it's tempting to rely on a trusted advisor—a confidant—who can help us make the right decisions. But what can we do to prevent our advisor from influencing us in the wrong direction, and how do we know if our confidant is actually becoming dangerous to us? Let's start off by considering what we can do in the form of preventive medicine that helps an advisor give us proper advice [30].

- Be a good communicator. Don't tell lengthy stories to your advisor that are so detailed and boring that your advisor starts to tune you out. Be concise and truthful. Give your advisor just enough information about your problem.
- Don't place artificial boundaries around data. If it's potentially important, tell it to your advisor because excluding important information will skew your advisor's assessment.
- Don't escalate your own opinion as being more important than your advisor's, especially if the advisor has unique expertise that you don't have. Adolf

Hitler was never a brilliant field commander, yet at times he valued his own opinion over those of his professional military advisors.

■ Do not value the advice of a confidant over that of an expert. Confidence does not equate to validity.

■ Look for advice you can act on. Advice such as "improve your presentations," without being more specific, is just poor advice.

■ Be part of the discussion and ask questions about costs, benefits, rationale, and so forth. Offer up your own thoughts.

Now let's look at some action items that can be used to prevent leaders from being misled by their followers or advisors [31]:

1. *Keep vision and values front and center.* In Chapter 1, we discussed the importance of an organization having a clear mission and vision. The mission delineates the reason for the organization's existence, whereas the vision sets a general direction for the foreseeable future. If your advisors are suggesting you go outside of your organization's vision, you should think very carefully before acting. For example, a university-based animal facility is usually not in the business of performing contract research for private organizations, but here and there it does occur. If the vision statement of your animal facility is to decrease the university's financial subsidy from 40% to 25% in two years, increasing the amount of contract research should be considered by you as a leader and manager, even if your senior staff (i.e., your advisors) argue that the lofty research ideals of the university would be compromised. You will have to decide if contract research is within the vision and mission of the university and the animal facility, even if it isn't something that's often done. You may have to set a new direction.

2. *Make sure people feel free to disagree. Cultivate truth tellers.* This brings us back to the concept of positive conflict in the organization, which was discussed in Chapter 3 under the heading of "Morale and Organizational Conflict." In a nutshell, the argument was made that a manager can't be surrounded by yes men or by people who are afraid to present disagreements. Thoughtful disagreement can often help a leader see through the fog. A good manager or leader will be willing to give up a tightly held belief if there is a good reason to do so. Most every animal facility, like most businesses, will have some person who specializes in finding the downside of every new project. Rather than putting this person's opinions aside, perhaps we can use him or her to find actual flaws in our plans.

3. *Do unto others as you would have others do unto to you.* The ethical climate that you set for your animal facility should be the same one in which you would like to work if you were not the leader. If people think they can cross the ethical line, some of them will do so, and they may either directly or indirectly try to influence you to do the same. Just think about the IACUC and how some people cross federal boundaries to advance their own needs. They would be

perfectly happy if you looked the other way. If, as a leader, you advocate and demonstrate an ethical and positive corporate culture, chances are strong that your followers will also act in an ethical, positive manner.

4. *Honor your intuition.* Here I will simply quote Offermann [31]: "If you think you're being manipulated, you're probably right."

Recognizing a Bad Advisor

The suggestions from the last section are more aligned to establishing a managerial style and business culture that helps prevent a leader from being adversely manipulated by her followers or advisors [31]. Sometimes, in spite of your best intent, you may rely more and more on the advice of one or two trusted advisors who, unknown to you, cannot be trusted to give you the wise counsel you need. Therefore, we'll look next at findings focusing on five specific warning signs that suggest a dangerous relationship with a trusted confidant is developing or has already occurred [29].

1. *People complain that you're inaccessible.* Many of us have seen leaders and managers who send a clear message that they are far too important to be bothered by the masses. However, an "important" advisor has ready access to the boss. You may have to alter your own leadership style and, at the same time, rid yourself of an advisor who supports this aloofness.

2. *You feel that only your confidant understands you.* Be careful and don't give excessive weight to any one person's opinion. Your trusted advisor may find it easier to conform than to voice objections [32].

3. *Your close advisor discourages you from seeking the opinions of others.* Why? Perhaps he wants status or perhaps he wants your job.

4. *Your advisor starts becoming the boss.* When a trusted lieutenant almost invariably speaks for you—the brigadier general—people are going to start wondering if it's the general or the lieutenant who is in charge.

5. *You get too much praise from your confidant.* You don't need and shouldn't want yes men surrounding you. If you only hear the good news, and none of the bad, then maybe it's time for a new advisor.

David DeSteno [12] wrote that "decades of scientific research show that people's accuracy in deciding if another can be trusted tends to be only slightly better than chance." Other research has found that people with power lie more easily and effectively than those without power [33]. It can be hard to accept that people we lead or manage, a person we trust as a confidant, or a person we report to can lead us down the wrong path. Sometimes, this is unintentional, and others times, it is calculated. But just as we don't want to surround ourselves with yes men, we can't be so pigheaded as to reject the concept that we might be used by others for their own benefit. We have to be aware of, and alert to, factors that might negatively impact our leadership.

Leadership Effectiveness

Chapter 1 and Appendix 1 discuss measurements of managerial effectiveness, and one of the take-home messages is that making these measurements isn't as easy as you might think. It takes a lot of hard work to determine how you are progressing toward a goal. We try to include as many numerical measurements as possible, and we try to include both financial and nonfinancial measurements. For example, a nonfinancial measurement that was mentioned in Chapter 1 was a customer satisfaction survey. Leadership should also be measured because the effectiveness of leaders is based on their accomplishments, not personality or other traits. Therefore, we might consider a survey of the people who are following the leader, asking about their job satisfaction and their motivation. We can also look at financial measures, such as growth in research funding or income from supplementary services. In our animal facilities, we can survey the people who directly report to the leader and look at changes in employee retention, the ability to attract needed equipment, the number of union grievances, completed renovations, progress toward accomplishing the vision, and any other relevant considerations that focus at least as much on leadership as on management.

Interestingly, some initial research has given birth to a theory that suggests that the means of evaluating the effectiveness of a leader may depend on the specific situation. It was found that when it was very clear that an organization was performing well or poorly, observers (such as you or me) focused on the leader's skills when evaluating him or her. However, when an organization's performance was not particularly clear, the evaluation of a leader's performance was based on the amount of charisma (i.e., personality or charm) that was attributed to the leader [34]. In a real-world context, a determination of whether your animal facility is clearly doing well or not will depend on measureable criteria, which may include finances, the number of complaints reaching upper management, and the speed of reviewing animal use protocols. It's only after those measurements are made (or perceived) by others that there will be a determination of your effectiveness.

No matter what criteria are used for evaluating a leader, money is often part of the equation. Most of us know that if a for-profit business leader cannot lead the organization's financial growth, that person will not be with the company for long. The same can be said for the president of a not-for-profit university, because if the school's finances don't hold up, the academic mission will eventually suffer or fail. Financial success is important, but it is not the only measure of leadership effectiveness. Look at the financial and nonfinancial accomplishments of Wayne Pacelle, the president of the Humane Society of the United States since 2004:

> Pacelle ... was the [Humane Society's] top lobbyist before taking over 2½ years ago. He quickly consolidated his power base by merging with two other animal-protection organizations—the Fund for Animals and the Doris Day Animal League. He also formed a political affiliate, the

Humane Society Legislative Fund, which spent $500,000 on key races in last year's elections, and established his own political action committee called Humane USA, which funneled an additional $300,000 to pro-animal candidates. Pacelle's goal is to double the amount of his electoral giving and the extent of the society's lobbying over the next two years. [35]

Pacelle is quoted as saying that his goal is to continue to expand the Humane Society by joining forces with like-minded organizations, capitalize on their strengths and eliminate duplicated efforts by having them under the umbrella of his organization. [36]

Whether or not you agree with Mr. Pacelle's goals, there is little doubt that he has shown strong and apparently effective leadership for his organization. But let's think hypothetically and try to imagine what would happen if Pacelle's mergers and other activities distance him from his staff and are contrary to the desires of many of the society's members. It's likely that nothing good will result, and it may, in the long run, harm the organization. If a leader (or at least a presumed leader) is a financial whiz kid but alienates half of the organization, there's a strong possibility that person will be terminated. I grant you that termination may take longer if the organization's finances are still strong, but no organization will survive with high employee turnover, low employee morale, and growing customer dissatisfaction. The same can be said for the leader of an animal facility. You can turn around the facility's finances, but if you continually alienate every researcher, you're history. No vice president for research will sit idly by and listen to day-in and day-out complaints about the director of the animal facility. This director has failed in communication, trust, socialization, and probably many other areas. Maybe he really wasn't a leader.

There's an interesting footnote to the activities of Wayne Pacelle. In a 2011 article [37], he reminisced about how political lobbyists undermined a battle he thought he had won concerning an attempt by the Humane Society to ban the use of "downer" cows (injured or sick cows that can barely stand) for human consumption. He concluded that he declared "mission accomplished" too early. "The slap-down from the conference committee taught me that you have to know who wields the power." "When you win by overpowering your opponents, they may try to find ways to subvert your goal. Lasting victory comes when you find common ground." That's certainly good advice, yet it was unexpected, coming from a person who himself had spent 10 years as a political lobbyist for the Humane Society.

Another way of approaching the evaluation of a leader's effectiveness is by assessing the quality of the people he has hired and trained. This will not come as much of a surprise to those readers who work in research universities because like good leaders, good researchers try to attract the brightest graduate students and postdoctoral fellows, and then try to train them to be high-quality scientists.

In fact, researchers take great pride in telling others where their grad students and postdocs are working and what they are researching. Likewise, the exceptional animal facility leader understands that part of his success is predicated by the quality of people he attracts, how he helps them to develop, and eventually, where they go. Now it may seem odd to plan on having people leave, but we all know that very few people stay in the same job forever. So if somebody who is doing a fine job is nevertheless going to leave, why not work with them, rather than against them? You're building goodwill that certainly may help you in the future. Admittedly, I'll do my best to keep my core coworkers financially and otherwise satisfied because we are mutually dependent, but if I know that somebody truly wants to leave, I'll discuss that with the person and do my best to help him or her obtain a new position. It's rare that you can keep a person happy by money alone, so if someone is looking to move up to the next step on the career ladder, let's hold the ladder for her. Who knows, someday we may be working together once again.

Moving Up the Leadership Ladder

With all of this talk about leadership, you may be wondering if, after a period of time as a successful manager, you will be recognized for your skills and promoted into a higher leadership position. Yes, this can happen, but it can also happen that the skills you have as a manager may not be the skills needed for a more advanced leadership position or, as often happens, your leadership potential is not recognized by your organization's upper management. Parts of this chapter describe the differences between a manager and a leader, and you should review them to make sure that you truly want to be in a leadership or a senior management position. Even if do you want such a position, you have to realize that there are likely other members of your department, your institution, or even outside of your institution who may be vying for the same position. Let me quote a fitting passage from the book *Executive Warfare* [38].

> Middle managers get ahead by being smart, hard-working, and showing results. But at the top of an organization, that's suddenly not enough, because everyone is smart, hard-working, and able to show results.

The take-home message is that to advance into upper levels of management, you have to combine the managerial and leadership skills you previously demonstrated in the workplace at a level that is at least as high as that of your peers and hopefully somewhat higher. But people have to know who you are and what you can do. They have to perceive you as being competent, trustworthy, and fair. Therefore, the keys to advancement are to make yourself and your skills known to senior-level leaders and to project an image of leadership potential. What follows are thoughts on how to accomplish this.

First and foremost, practice honesty. The need for honesty has been reiterated throughout this book and by others [38,39] as one of the keys to successful management. If your boss or others in positions of power find that you have intentionally twisted facts or have been dishonest with them or others, your chances of being promoted are near zero. You will have lost their trust, and anything you did or said in the past, or will do or say in the future, will be open to question. A simple example of workplace dishonesty is stretching the truth or adding outright fabrications on your résumé. According to a CareerBuilder survey, 58% of employers have found a lie on an applicant's résumé [40]. Don't help raise the percentage any higher.

Perhaps the second most important item, in my opinion, is how you present yourself. For example, does the way you dress fit the corporate culture? When you make a presentation at a meeting, are you well prepared, articulate, convincing, and able to adequately answer questions? Can you do the same during an informal conversation in your office? When meeting a person, do you make eye contact and smile? After work, when at public places such as a gym or a restaurant, are you composed or raucous (you never know who is observing you)? Do you act with civility to all people you meet, in and out of the workplace? At meetings, do you contribute in a positive manner or sit off in a corner where you think you will be safe? I used to make it my business to arrive early at meetings and take the chair immediately to the right of where the person who would be leading the meeting would sit. Say what you wish about doing that, but it ensured that I was noticed. All of these items reflect on you as much as does the quality of your daily work output.

In Chapter 3, I provided some thoughts about communicating with your own supervisor, and I said that it was both important and good politics to keep her aware of any information that could affect her work. As you probably know, as you climb the corporate ladder business, politics becomes progressively more important (and more frequent), and the savvy manager will always support his own supervisor and his supervisor's decisions once they have been finalized. That's part of your job. It also should be part of your job, particularly if you want to be noticed, to go beyond simple support and try your best to actively move any of your boss's decisions from the talking phase to the action phase. If a project is of particular importance to your boss or another leader, consider copying her on pertinent emails, but don't waste her time by sending her copies of every email that's even peripherally related to the project.

What else can the shrewd manager do to become more visible to upper-level leaders? One thing many managers have learned is to be friendly to leaders' administrative assistants, chiefs of staff, or people with similar titles. These people are often confidants of the leaders you are trying to influence, and a good word from them may be worth more to you than you can imagine. Being friendly means just that; be polite, chat with them, and show them respect for what they do. It doesn't mean take them out to lunch.

Consider volunteering to participate on high-profile committees or projects. If you are selected, you may find yourself working with some of your organization's movers and shakers. If there is no room on a particular committee, ask if you can attend its meetings as an observer. Your visibility, even if you just sit silently on the sideline, will help you. You may find that the committee may occasionally ask for your opinion, and if in the future a committee opening occurs, you may be considered to fill it as you have already demonstrated interest in the committee's work.

Don't "trash talk" behind the scenes. If there is a person or an activity that makes you unhappy, you can always express your opinion in a relatively neutral manner. For example, if you disagree with a school policy and feel compelled to say something, I suggest limiting your displeasure to commenting that you don't think the policy is in everybody's best interest, but you can appreciate that others have a different viewpoint. Saying that the policy is a disaster waiting to happen and those that support it are crazy can find its way back to those who made or support the policy and become a disaster for you.

Here are three final suggestions. First, hire people carefully, because hiring the wrong person can be seen as a reflection on your abilities. Next, if you do make a mistake (in hiring or otherwise), admit to it. People can forgive a small number of mistakes, but it's easier for them to forgive you when they know you have recognized the error and you are doing what you can to prevent it from recurring. And lastly, meet people. The more people you know in your organization, the more your name will come up in conversations and the more likely it is that upper leadership will think of you when the opportunity for a promotion arises.

With all of these suggestions, you may have noticed that I did not include knowledge of laboratory animal science as being a prime necessity for moving up the leadership ladder. That omission was intentional because you have already established yourself as a reasonably competent manager, a feat that for most of us includes having a reasonably in-depth knowledge of laboratory animal science, and for veterinarians, it also includes laboratory animal medicine. You should have a few areas of outstanding strength to make yourself more visible to those who have a certain amount of control over your career [41]. However, even if we are generally competent and knowledgeable managers, we all have pockets of weakness. Mine was waiting too long to let go of consistently underperforming coworkers. Perhaps yours is a lack of clear verbal communications or poor writing skills. Whatever your weakness, even if it is not a major problem, that is where you have to work on improvement. It makes little sense to improve a managerial strength you already have; that's almost wasted time. Spend your efforts on turning weak areas into strengths. It's the same old story about turning lemons into lemonade.

To help advance your career, you should have a few areas of managerial excellence while working to improve any areas of managerial weakness.

As a potential leader, you will likely need a different set of skills than what you have now. You require the skills to lead people to places where they have not been (paraphrasing the quote by Henry Kissinger at the beginning of this chapter). The people who report to you, directly or indirectly, are those who may be required to have more technical or medical skills than you have, but your leadership position does not require you to be the smartest doctor or most competent technician. Yes, you have to remain reasonably astute and speak the same technical language used by your staff and researchers, and you must be able to interpret new findings in lab animal science, but now your main responsibilities are to move your division forward whenever possible, ensure that the daily problems facing animal facilities are dealt with in a timely manner, and at the same time, maintain the provision of high-quality service for your customers. Good management takes significant effort, and good leadership demands the same.

Leadership Succession

Without doubt, you have to be competent and able to speak for yourself to rise through the leadership ranks, but once you are a leader, especially if you are an animal facility director or other high-level leader, you have to start thinking about identifying a person within your organization who will be able to step into your shoes should you retire or move on. When I first came to the University of Massachusetts Medical School, my retirement was at least 10 years away, and therefore I told the school that my intent was to stay on the job for at least 10 more years. However, one of my first priorities was to consider if anyone on the existing staff might be able to replace me when I did retire. But what does "be able" encompass? Certainly, there were people present who were competent in their current position, but did any of them desire to move up into a higher management position? Did any of them want to learn more about management? Who appeared to be a quick learner? Who could keep calm and collected under pressure and be able to make the right decisions at the right times? Who was willing to do more than their job required? Who had a vision for the future of the department? Was anybody already considered a leader by their peers? Fortunately, there was such a person.

These are some of the key questions a leader has to ask when planning for his or her own succession. This can be a challenge because there may not be any one person who can fulfill all of the leadership requirements now, but with training and mentoring, there may be one or more people who can successfully climb the management ladder. It is also challenging because the work in an animal facility often does not allow people to rotate through many different positions where they can learn different skills and demonstrate their existing management skills. Still, it is possible to give people increasing levels of responsibility and evaluate their adaptability and leadership potential.

In most instances, the person who is serving in an animal facility leadership position does not have the authority to appoint her own successor. But that isn't a death blow. If you are the incumbent person, you have the responsibility to mentor and promote the career of the person you believe is qualified to become your successor. You have to make sure that person serves on committees, meets with your own boss, and in general, is given strong visibility within and outside of the animal facility. Of course, if you do not believe there is any internal candidate who can grow to become your successor, then you should not be advocating for anybody. You want to leave on a high note, and the person you have mentored and recommended to be your successor should be at least as competent as you are and will be able to use the base you laid to build an even stronger animal facility.

Tying Together Leadership and Management to Accomplish Goals

As we approach the end of this book, I hope you've gained some valuable information about management and leadership. One of those lessons is that a manager uses resources to reach goals, and leaders and managers attempt to motivate people (a resource) to reach those goals. We've also discussed the basic needs of people (e.g., fair salary, good supervision, good working conditions, recognition, and a feeling of accomplishment), along with the need for open communication and trust. We said that leaders are ultimately responsible for developing a vision statement (a written statement of a major direction or action the organization will be taking over a defined time period). The leader has to work with his or her coworkers and develop goals and strategies to accomplish the vision. Now, let's take a look and see just how a leader gets things going. The vision we'll use is one we've already discussed: decreasing the animal facility financial subsidy from 40% to 25% in two years. This is a very narrow vision statement, but let's use it as a simple example (Table 7.2). We'll make one change. In Appendix 1, I said that the specific goal of decreasing the subsidy was dictated by upper organizational management. For now, we'll assume that it was the animal facility's management that recognized grant money for animal research was limited due to federal cutbacks in research funding, and that the facility had to be proactive about rationing its finances before the school made even more drastic cuts to its budget.

Rob Rubino, our exemplary leader, demonstrated a combination of leadership and management skills. He wasn't born with those skills; he learned them through readings, meetings, the mentoring of others, and much on-the-job practice. He worked hard every day to implement what he had learned. Nothing magical was involved. He had the same frustrations, successes, and failures that you have had or will have, but he persevered. His inner motivation to be a better manager and leader today than he was yesterday kept him going, and eventually he climbed to the top

Table 7.2 Path to Leadership

Action	Commentary
1. Rob Rubino was a seasoned laboratory animal veterinarian but new to Great Eastern University, where he now directed the animal facility. He quickly discerned that the 40% subsidy that supported the animal facility was an unnecessary financial drain on the school because the facility could be run more efficiently and effectively. He also knew, from meetings and what he read, that federal funding for research grants was being curtailed. This would further impact the school's and animal facility's finances.	1. Rubino evaluated the external political and economic environment (the federal government) and the internal environment of his school. He used information resources to determine there was a looming financial threat to the animal facility, and he recognized an opportunity to overcome it.
2. Rubino met with his managers and asked for their opinions. He was calm, collected, and business-like. Did they think this was a real problem? What might happen if they did nothing? What might happen if they did something? They were pleased that Rubino was including them in the discussion.	2. Rubino showed leadership, communication, and trust for the opinions of his managers. He organized and planned. He was self-regulated. He tried to align his management team while adapting to the needs of the school and his department.
3. Rubino listened to the managers' input. Based on what he heard, he suggested that they cut the subsidy from 40% to 25% in two years. Then he asked for more feedback. Could it be done? Would another figure be more realistic? He was enthusiastic. His managers were accepting the developing vision.	3. Good communication and good leadership. At this point, he suggested but did not demand. He was motivated. He was building a coalition of support.
4. Rubino and his most senior managers developed specific goals that had to be reached in order to fulfill the vision (these are detailed in Appendix 1). Feedback from technicians and other coworkers was solicited and evaluated. The managers then delegated to others in the animal facility the responsibility for developing strategies to reach the goals. Other managers were put in charge of developing means to measure progress toward the goal.	4. Delegation is a proper use of time resources and demonstrates trust. Rubino has to make available any needed resources. The vision was clear, not complicated. The best goals are those for which progress can be measured.

Table 7.2 (Continued)   Path to Leadership

Action	*Commentary*
5. The vision to decrease the subsidy to 25% in two years was communicated verbally to everybody, reinforced at meetings, and posted on bulletin boards. Rubino himself was the bellwether. He encouraged people to take reasonable risks to meet their goals.	5. For a vision to remain viable, it has to be reinforced, especially by the leadership. This entails communication and motivation.
6. There were biweekly meetings to report progress. Charts were publicly displayed showing progress toward reaching specific goals.	6. This reinforces goal fulfillment, gives people a sense of accomplishment, and allows for strategy corrections as needed.
7. Rubino continued his support of the program and made sure those groups or individuals making unique progress were rewarded. He would chat with people in hallways and keep his door open for them to chat with him. He occasionally showed up at group meetings to lend his support and make sure the group had what it needed.	7. Rewards, when used carefully, can be motivators. Social skills are important in leadership. Rubino was empathetic and supportive, but did not micromanage.
8. Rubino quietly ensured that his own boss and other key persons were aware of the efforts and progress in the animal facility. He gave the credit to his coworkers.	8. Good politics. You will always get part of the credit for a job well done. Keep your boss informed about important happenings.

of his profession. His work will not be done until the day he retires, but while he is there, he wants to be the best. And he wants to teach the person who follows him to be even better.

References

1. Ramaswamy, K., and Youngdahl, W. 2013. Are you your employees' worst enemy? *Strategy+Business* 73: Winter. www.strategy-business.com/article/00222?pg=all& tid=27782251 (accessed March 17, 2015).
2. Kotter, J.P. 1990. What leaders really do. *Harvard Bus. Rev.* 68(3): 103.

3. Clark, D.R. Concepts of leadership. Performance Juxtaposition Site. http://www.nwlink.com/~donclark/leader/leadcon.html (accessed March 7, 2016).

4. Howell, J.P., and Costley, D.L. 2006. *Understanding Behaviors for Effective Leadership.* 2nd ed. Upper Saddle River, NJ: Pearson Prentice Hall.

5. Kouzes, J.M., and Posner, B.Z. 2008. *The Leadership Challenge.* 4th ed. Hoboken, NJ: John Wiley & Sons.

6. Kotter, J.P. 2007. Leading change: Why transformation efforts fail. *Harvard Bus. Rev.* 1: 4.

7. Cuno, S. 2008. *Prove It before You Promote It: How to Take the Guesswork Out of Marketing.* Hoboken, NJ: John Wiley & Sons.

8. deVries, R.E., Bakker-Pieper, A., and Oostenveld, W. 2010. Leadership = communication? The relations of leaders' communication styles with leadership styles, knowledge sharing and leadership outcomes. *J. Bus. Psychol.* 25: 367.

9. Bolman, L.G., and Deal, T.E. 2014. *How Great Leaders Think: The Art of Reframing.* Hoboken, NJ: John Wiley & Sons.

10. Denning, P.J., and Dunham, R. 2010. *The Innovator's Way: Essential Practices for Successful Innovation.* Cambridge, MA: MIT Press.

11. Buckingham, M. 2005. What great managers do. *Harvard Bus. Rev.* 83(3): 70.

12. DeSteno, D. 2014. Who can you trust? *Harvard Bus. Rev.* 92(12): 112.

13. Maxwell, J. 2005. *The 360° Leader: Developing Your Influence from Anywhere in the Organization.* Peabody, MA: Thomas Nelson Publishers.

14. Daniels, A.C., and Daniels, J.E. 2006. *Measure of a Leader: The Legendary Leadership Formula for Producing Exceptional Performers and Outstanding Results.* New York: McGraw-Hill.

15. Blickle, G., Kane-Frieder, R.E., Oerder, K., Wihler, A., von Below, A., Schutte, N., Matanovic, A., Mudlagk, D., Kokudeva, T., and Ferris, G.R. 2013. Leader behaviors and mediators of the leader characteristics—Follower satisfaction relationship. *Group Organ. Manage.* 38: 601.

16. Goleman, D. 2004. What makes a leader? *Harvard Bus. Rev.* 82(1): 1.

17. Zenger, J.H., Folkman, J.R., Sherwin, R.H., and Steel, B.A. 2012. *How to Be Exceptional: Drive Leadership Success by Magnifying Your Strengths.* New York: McGraw-Hill.

18. Anonymous. 2006. A conversation with … Warren Bennis. *Harvard Manage. Update.* 11(12): 9.

19. Kramer, R.M. 2006. The great intimidators. *Harvard Bus. Rev.* 84(2): 88.

20. Cross, R., and Thomas, R. 2011. A smarter way to network. *Harvard Bus. Rev.* 89(7/8): 149.

21. Kramer, R.M. 2002. When paranoia makes sense. *Harvard Bus. Rev.* 80(7): 62, 124.

22. Kramer, R.M. 2003. The harder they fall. *Harvard Bus. Rev.* 81(10): 59, 136.

23. Maxwell, J.C. 2007. *The 21 Irrefutable Laws of Leadership: Follow Them and People Will Follow You.* Nashville: Thomas Nelson.

24. Nisbet, M.C., and Scheufele, D.A. 2007. The future of public engagement. *The Scientist.* http://www.the-scientist.com/article/home/53611/ (accessed March 7, 2016).

25. Farkas, C.M., and Wetlaufer, S. 1996. The ways chief executive officers lead. *Harvard Bus. Rev.* 74(3): 110.

26. Collins, J. 2005. Level 5 leadership: The triumph of humility and fierce resolve. *Harvard Bus. Rev.* 83(7): 136.

27. Surowiecki, J. 2005. *The Wisdom of Crowds.* New York: Anchor.

28. Ciampa, D. 2006. Getting good advice. *Harvard Manage. Update.* 11(12): 1.

29. Sulkowicz, J.F. 2004. Worse than enemies: The CEO's destructive confidant. *Harvard Bus. Rev.* 82(2): 64.

30. Garvin, D.A., and Margolis J.D. 2015. The art of giving and receiving advice. *Harvard Bus. Rev.* 93(1–2): 61.

31. Offermann, L.R. 2004. When followers become toxic. *Harvard Bus. Rev.* 82(1): 54.

32. Soyer, E., and Hogarth, R.M. 2015. Fooled by experience. *Harvard Bus. Rev.* 93(5): 73.

33. Anonymous. 2010. Defend your research: Powerful people are better liars. *Harvard Bus. Rev.* 88(5): 32. Also see Carney, D.R., Dubois, D., Nichiporuk, N., ten Brinke, L., Rucker, D.D., and Galinsky, A.D. The deception equilibrium: The powerful are better liars but the powerless are better lie detectors. http://faculty.haas.berkeley.edu/dana_carney/deception.equillibrium.ms.and.ols.pdf (accessed March 7, 2016).

34. Jacquart, P., and Antonakis, J. 2014. When does charisma matter for top-level leaders? Effect of attributional ambiguity. *Acad. Manage. J.* 58: 1051.

35. Birnbaum, J.H. 2007. The Humane Society becomes a political animal. *Washington Post*, January 30, p. A15.

36. Blum, D.E. 2007. Animal charity extends its reach by joining forces with others. *Chronicle of Philantrophy* 20(2): 50.

37. Pacelle, W. 2011. I hadn't factored in the pressure from the lobbyists. *Harvard Bus. Rev.* 89(4): 98.

38. D'Alessandro, D.F., and Owens, M. 2008. *Executive Warfare: 10 Rules of Engagement for Winning Your War for Success.* New York: McGraw-Hill.

39. Garfinkle, J.A. 2011. *Getting Ahead: Three Steps to Take Your Career to the Next Level.* Hoboken, NJ: John Wiley & Sons.

40. CareerBuilder. 2014. http://www.careerbuilder.com/share/aboutus/pressreleasesdetail.aspx?sd=8%2F7%2F2014&id=pr837&ed=12%2F31%2F2014 (accessed May 21, 2015).

41. Zenger, J.H., Folkman, J.R., and Edinger, S.K. 2011. Making yourself indispensable. *Harvard Bus. Rev.* 89(10): 85.

Appendix 1: Productivity Goals and Measurements

> The productivity of work is not the responsibility of the worker but of the manager.
>
> **Peter Drucker**

> If a manager's expectations are high, productivity is likely to be excellent. If his expectations are low, productivity is likely to be poor.
>
> **J. Sterling Livingston**

In 2008, it was estimated that actively disengaged employees (i.e., those that are unmotivated) within the U.S. workforce cost more than $300 billion in lost productivity [1]. That's a lot of money. In 2012, it was found that only 35% of the 32,000 people in a worldwide survey were highly engaged in their work [2], and in 2013, a Gallup study found that only about 30% of American workers were engaged in their work [3]. Now, consider that most animal facilities require a monetary subsidy from their parent organization in order to break even financially, and that subsidy is needed whether or not the workforce is motivated to work efficiently and effectively. The actual dollar amount of the subsidy is related to productivity, with lower employee dedication to their work leading to poorer productivity and a higher subsidy. The opposite statement is that good productivity requires a motivated, engaged workforce and can lead to a lower institutional subsidy. (See the discussion of motivation in Chapter 3.)

Characteristics of a Productivity Measurement System

Laboratory animal facilities perform an important service for their parent organizations. This holds true for animal facilities in profit-making and not-for-profit organizations. At best, most animal facilities try to run efficiently and break even financially (with or without a financial subsidy from the parent organization). Nevertheless, one reality about running an animal facility is that there has to be a balance between operating efficiently (getting the most output from the least input) and operating effectively (having a high-quality output). There are financial and nonfinancial aspects of efficiency and effectiveness.

> *Consider an animal facility director who buys 1000 mouse cages from Vendor A, saving about $10 a cage compared with purchasing the same-size mouse cage from Vendor B. On the surface, she's being efficient because she saved $10,000. There was more output (cages) for less input (money). In fact, if the cages from both vendors were made of the same material, were of the same quality, and met the needs of the animal facility, then indeed she was an efficient manager. But if Vendor A provided standard polycarbonate cages and Vendor B provided high-temperature polysulfone cages, and if the facility needed to repeatedly autoclave the cages, then purchasing the polycarbonate cages wasn't such a bargain. They will likely break down more quickly than the polysulfone ones and therefore have to be replaced sooner. The cost efficiency may have evaporated. The director should have balanced efficiency against effectiveness.*

Now, let's take a look at a nonfinancial example of the balance between efficiency and effectiveness. If we can put 500 cages an hour through a tunnel-type cage washer, most of us would agree that's pretty efficient by today's standards. Of course, if most of the cages come out dirty, we would also agree that we're not being particularly effective. But, if we slow down the cage washing process to the initial point where essentially every cage comes out clean (perhaps 250 cages an hour), then we've reached the proper balance between efficiency and effectiveness, at least with the current cage washers.

You can probably think of other examples, but it's fair to say that animal facilities have to operate their financial and nonfinancial activities both efficiently and effectively. Taken together, determining how efficiently and effectively an organization uses all of its financial and nonfinancial resources to reach its goals is termed *productivity* [4]. A highly productive animal facility is highly efficient and effective.

An organization with high productivity is highly efficient and effective.

You can visualize the connection between efficiency and effectiveness from the following formula:

$$\text{Productivity} = \text{Efficiency} + \text{Effectiveness}$$

The formula simply says that as efficiency and effectiveness increase, so does productivity, and of course, if efficiency or effectiveness decreases, so does productivity. Let's take this formula a step further and arbitrarily assign numbers to efficiency and effectiveness. A zero could indicate a very inefficient (or ineffective) process, and we'll say that a 50 indicates a very efficient (or effective) process. Therefore, there can be a maximum productivity score of 100 (50 for efficiency + 50 for effectiveness). If either efficiency alone or effectiveness alone were to drop below about 25 (which we can arbitrarily define as being average efficiency or effectiveness), it would suggest a need for improvement because we don't want good efficiency but poor effectiveness or good effectiveness with poor efficiency. In the best of all worlds, we will strive for a score of 100, but that's usually unrealistic. Nevertheless, we would never want to see either efficiency or effectiveness fall below 25.

Is measuring productivity common in animal facilities? The reality is that most laboratory animal facilities do not measure productivity numerically, although it's not that hard to do. The key is to develop a goal and then measure progress toward the goal, as described elsewhere in this appendix. Here's a simple example: Let's say you have one cabinet-style cage washer and your goal is to wash 10 full loads of cages a day. You want 100% of the cages to come out clean. If you wash eight loads a day, you are at 80% efficiency toward reaching your goal. If all of those cages are clean, you have 100% effectiveness. This is an easy-to-understand example, but as you read further, you will find that at times, measuring productivity isn't quite that easy.

Historically, most organizations (including animal facilities) have relied primarily on financial measures to determine the overall productivity of its operations. I believe that doing this was a holdover from a corporate accounting mentality where the "bottom line" was of primary importance. However, in service organizations, such as our animal facilities, the quality of the service (effectiveness) is a major factor to be evaluated when considering overall productivity, perhaps more so than efficiency. We all know it's easier to count the number of cages coming out of the cage washer than it is to determine if the cages are clean. Yet, eventually we have to come to grips with the problem and determine what the word *clean* really means. Is it a matter of just looking at a cage, or do we have to count bacterial colonies cultured from its inner surface or use an adenosine triphosphate (ATP) detector as a surrogate?

Now, let's look at some trickier measurements. How do you measure the quality (effectiveness) of a veterinarian's output? Or, if you have a manager who is developing a long-range plan, is there a good way to measure that person's output? What we need in our laboratory animal facilities is a system to measure progress toward our productivity goals that incorporates important financial *and* nonfinancial parameters, and perhaps gives a little more weight to effectiveness over efficiency. If I had to make an educated guess, I would suggest that our measurements are about 60% focused on effectiveness and 40% focused on efficiency. But even more important is to be able to ensure that everybody is in agreement on how productivity will be measured. I'll say more about this in just a moment, but for right now, I will point out that once agreement is reached on how productivity will be measured, peer pressure can help spread best practices toward reaching whatever goals are set [5].

A productivity measurement system is part of an overall management control system in which we develop important goals, develop the strategies to be used to reach those goals, and of course, use the measurements to gauge our progress toward reaching those goals. Yet, it takes more than mere words to measure productivity; it takes commitment. As managers, we have to be self-motivated and continually promote productivity; otherwise, whatever efforts we expend will most likely lead to failure. We have to believe that the performance goals we set are important to us and our animal facility, and we are going to provide as much support as possible in achieving those goals. Productivity goals require the support of the animal facility's upper management to give the goals and their measurement the organizational backing needed to implement them.

Productivity measurement is part of a management control system that managers use to gauge the progress toward reaching goals.

Measuring productivity is not a passing fad. Every manager has to do this to determine if his business is progressing toward its goals. Basically, productivity measurements "help managers translate strategies into measurable actions and meaningful business results" [6]. As a practical matter, a productivity measurement system should have six important features. Based on the earlier discussion in Chapter 1, some of this will be a review.

1. *The goals to be achieved will be aligned with the mission of the animal facility and its parent organization.* A good mission statement outlines what your organization does and who its main customers are. For a laboratory animal facility, we don't want to waste our time with measurements that are not aligned with our animal care mission and the mission of our parent organization. These measurements eventually should bring added value to either or both of our two key customers, animals and investigators. For instance, it might be valuable to measure our progress toward decreasing the number of investigator complaints because that is obviously aligned with our animal care mission, but it would probably serve no useful purpose to have goals and

related measurements for increasing the number of white versus black mice in the animal facility.

2. *The goals will be specific.* If we don't have clear goals, we don't know what we're trying to achieve, and it makes little sense to try to measure progress toward reaching unclear goals. A goal such as being the "best in the business" is nice, but not particularly specific. On the other hand, a goal such as "decreasing our annual personnel turnover by 20% over the next two years" is quite specific.

   Goals are not items that can be accomplished overnight; they usually have to be reached over time. For example, changing the temperature setting on a cage washer can be done almost immediately by pushing some buttons. Yes, it's a goal, but it's not much of a goal. On the other hand, decreasing annual personnel turnover by 20% will take some time and may be a legitimate goal for many animal facilities.

3. *There should be clearly defined strategies for reaching the goal.* There will probably be different strategies and goals for different working groups within the animal facility, but when we put everything together, the various strategies should result in reaching the overall goal of the animal facility. Some examples are provided later in this appendix.

4. *There will be a limited number of items we have to measure.* We want to evaluate only those activities that are really central to enhancing productivity and provide value to our key customers. We don't want to waste time measuring interesting but nonessential items. For example, let's say you have goals of cutting personnel turnover from 25% a year to 10% a year, decreasing overtime by 20%, increasing American Association for Laboratory Animal Science (AALAS) certification by 20%, starting a section of comparative medicine, and repairing the ceilings in 10 animal holding rooms. Those are a lot of important goals, but which ones are really the most important? Which ones enhance productivity and provide value to our customers? You have to prioritize the importance of your goals and develop measurements for a few (perhaps your top three), not each and every one.

5. *A good productivity measurement system will be numerical.* A little earlier in this appendix, I mentioned that it is important to have everybody in agreement on how productivity measurements will be made. It is obviously more accurate to be able to count bacterial colonies or interpret a readout on an ATP detector than to look at a cage and try to guess how many colonies survived going through the cage washer. For ease of understanding, numbers on a graph or table make our comprehension much easier. Measuring outcomes in numerical terms can be difficult, particularly when evaluating both financial and nonfinancial goals. This is because many of the nonfinancial goals are quality, not quantity, oriented.

   Here are just a few examples of laboratory animal facility activities that are amenable to numerical measurements. Some are primarily measurements of efficiency, and others are primarily effectiveness measures.

Activity	*Measurement(s)*
• Time from placing an animal order to the time the animals are received	• Efficiency
• Time to take an animal census	• Efficiency
• Number of census errors	• Effectiveness
• Time from taking the animal census to the time bills are sent out	• Efficiency
• Number of *per diem* billing errors	• Effectiveness
• Time from placing a request for veterinary services to the time the service is provided	• Efficiency
• Time needed to clean a given number of mouse cages	• Efficiency
• Number of mouse cages that need recleaning	• Effectiveness
• Reduction in the number of cracked or discolored cages in use	• Effectiveness
• Customer satisfaction with veterinary services (via customer survey)	• Effectiveness
• Customer satisfaction with front office services (via customer survey)	• Effectiveness
• Employee satisfaction with specific departmental personnel practices (via survey) [7]	• Effectiveness
• Number of publications during the past year	• Efficiency
• Impact quality of journals in which publications appear (define what impact quality number is desired from published lists)	• Effectiveness
• Number of sick days used	• Efficiency • Effectiveness
• Number of investigator complaints	• Effectiveness
• Number of negative findings arising from U.S. Department of Agriculture (USDA) or Institutional Animal Care and Use Committee (IACUC) inspections	• Effectiveness
• Dollar amount of a budget subsidy from the parent institution	• Effectiveness
• Total income versus budgeted income; total expenses versus budgeted expenses	• Effectiveness • Efficiency

Activity	Measurement(s)
• Number of people taking an AALAS certification exam	• Effectiveness
• Number of new AALAS certifications versus the number of people taking the certification exams	• Effectiveness
• Number of on-the-job accidents	• Effectiveness
• Number of unneeded inventory items	• Efficiency • Effectiveness

6. *The system should be able to use past data (what we have just measured) as a reasonable indicator of future productivity.* Collecting data is interesting, but if it can't predict anything about the future, it will have limited importance to us. For instance, let's say that last year the average number of mouse cages that a person was able to change both efficiently and effectively was 300 cages per day. This year, we purchased 10 new cage racks and found that using the new racks, an average person changed 350 cages per day. We can say that the data we collected from the new racks predicted that they could lead to changing 50 additional cages per day. In this example, a simple count of the number of cages changed was able to predict future productivity. That's all fine and good, but before you run out and buy more of the new cage racks, consider that most animal facilities only compare their productivity changes to what was measured in their own animal facility. Three hundred fifty cages per day is certainly better than 300 cages per day; however, if employees at a nearby animal facility that has the same cage racks and the same general room conditions that you have routinely change 450 cages per person per day, then your facility's productivity still may not be up to snuff. The new racks may have helped, but you may want to talk to your colleagues at the nearby facility and see if you can determine why they are more productive than you are. Perhaps better training will be found to be more important than the number of the cage racks. As a manager, you should consider benchmarking productivity outside of your own vivarium [8].

As you might suspect, people have been trying to measure productivity for years. One of the most recent iterations of productivity measurements, the Balanced Scorecard [9], began in the 1990s and its still going strong. At its bare bones, the Balanced Scorecard is a management control system because it allows a manager to maintain oversight of activities as they progress. Over the years, it has morphed from a measurement system into a combination of a measurement and management control system [10]. We can incorporate many of the concepts of the Balanced Scorecard and other measurement systems to develop a productivity measurement and enhancement system for laboratory animal facilities. Similarly, *Lean management* is beginning to be incorporated into laboratory animal facilities (and is discussed later in this appendix), but in reality, these systems still require basic, good, day-to-day management skills that have been reworked into a system of measuring productivity.

Summary of Important Features of Productivity Measurement

- **Goals to be measured are aligned with your mission statement and will bring value to your customers.**
- **The goals to be measured are specific.**
- **Each goal is associated with clearly defined strategies that will be used to help reach the goal.**
- **The number of items to be measured is limited to those that are central to enhancing productivity.**
- **Measurements are numerical.**
- **Collected data will help predict future outcomes.**

Linking Productivity Measurement to Specific Goals and Strategies

We can't escape the management adage "Where are we now? Where do we want to be? How are we going to get there?" The same adage applies to productivity measurements. If we don't know what our goal is (i.e., where we want to be), then how can we measure progress toward that goal? It doesn't matter how much money we have available, how many people are available, or how much fancy equipment or space is available; we still have to know what our goal is before we can develop and evaluate the strategies and resources needed to reach the goal. And then, of course, we have to be able to measure progress toward reaching the goal.

We'll use as an example an animal facility that historically gets a 40% revenue supplement from its parent organization in order to balance its annual revenues against its annual expenses. That is, if it wasn't for the 40% financial subsidy, the animal facility would be losing money. Responding to a decision by corporate management, the animal facility's management tells the staff that they have to cut the subsidy down to 25% in two years. Now, we can argue at some other time if upper management should have taken the lead in making the decision to set that goal, or if the animal facility's staff should have set a goal of reducing the subsidy, but let's not face that issue now. The bottom line is that a decision was made to cut the subsidy to 25% in two years. That's our goal because the big bosses said so. It's pretty easy to measure progress toward that goal, isn't it? Our starting point is the current 40% subsidy, so it is an obvious financial-based measurement if we can show that it decreased to 30% in the first year. Thirty percent might be a pretty good first-year goal. If, by the second year, the animal facility manages to reduce the subsidy to 25%, it met its goal. Good for the animal facility; it was very efficient and we were readily able to measure that efficiency. But how did this happen? What were the specific strategies the animal facility had to take to reduce the subsidy? We know it was efficient, but was it effective in maintaining the mission of the facility? Did the quality of our output suffer? How was effectiveness measured?

We'll see now if we can integrate the six ideal productivity measurement parameters described previously into an animal facility's goals. Let's use the example given above, which was to decrease the animal facility's annual operating subsidy from 40% to 25%. If the animal facility's total annual expenses are $100,000 and its revenues are $60,000, then the parent organization has to provide $40,000 as a revenue subsidy to balance revenues against expenses ($100,000 expenses – $60,000 revenues = $40,000 needed revenue subsidy). Since the goal is to lower the subsidy to $25,000, we have to either lower our expenses, increase our revenues, or both. Although the end result is going to be financial in nature, we have to balance financial and nonfinancial productivity goals. If we didn't have to do that, we could be callous, lay off a person, and save the cost of his or her salary and benefits. That would be an efficient way to save money. But, if that person is truly needed to care for animals, it would not be a particularly effective strategy because quality would suffer. We need balance.

> *Great Eastern University is a large, multidisciplinary academic institution with major strengths in business administration and biomedical research. The university's stated five-year goal is to become the generally recognized leader in academic financial efficiency. The medical and veterinary schools, along with all the other schools within the university, are included in this goal.*

This so-called goal of the university is really a vision statement, and we've discussed vision statements in Chapter 1. If Great Eastern wants to call it a goal, that's just fine with us, but we know it's really a vision statement because it gives us the general direction the university wants to take over the next five years. To get there, it will require all the schools of the university to have their own goals, which are needed to fulfill the Great Eastern vision. From this university-wide financial goal, we can begin to understand why the animal facility was given a specific goal to lower its operating subsidy down to 25%. We can also see that the university's goal didn't say anything about maintaining effectiveness, probably because it was assumed that effectiveness had to be maintained. Therefore, we'll assume that effectiveness is important. It also was not as specific as a good goal should be (what does a "generally recognized leader" really mean?), but that doesn't suggest that the animal facility's goal has to be equally nebulous. In fact, I would say that lowering the subsidy to 25% is pretty specific. Take a look at Figure A1.1, which shows productivity relationships within the Great Eastern University laboratory animal facility.

Figure A1.1 shows that Great Eastern University has a specific expectation (goal) for the animal facility (which, as we now know, is to have a 25% financial subsidy). The university also has a more general expectation to be a national model of financial efficiency. Because the vertical arrows point in two directions, the

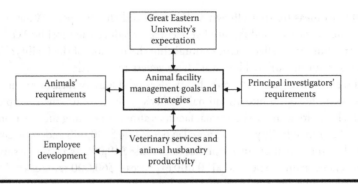

Figure A1.1 Productivity relationships. (*From* Chang, R.Y., and Morgan, M.W., *Performance Scorecards,* San Francisco: Jossey-Bass, 2000. Modified with permission.)

diagram suggests that the goals of the animal facility and the university have to be in concordance. That is, the animal facility must have a mission, vision, and goals that mesh with the university's mission, vision, and goals, and the animal facility must develop a financial goal that fits with Great Eastern's financial goal. Likewise, the goals of the animal husbandry and veterinary services groups have to be aligned with the overall goals of the animal facility. We also see two-way horizontal arrows between the animal facility's goals and strategies and the facility's two key customers, which are the principal investigators and the animals. Those two-way arrows suggest that the goals and strategies of the animal facility are tightly linked to the needs of the investigators and animals. Finally, the diagram shows that employee development (such as education, training, and having a career ladder) is important to achieve high levels of animal husbandry and veterinary services productivity.

To summarize the concepts shown above, we can say that productivity goals have to be viewed vertically and horizontally. Vertically, the productivity goals of the university, animal facility, veterinary services, and animal husbandry have to be aligned so that all the goals lead to a reduction of the animal facility's subsidy to 25%. Horizontally, the animal facility can't focus on saving money to the exclusion of serving its two customers, animals and investigators.

Going back to the six areas of importance that were previously described, we will ask the following questions that are central to productivity and its measurement:

1. Does the goal to be measured (reducing the school's subsidy from 40% to 25%) align with the mission and vision of the animal facility and with the mission and vision of Great Eastern University? This is what you might call a baseline question, because if the answer is no, it has to be rectified and changed to yes before moving forward with the measurement of progress toward specific goals. In this case, the answer is probably yes, although we never stated the specific mission of Great Eastern University and its animal facility. Even so, the mission of every university includes the service of its

faculty (to the school and community) and teaching. Likewise, the mission of every animal facility includes the provision of service to the animals and investigators of the school, and ultimately to humans and other animals. Very often, the animal facility has to take a role in teaching investigators and other employees about animal care and use issues. Therefore, there is reasonable concordance between the missions of the university and its animal facility. If somebody suggested that the animal facility should make and sell pizzas on the weekends, we would have a lot of trouble integrating that idea into the educational and service missions of the school and the animal facility. I hope we can all agree that the financial goal of the animal facility (the reduction of the subsidy to 25%) meshes with the financial vision of Great Eastern (to become a leader in academic financial efficiency).

2. *Are there specific goals to be reached, so we know what parameters we have to measure on route to reaching that goal?* Here again, the answer is yes. Great Eastern has set a goal of becoming a generally recognized leader in academic financial efficiency. Given Great Eastern's strength in business administration, we'll accept it as being a reasonable, albeit somewhat nebulous, goal. The animal facility has been given a related but more specific goal: to decrease its financial subsidy to 25% of its total operating funds within two years. Each subdivision (animal husbandry, veterinary services, and administration) will have somewhat different but related goals that, when taken together, hopefully will allow the animal facility to reach that 25% mark.

In the goals I will describe below, you will see that I grouped all of the animal care technicians together as if they were one person, and I did the same for all of the veterinary technicians. A particular working group of people can be considered one person if the people in the group all do pretty much the same work (such as animal care technicians who care for the mice in a particular building). You will also notice that specific goals are linked to specific people. This is an important concept because if individuals are not assigned specific goals, then nobody is really responsible for those goals, and there's a good chance that they will never be properly addressed. This starts a dangerous slide downward because as you know, we are evaluated by the results we generate, not by lofty goals or strategies. If we don't produce acceptable results, we have failed.

Specific individuals should be associated with specific goals.

Here is a listing of some potential specific goals to reduce the subsidy.
- Goal for the animal facility as a whole
  - Decrease the school subsidy to 30% in the first year and 25% in two years.

- Goals for the animal facility director
  - Decrease overtime by 20%.
  - Increase revenues by 5%.
- Goals for the animal care division supervisor
  - Decrease overtime by 20%.
  - Increase number of cages washed per day by 10%.
  - Have no more than 1% of cages in need of rewashing.
- Goal for the animal care technicians in Building A
  - Get all dirty cages to the cage wash room before noon.
- Goals for the veterinary services division supervisor
  - Decrease overtime by 20%.
  - Increase fee-for-service income by 10% in one year.
- Goal for the veterinary technicians in Building B
  - Respond to all morning sick animal reports by 11:00 a.m.

These are interesting goals, but how do we know that they are realistic? How do we know, for example, that having less than 1% of cages in need of being rewashed is achievable or even important? Is it really possible to have the veterinary technicians respond to all sick animal reports before 11:00 a.m.? Even if we assume that these goals are realistic, we have to be sure that they relate to the specific goals of other people or groups above and below in the organizational chain of command. In other words, will responding to all sick animal reports by 11:00 a.m. help the veterinary services supervisor with his or her goal to decrease overtime by 20%? In turn, will decreasing overtime by 20% have any significant impact on the animal facility director's goal of decreasing total overtime by 20%?

Adapting the thoughts of Chang and Morgan [6] to laboratory animal science, there are at least four sources that we can use to help us set realistic performance goals:

- Statistical and historical data (e.g., information from laboratory animal science publications or our own institution)
- Customer requirements (i.e., requirements of the animals and investigators)
- Generally known, although not necessarily published, standards within the laboratory animal science field (such as the number of isolator-style cages one technician typically can change in one day)
- What your boss wants (a political reality we have to live with)

3. *Are there clearly defined strategies for reaching the goals?* Having strategies is critical. Goals without strategies are like plants without water; eventually, they wither and die. Since this is a hypothetical (although plausible) scenario, we can suggest some strategies for reaching the goals shown above. Here are just two examples.

- Strategy for the animal care division supervisor to reach the goal of decreasing overtime pay by 20%

- Have a split-shift day. Observe people working. Do they have the cages, equipment, and supplies that they need and when they need them? If not, then changes in procedures are needed. If they do not have what they need, is the problem that the cage washers cannot process cages fast enough? If that is the problem, then a split shift might be the answer. Some people work from 7:00 a.m. to 3:30 p.m., while others work from 9:00 a.m. to 5:30 p.m. This keeps the machines working for an extra two hours per day without incurring overtime.

- Strategies for the veterinary services division supervisor to increase fee-for-service income by 10% in one year
  - Survey investigators about possible service needs.
  - Advertise monthly in the school newspaper about the availability of service.
  - Send a notice of service availability along with the monthly animal care bills.
  - Send emails to principal investigators about service availability.
  - Ensure that all technicians have appropriate skills to provide the extra services.
  - Increase the existing fee schedule by 2%.

You can see that many of these strategies don't require much detailed planning, although there will always be some effort involved. For example, you don't have to make grandiose plans to increase the existing fee schedule by 2%, but you do have to consider the timing of this increase and how it might affect the researchers. Likewise, it may be fairly easy to initiate split work shifts, but timing, having enough people to operate the cage wash facility, the possible need for supervisors to also have a split shift, union agreements, and other factors may have to be taken into consideration before you actually implement this strategy.

4. *Are we limiting the number of items we are going to measure to only those that are really central to productivity?* This is always a difficult question. We all have our own prejudices, so you can expect that there will be a lot of discussion needed to agree on the goals that we will eventually measure, which in turn will lead to reaching the overall goal of decreasing the subsidy to 25%. For any one group of people in our example (animal care technicians, veterinary technicians, or managers), it's probably not realistic to have more than three or four specific goals. Each goal can have one or more associated strategies, so the more goals you have, the more work you have to do to reach those goals, and pretty soon, you get so wrapped up in meetings and measuring your progress toward your goals that you don't have time for your regular work. So, my advice is to go easy on the number of goals you select.

In the example that I've been using, I've shown only a limited number of goals for each of the major divisions of the animal facility. Here again are the three goals described earlier for the animal care division supervisor:

- Goals for the animal care division supervisor
  - Decrease overtime by 20%.
  - Increase the number of cages washed per day by 10%.
  - Have no more than 1% of cages needing rewashing.

The question we are now asking is, are these three goals really important to the overall goal of decreasing the subsidy to 25%? The first goal, decreasing overtime, will almost undoubtedly save money, but the animal care supervisors and others will have to determine if the amount to be saved is of any real significance. If the facility is spending $150,000 a year on overtime, then a 20% decrease in overtime will save $30,000 a year. Depending on the cost of all salaries and benefits, that $30,000 might be a substantial savings. On the other hand, if the total overtime cost for the year was about $1500, then a 20% decrease would lead to savings of only $300. This would not have much of an impact for most animal facilities. Therefore, you can see that the concept of having a significant goal to increase productivity must be viewed in the context of your own animal facility.

An efficiency goal that is related to decreasing overtime is to increase the number of cages washed each day by 10%. This is linked to the strategy of having a split work shift. The more cages that can be washed per day without overtime, the more money that will be saved. But once again, human beings, not machines, will have to decide if the goal is really important. It will be interesting to see the strategy that the cage wash team develops to increase productivity. Maybe they'll figure out a way to get more cages through the cage washer in a shorter time, yet maintain the needed level of sanitation.

The final goal, having no more than 1% of cages in need of rewashing, is clearly based on effectiveness. But there's also an efficiency component, because the fewer cages we rewash, the less time it takes, and the animal facility will save the labor cost, chemical cost, water cost, and so forth. But what if normally there are no more than 1.5% of the cages needing to be rewashed? Would it make sense to try to lower that number by another half percent? Probably not, unless the total number of cages washed per day was extraordinarily large. Perhaps this goal is not of much significance and should be discarded.

The bottom line is that people have to make a decision about each strategy and its importance to the overall goal or goals. There's no way around this.

5. *In order to easily visualize progress toward our goal, can we make our measurements numerical?* This is not always easy, but almost always possible. Different work groups or different individuals may have to use some different measurements, but they are all related to the same goal, so here are a few possible measurements. Some are the same as those presented earlier in this appendix.
   - Let's look first at some possible measurements used by the *animal facility director* to gauge progress toward the goal of cutting the total animal

facility subsidy to 25%. The director has to see the overall picture, so his key measurements are going to differ in scope from the measurements of the people below him in the animal facility hierarchy. He will measure

- Total monthly and year-to-date income (a financial efficiency measurement)
- Total monthly and year-to-date expenses (a financial efficiency measurement)
- Total salary versus nonsalary income and expenses, evaluated monthly and year-to-date (a financial efficiency measurement)
- Results of an investigator satisfaction survey (a nonfinancial effectiveness measurement)
- Total employee turnover (a nonfinancial effectiveness measurement)
- Key personnel turnover (a nonfinancial effectiveness measurement)

– Next, we'll look at measurements used by the *animal care division supervisor* to gauge progress toward reaching the goals of cutting overtime by 20% and increasing the number of cages washed per day by 10%.

- Total overtime expense, monthly and year-to-date (a financial efficiency measurement)
- Number of cages washed per week (a nonfinancial efficiency measurement)
- Number of union grievances filed (a nonfinancial effectiveness measurement)

Lastly, let's look at the measurements used by the *veterinary services division supervisor* to measure progress toward reaching the goals of decreasing overtime by 20% and increasing fee-for-service income by 10%.

- Total overtime expense, monthly and year-to-date (a financial efficiency measurement)
- Total fee-for-service income, monthly and year-to-date (a financial efficiency measurement)
- Number of investigator requests for fee-for-service work (a nonfinancial effectiveness and efficiency measurement)
- Number of investigator complaints about the quality of the fee-for-service work (a nonfinancial effectiveness measurement)

6. *Are the data we are collecting reasonable indicators of future productivity (i.e., of future efficiency and effectiveness)?* Just because we *measure* productivity doesn't mean that there is a *correlation* with productivity. If productivity measurements were always strongly correlated with actual productivity, then we would not need as many measurements as we use [11]. We will certainly try to develop measurements that have the ability to help predict future productivity, but it may take a while to see if there is a correlation. The various financial measurements used in the above examples would appear to have the ability to do so, but only time will tell for sure. For example, we would like to be able to show that as total overtime expenses decrease, there is a direct correlation with a decreasing

total subsidy from Great Eastern University. Likewise, it would be nice if we can link a decrease in the number of union grievances filed to an increase in the number of cages washed per day. If, after a period of time, we cannot detect any useful correlation between our measurements and a desired outcome, it will probably be wise to look for a different measurable goal that lends itself to being able to correlate with our needs. We simply have to carefully choose our goals and how we measure progress toward fulfilling those goals.

How Does Lean Management Fit into the Productivity Picture?

Lean management [12] is a term that's frequently used in business circles and is finding its way into laboratory animal facility management. In its most basic form, it is a combination of management practices that are designed to get the most effective use from existing resources while using the fewest resources possible (i.e., becoming an efficient and effective vivarium). Lean management emphasizes the importance of all people who are involved with resource use (not only managers) by giving employees the authority to solve many of the problems that occur in their area of work. By doing this, the employee recognizes the importance of his or her work (which we know is an important motivator), is informed that his or her opinion is valued (which we also know is an important motivator of people), and saves time since the resolution of a problem may not need to go up and then back down the management ladder. Charron et al. [13] summed this up by writing that "Lean is a process management system that is controlled by the workers." A true Lean management program will start at the top of an organization (e.g., with the support of the CEO of your institution) and encompass everybody, down to the most recently hired person. If you are wondering if it's possible to have a Lean management program in the animal facility if it doesn't exist throughout the organization, the answer is yes, you can. Yet, it is unlikely that it will be as effective as when the entire organization participates in the program.

There are three basic pillars upon which Lean management is built: The first is continuous improvement of a service or product (this is known as *kaizen*). The second pillar is the elimination of waste (known as *muda*). Waste can be unnecessary movements, unneeded inventory, excessive time to complete a task, time spent waiting for something to happen, overly strenuous work, work processes that vary significantly on different days, and so forth. The final pillar is that any process must be of value to the end customer (such as animals or investigators). If you keep these three pillars of Lean management in mind, you will find it much easier to implement a Lean management system in your animal facility. There is one important caveat I'll give you about Lean management: it is not a magic bullet that will cure all animal facility problems. All of the basic management concepts discussed in this book hold true for Lean management. At its core, Lean management just puts these same concepts into a different terminology, but it is still just a basic good management program. I've often told people that Lean management is mostly the same good management practices that we've been teaching for years, but it's packaged differently. If you are interested in implementing Lean management in your animal facility, I strongly recommend that at the very least, you first read a few books on Lean basics (e.g., Charron et al. [13]) and consider using a Lean coach to get your program going. If you already have a well-run animal facility, you will be pleasantly surprised to find that you are already practicing many, perhaps most, of the ideas behind Lean management.

Lean management is a combination of management practices that are based on continuous improvement of the product or service, minimization of wasted resources, and providing value to the customer.

Lean management and high productivity are very similar concepts because we defined high productivity as being both efficient and effective in using financial and nonfinancial resources to reach a goal. Lean incorporates all of the basic management principles we've discussed, such as planning, goal setting, making decisions, organizing, directing, and establishing controls, in order to eliminate waste and any activity that does not provide value to the customer. You or your management team do this by examining all of your work processes for efficiency and effectiveness. For example, consider an animal facility that performs an animal census once a month by manually counting all its cages. Lean management would ask, where are we now? The answer might be that we do a monthly manual census that takes two hours of most every technician's time, and it is about 80% accurate. When errors are found, our customers (the researchers) are usually not happy. The next question is, where do we want to be? Here, the answer might be that we want to do a twice-weekly census in two hours per technician that is 99% accurate and may bring in more *per diem* money to the animal facility. Even if the twice-a-week cage census takes more total time, does the resulting income and researcher satisfaction more than offset the added labor? That's the question Lean management is asking, and subsequently, a strategic plan will be developed to reach the 99% goal. First, the management team might query other animal facilities to determine if they have experience with more frequent and more accurate means of doing a census. This is just basic background work. If the idea seems feasible, it is likely that the plan that will eventually be used will incorporate an electronic census method, such as bar code reading.

Now let's return to the question of whether taking an animal census twice a week, or even every day of the week, will result in more labor costs than will be recovered through more accurately recorded *per diem* charges. If it does lead to additional labor costs, then it would make sense to reconsider and perhaps modify the plan. At the same time, Lean management will not only focus on the financial result (the money an additional census may bring to the animal facility), but also consider the impact it has on the researchers.

Lean management uses the notion of kaizen, which in addition to the basic concept of continuous improvement noted earlier, emphasizes the *measurement* of continuous improvements in activities. This is no different than the continuing measurements of productivity that were described earlier in this appendix. Kaizen emphasizes the use of standardized activities (such as standard problem-solving methods or the use of standard operating procedures) and a measurement of those activities (e.g., How long does the activity take? How many animal health exams are performed per day? How many cages are discarded per day?) against the goal you have set, and then makes any needed changes to the activities in order to reach the goal. Once the goal is reached, the process begins anew to make an even more efficient goal that further reduces any residual wasted resources or activities. With kaizen, as with any good management paradigm, there is no such thing as good enough; there is a continual quest to do things better.

One interesting and useful aspect of Lean management is the use of just-in-time activities by some companies, particularly those that manufacture products. Just-in-time refers to keeping a minimal inventory of extra supplies or products, and therefore reducing the cost of buying or storing rarely needed inventory (this is reducing waste, or *muda*). In an animal facility, that may mean keeping a minimal inventory of excess clean mouse cages. When a customer (such as an animal care technician) puts in an order (e.g., asks for clean mouse cages), the company (i.e., the cage cleaning crew) will have the right number of needed supplies and will quickly fill the order. But just-in-time has a different benefit to the laboratory animal community; it allows us to investigate flaws that may be inherent in certain of our daily activities that are rarely analyzed for efficiency and effectiveness. Consider an investigator who had forgotten to send in a request to the animal facility to ship animals to a colleague across the country. The investigator asks if you can do him a favor and expedite the shipment because his colleague really needs the animals to finish his research. Normally, this process would take you at least a week (and probably longer) because the receiving institution has to be notified of the shipment, the receiving institution will likely want an animal health report for the room from which the animals are coming, you have to prepare the report and get permission from the receiving institution to ship the animals, material transfer agreements may need to be executed, shipping labels must be prepared, technicians must be informed to remove the animals (usually mice) from the specified cages and put them in shipping containers, you have to make arrangements to get the animals to the airport, and so forth. The investigator doesn't know this; he thinks you just put the mice in a box with some food and water and ship them off. But if you were to closely look at the entire process and ask for (and receive) feedback from all concerned parties with the goal of expediting the whole animal export process, perhaps you could eliminate some steps that would save you time and labor whenever you ship mice, not just when an emergency call comes in. You really didn't need the just-in-time call from the investigator to start the ball rolling, however, you may never had recognized a need to do things differently if that push hadn't come and if you were not always thinking about "How can I do this better?"

There are some aspects of Lean management that differ a little from the more traditional, or Western, style of management, and I'll just mention one that I see as being of importance to animal facilities. Earlier in this book, when discussing decision making, I wrote that one commonly used method of making a decision is to begin by brainstorming, that is, listing all possible reasons and solutions to a problem and then eliminating those that are most unlikely and focusing on those that have the most relevance. I also wrote that a final decision about the solution to a problem is rarely made in a formal manner, and it is more likely that the person who is highest on the chain of command and responsible for making or approving the decision often considers the problem in a less than formal manner. That means possible decisions get discussed over coffee with some trusted associates, or are considered in the shower, on a ski slope, and so forth. A clear and defined process for decision making may not have even been considered. Not so with Lean management. In Lean management, problems are addressed in a more formal and defined process, with much emphasis being placed on incorporating the knowledge of the people most closely associated with the problem into a final decision. This can mean that a veterinary technician who has been treating a mouse for dermatitis will meet with the attending veterinarian and clinical veterinarians. The conversation does not always start with brainstorming, but it can. The goal is to break the problem into its component parts to make it easier to address.

Let's take a look at an example of how a Lean manager might approach a common problem in animal facilities: not enough clean cages are available for the animal care technician to use during cage changing.

> *The cage-changing problem wasn't new. At least once a week, Joan White complained to her supervisor that there were not enough clean and bedded cages for her to use. White's supervisor would go to the cage wash area and ask the wash crew to work a little faster. The problem would be resolved for a few days, and then it would surface again.*
>
> *The problem continued until a new supervisor took over when the original one left the university. The new supervisor looked first at the end user (Joan White) and then worked backwards to find where the problem may have originated. He had four major considerations that were considered in a formal, logical process [14]:*
>
> 1. *The manpower involved (such as Joan White's efficiency, the cage wash crew's activities, and the cage wash supervisor)*

2. *The machinery involved (the cage washer and bedding filler)*
3. *The materials involved (cages, caging material, and bedding material used)*
4. *The method being used (when and how cages reached the cage wash team, and when and how the cages were washed and filled and made ready for Joan White's use)*

The new supervisor found that White's complaint was valid. She came to work on time and began work on time. There always was a stack of clean and bedded cages for her to use. She worked at a reasonable pace until she ran out of cages. Therefore, the new supervisor was able to determine that Joan White was not the problem, nor was the initial supply of cages part of the problem. The supervisor kept working backwards and turned his attention to the cage wash area and the people working there. He determined that the cage wash team worked efficiently and effectively as long as they had cages to wash. Looking at the machinery function and talking to the cage wash team made it clear that the machinery was working properly. However, the cage washers and the cage wash supervisor pointed out to the new supervisor that most of the plastic cages being used were more than five years old and constantly breaking as a consequence of frequent washing and autoclaving. This slowed down the processing of cages, and once or twice a week, it slowed it down so much that there weren't enough cages for the animal care technicians who were changing the cages. When that happened, the cage wash supervisor would add some brand new cages, and everything worked well for a few more days.

By methodically working backward and focusing on manpower, machines, methods, and materials, the new supervisor was able to analyze the individual parts of the problem, and by doing so, he determined that there were two related problems. The first was materials (the cages themselves), and the second was the method (cages were not being replaced on an annual basis to account for the usable life span of plastic animal cages). With this information in hand, the new supervisor was able to make a strong recommendation to the facility director to consider an annual replacement of about one-fourth of the plastic cages every year, since the typical life span of a polysulfone plastic cage is about four years. Once the problem was defined but not yet resolved, the director may have wanted to consider alternative solutions and may have had a brainstorming session with various people to eliminate some obviously weak solutions and focus on some solutions that seemed stronger. The problem-solving team may have looked at these possible solutions and started to dissect them in detail. For example, if certain assumptions were required for a potential solution to work, and any one of those assumptions were either false or not possible, then that idea would be discarded [15]. The take-home message from problem solving using Lean management methods is that logical thinking and a clear procedure for addressing a problem are worthwhile considerations for laboratory animal managers.

Of course, the example I used was a simple one, but even for more complicated problems, the lesson we learn from Lean management is a good one. When you are investigating a problem, try to get to its root cause by working backward to determine where the problem started. Is the problem caused by manpower, methods, materials, or machinery?

In summary, Lean management is not a magic bullet that makes an animal facility work more efficiently. It is a worthwhile management method that uses all the basic management skills and resources that are at your disposal. It is best to think of Lean management as a disciplined way of performing the daily activities of managers and nonmanagers. It has its place in animal facility management, but it still incorporates the basic management skills that are the focus of this book.

Getting the Productivity Measurement Process Moving

Hopefully, everything I have written above is interesting to you and makes intuitive sense, but a key question is, how do we get the productivity measurement process moving? Who takes the lead? In general, the leader can be within your organization if he or she is familiar with techniques such as the Balanced Scorecard or Lean management approaches, mentioned earlier, or you can use an outside consultant (the latter is a potentially costly but worthwhile approach). Yes, you can muddle through it yourself, but if you choose to do so, I strongly suggest practicing with one or two trial groups of people before you do it for real. It's akin to doing a pilot study before doing the full study. There is a learning curve, and jumping in without proper preparation is a recipe for disaster.

Now let's have a hypothetical meeting in the laboratory animal facility. In it, we'll see some typical comments about setting and quantifying goals. To make life a little easier for us, let's assume that John, the animal facility director, has experience in setting and quantifying goals, and therefore he is leading the discussion. It's also important to have a senior manager as part of the evaluation process because that person has the overview of the animal facilities resources and vision. The discussion may go something like this:

John: Colleagues, we have a challenge ahead of us to decrease our operating subsidy from 40% to 25% while maintaining or even increasing our productivity. Can we do it?

Sheila: The only way I know how to save money is either to cut back on services or increase our charges.

John: Do you think we're already operating at 100% efficiency?

Sheila: No, I don't know of any animal facility that runs at 100% efficiency, but we can't stand over everybody with a whip and make them work faster and harder. If we do that, you can be sure that we're going to make mistakes and alienate people.

Lannie: Well, here in the front office, we could get our work done in half the time if we had faster computers and if we had a computerized financial management system. It would also cut back on all the transcription errors we make. Every year, we talk about updating, so maybe this is a good time to actually do it.

John: If we save time and make fewer errors, how will that benefit either animals or investigators, and how will the cost be recovered within the two years that we have to decrease our subsidy?

Lannie: It would save money in the long run and cut down on investigator complaints and the need to redo an animal census.

John: How long is the long run? We have to focus on two years, which isn't a lot of time. By the time of our next meeting, can you research the payback period in terms of dollars to purchase a typical system and the time spent learning how to use the system? Also, can you come back to us with details on how this will save us money? For example, will we save enough time that we'll need fewer people in the front office?

Lannie: I can try to do that if nobody kills me first for making the suggestion.

Ethan: Well, for the short range, I have some ideas. To begin with, instead of paying us time and a half for compulsory overtime for cage washing, why not just have a split shift during the working day, with some people working 7:00 a.m. to 3:30 p.m., and others from 9:00 a.m. to 5:30 p.m.? A few of us would like to start a little later, and it would keep the machines working for the same length of time. It would save money and have no negative effect on investigators or animals. I like the overtime money, but I don't like working late every other day. I think it's a win–win deal.

Carrie: I agree with Ethan. If the cage wash team wants to cut back on overtime and the number of cages washed doesn't change, why should that be a problem?

John: OK. We'll consider that as long as splitting the shift doesn't mean we have to hire more people to be present in the cage wash room. It's good because we can easily measure the amount of money saved, and it will have no influence on effectiveness because the cages are going through the same machines in the same time. Right now, overtime work costs us about $150,000 a year, so your idea has the potential of making a big difference.

Ethan: Here's another idea. Maybe using disposable cages will save us time and money?

John: Fair enough. Ethan, will you work with Bob in the front office to find out if this is a viable option for us? We'll need that information by the end of this month.

Ethan: OK, I'll do that, but I have another idea. There are a lot of investigators who use one of their own technicians to manage their mouse breeding colonies, but it's not a full-time job for any one person and the principal investigators would really rather have them in the lab. Why don't we hire one or two additional veterinary technicians and have them cover the breeding colonies of four or five researchers and we can charge them for the service? We have plenty of seating room for two more technicians, and we can set a fee that will cover their salaries and benefits and still have money to spare. The extra income will help decrease the subsidy, benefit the researchers, and give more consistent care to the animals.

John: That's another good idea, assuming that we can quickly and properly train the new technicians. We can easily measure the breeding and weaning success of the colonies we would care for to see if they can be used as indicators of increased success compared with the present breeding and weaning figures. If the breeding figures are better with our technicians than with the investigators' technicians, we can use those numbers as a selling point to provide even more services to the research community. Of course, it's easy to determine if the cost of the service will bring us surplus income. But first, please work with Sue and Jack and formally survey our investigators and let us know by the end of the month if they are really willing to commit themselves to having our people take care of their colonies. I don't want to hire people and then find out that nobody wants to use them. Let's take a break until next week and meet again at the same time.

The key points in the above discussion were as follows:

■ John led the discussion, but he didn't dominate the discussion and he didn't tell people what goals to set. He did state the overall goal for the animal facility, but then he let others think about the specific subgoals and the strategies needed to reach those goals. This certainly may help motivate people to reach goals that they themselves helped establish. John also made sure that specific people had specific tasks to do. That way, there was accountability for actions. Finally, he set a time for the next follow-up meeting.

■ Overall, people were thinking "Lean." Some of the best suggestions came from Ethan, who appears to be an animal care technician.

■ The specific strategies proposed by Ethan were focused on the goal of decreasing the subsidy (increasing efficiency) without having a negative effect on the animal facility's effectiveness.

■ All of the proposed strategies were measurable.

- The strategies were in line with the missions of the university and animal facility. Nobody suggested opening a pizzeria to make extra money.
- The measures for the proposed strategies were balanced; that is, there were some financial measures and some nonfinancial measures.
- There was the potential of using collected data to make future predictions about financial success.

One additional point: Although I am suggesting that we often need measurements to help drive change (to show people where they are relative to where they want to be), there is no one group of measurements that will fit the goals of everybody in the animal facility. I'm the director of an animal facility, and it's part of my job to ensure efficiency and effectiveness. On a day-to-day basis, I have to know about overall budget integrity, investigator satisfaction, staff satisfaction, needed staffing levels, equipment needs, and many other general issues. Therefore, if I were the facility director in the example I used above, I might be most concerned about the actual need for extra technicians to provide service to investigators and about the fees that I could reasonably charge. One way of measuring this would be to send a questionnaire to investigators to see what additional services they desired and what they would be willing to pay to obtain those services. I could set up a numerical scale of wanted services, or a more qualitative scale, but a measurement of some sort could be developed. Similarly, I could very easily get monthly financial reports of overtime costs, and a monthly "best-guess" estimate of what our annual subsidy from the university will be.

It would not be unusual for the people in charge of overseeing the cage washer operation to have different goals than mine, although we are all shooting at the same target of decreasing the subsidy to 25%. They have to be able to measure the effectiveness and efficiency of having a split shift. To that end, they might measure (i.e., count) the total number of cages processed per week to see if there is any difference between the overtime system and the split-shift system. These measurements have to continue so we can see, at our monthly meetings, the progress being made toward reaching the goal. The cage wash group would also have to report back to us the amount of money being saved by using the new system. The veterinary services supervisor would have to monitor investigator satisfaction with the new breeding colony service (e.g., by periodic satisfaction surveys) and also the income from breeding colony activities versus the cost of salaries and benefits. This will be reported at our monthly meetings, which will also include feedback about the entire goals, strategies, and measurement process. Every supervisor will have different specific goals and strategies, but all must relate to the goals of people above and below them in the animal facility hierarchy and, of course, to the larger goal of the university.

What happens if everything described above falls into place and the animal facility staff meets their two-year goal of decreasing the subsidy to 25%? Do they pat themselves on the back and focus on another goal, or do they do something more tangible, such as throwing themselves a party or giving everybody a little financial bonus (obviously, not so much as to raise the subsidy)? Opinions will probably differ, but my vote is to show some form of appreciation, such as a bonus or a party attended by everyone, including senior management. If communicating appreciation for a job well done is an important part of a manager's job (which it is), and if we understand that people crave recognition for their work and accomplishments (which they do), then a somewhat tangible expression of appreciation goes a good deal further than a simple thank you. At the very least, don't forget to say thank you.

I'll conclude this discussion by pointing out that measuring progress toward a goal does not always have to be part of a big, department-wide plan. You can develop strategies and measurements for your personal goals. Think of a fairly simple problem, such as getting to work late because you have to take your child to school before your workday starts. You know your boss is unhappy about this. If you routinely get to work late three days a week (even if you stay late to make up the lost time), perhaps a reasonable goal for the first year would be to be late only one day a week. You can develop specific strategies, such as waking up earlier, getting your child's clothing laid out the night before, and preparing his lunch the night before. There is one specific and easy measurement that can be made, which is the number of days late per week or per month. However, you can also consider other measurements that may or may not forecast future success, such as the length of time it takes from waking up until you leave your house. The choice is yours, but as you can see, the process is basically the same one we used for defining a departmental goal, developing specific strategies, and measuring the success of those strategies toward reaching a goal. I can summarize this entire chapter with the diagram shown in Figure A1.2.

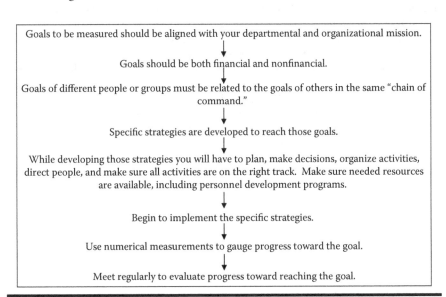

Figure A1.2 Quick review of developing measureable productivity goals and strategies.

References

1. Gallup Consulting. 2008. Employee engagement: What's your engagement ratio? http://mvonederland.nl/system/files/media/gallup_2008_pdf.pdf (accessed June 30, 2016).
2. Towers Watson. 2012. Global workforce study. https://www.towerswatson.com/Insights/IC-Types/Survey-Research-Results/2012/07/2012-Towers-Watson-Global-Workforce-Study (accessed May 15, 2015).
3. Gallup, Inc. 2015. State of the global workplace. http://www.gallup.com/poll/165269/worldwide-employees-engaged-work.aspx (accessed July 22, 2015).
4. Pritchard, R.D., Holling, H., Lammers, F., and Clark, B.D. (eds.). 2002. *Improving Organizational Performance with the Productivity Measurement and Enhancement System: An International Collaboration.* New York: Nova Science Publishers.
5. Lee, T. 2010. Turning doctors into leaders. *Harvard Bus. Rev.* 88(4): 50.
6. Chang, R.Y., and Morgan, M.W. 2000. *Performance Scorecards.* San Francisco: Jossey-Bass.
7. Bassi, L., and McMurrer, D. 2007. Maximizing your return on people. *Harvard Bus. Rev.* 85(3): 115.
8. Likierman, A. 2009. The five traps of performance measurement. *Harvard Bus. Rev.* 87(10): 96.
9. Kaplan, R.S., and Norton, D.P. 1996. *The Balanced Scorecard: Translating Strategy into Action.* Boston: Harvard Business School Press.
10. Niven, P.R. 2003. *Balanced Scorecard Step-by-Step for Government and Nonprofit Agencies.* Hoboken, NJ: John Wiley & Sons.
11. Meyer, M.V. 2002. *Rethinking Performance Measurement: Beyond the Balanced Scorecard.* Cambridge: Cambridge University Press.
12. Womack, J.P., and Jones, D.T. 2003. *Lean Thinking: Banish Waste and Create Wealth in Your Corporation, Revised and Updated.* 2nd ed. New York: Productivity Press.
13. Charron, R., Harrington, H.J., Voehl, F., and Wiggin, H. 2015. *The Lean Management Systems Handbook.* Boca Raton, FL: CRC Press.
14. Isaac-Lowry, J. 2013. Lean problem solving section 01: Why problem solving matters. https://www.youtube.com/watch?annotation_id=annotation_108550&feature=iv&src_vid=SrEgqwdz1Ps&v=eJpRsQYCJ5I (accessed April 28, 2015).
15. Rasiel, E.M., and Friga, P.N. 2001. *The McKinsey Mind: Understanding and Implementing the Problem-Solving Tools and Management Techniques of the World's Top Strategic Consulting Firm.* New York: McGraw Hill.

Appendix 2: Setting *Per Diem* Rates

Carpe *per diem*—seize the check.

Robin Williams

Basic Considerations

In Chapter 4, a *per diem* rate was calculated using one species (mice) and one service (animal husbandry). If that were always the situation, life would be a lot easier for us. Unfortunately, operating an animal facility with one species and one service is more the exception than the norm. In the real world, you will be caring for many species and, more likely than not, providing two or more services. The problem facing us is to establish *per diem* charges for each species. This is not as difficult as it sounds, and in fact, I'm going to make it even easier than in the last edition of this book. A step-by-step explanation of how these calculations are made is provided, with the understanding that it is just to help you understand the process. Even if you have a computer calculate your *per diem* charges, you should still understand the concepts. In all of the examples below, it makes no difference if your census is counted by cages (as most facilities do for rodents) or by individual animals. The concepts are exactly the same.

Let's consider some general rate-setting principles that must be used by academic institutions receiving grant or contract money from the National Institutes of Health (NIH), a major source of funding for research in the United States [1,2].

1. The *per diem* rates you develop must be based on the actual costs and actual usage of the services. That means you must develop *per diem* rates that include all of your federally allowable expenses, including the vivarium's direct costs,* such as animal care technician salaries and benefits; the cost of food and bedding, disposable clothing, and sentinel animals; and so forth. You are usually told by your institution's financial office which costs you should include in the *per diem* rate. You cannot develop a *per diem* rate that will intentionally lead to a profit for the animal facility. Here are just a few of the costs you *can include* in the *per diem* rate:
   - Labor (salaries and fringe benefits for the vivarium staff)
   - The costs directly associated with a preventive medicine program, such as costs involved with purchasing and testing sentinel animals
   - Service contracts for core equipment, such as cage washers
   - Operating supplies, such as animal food, bedding, and disposable outerwear
   - General maintenance costs (e.g., small repairs, rack filters, and cleaning supplies)
   - Professional meetings for the vivarium staff
   - Employee training
   - Depreciation and maintenance costs for equipment (such as a copy machine) that was not purchased with federal grant money *and is not included in your institution's facilities and administrative (F&A) (indirect) costs*

   Then there are certain costs that *cannot be included* in the *per diem* rate. Here are a few examples:
   - Alcoholic beverages
   - Bad debts owed to the animal facility
   - Contributions and donations made by the animal facility
   - Fund-raising expenses of the animal facility
   - Tickets to shows, sporting events, and so forth

2. The *per diem* rates you charge should be established using a documented method. For example, the NIH publishes a rate-setting guide for animal facilities [3] that can be used for calculating the *per diem* rate. The methodology you will be taught in this appendix also fulfills federal requirements. You should be consistent and use the same method every time you recalculate the *per diem* rate, not one method one year and another the next. Similarly, the

---

* Grants given by the federal government for biomedical research typically consist of two parts: (1) Direct costs are those grant funds that can be used for the salaries and fringe benefits of the investigators, and to purchase research supplies, for animal purchase and care, and other items directly benefiting the grant-supported project. (2) Facilities and administrative costs (also known as indirect costs or F&A costs) are negotiated between the federal government and your institution, and they provide support to your institution to help pay for heat, electricity, administrative offices (such as the dean's office), certain required regulatory activities (e.g., the IACUC), and many other allowable costs.

categories of items that are included in your *per diem* rate (such as salaries and benefits, food, and bedding) should not change from year to year.

3. Animal facility users whose research is supported by federal funds have to pay the same rate for the same service. That is, within your institution there can be no differences in the rates charged to investigators doing research that is supported by federal dollars. However, your institution can subsidize a lower *per diem* rate for a researcher if it does so by using nonfederal funds. On the other hand, you can charge an additional reasonable fee to an external user of your animal facility who is not supported by federal funds.

4. *Per diem* rates are to be reviewed and updated (if appropriate) at least every other year. When that is done, you must take into consideration any operating deficit or surplus from the previous *per diem* period. That is, you cannot "bank" any surplus funds from this year to pay for expenses that may be incurred in future years. If you do have a surplus (which would be unusual for most animal facilities), you must use that money to reduce the *per diem* rates charged.

5. If a federal grant or contract includes funds paid to your institution for indirect costs (such as heat and electricity), you cannot include those same costs as part of the *per diem* charge to investigators. That would be collecting the same cost twice (i.e., via the indirect cost allotment to your institution and via your *per diem* charge to the investigator). It's not legal to do that.

6. The information and methods used to calculate *per diems* must be made available to federal officials as required by 45 CFR 74.53 (as a rule of thumb, if you keep all of your *per diem* documentation for five years, you will be compliant with federal regulations).

Different institutions may include different items in their *per diem* rate calculation, and I only gave you a rough idea of what you can include. In fact, because institutions differ in what is included in the *per diem* rate, it's dangerous to compare *per diem* rates of one institution with those of another unless you know and appreciate what is or is not included in that rate. As a simple example, many institutions *do not* include all or part of their veterinarian's salary in the *per diem* rate. That salary comes from other sources. Likewise, in many institutions, the senior manager's salary is not included in the rate. At some institutions, the federally mandated Institutional Animal Care and Use Committee (IACUC) training costs are included in the institution's (not the animal facility's) indirect costs. On the other hand, where I work, my salary and those of the supervisors, the veterinary technicians, the animal care technicians, and all office staff; the cost of all supplies; and even capital equipment and building depreciation are included when I calculate our *per diem* cost for housing animals. I'm afraid that if you don't already know what your institution includes, you will just have to sit down with the powers-that-be and ask what they want included in the *per diem* rate.

As just noted, some *institutions* include certain indirect costs in their *per diem* rates, while others do not (for the former group, these are the so-called indirect or

F&A charges that were described in the footnote). These indirect costs can include part of the cost for heat, electricity, building depreciation, library use, the office of the institution's president, and so on. If you are including any indirect costs in your *animal facility's per diem* rate, you will have to be told by your financial office what guidelines to follow. Remember, though, that the more items that are included in the animal facility's *per diem* rate, whether direct or indirect, the higher the *per diem* charge to researchers will be unless your institution subsidizes the *per diem* rate with nonfederal dollars. The corollary is that researchers want as many costs as possible included with the *institution's* indirect costs, so that their *per diem* rates from the laboratory animal facility are (theoretically) lower. If you change jobs and begin work in a different laboratory animal facility, you may find that some items that are included or excluded in your new institution's *per diem* calculations differ from those of your old institution. As an example, the federal government allows a research institution to include in the *institutional* indirect costs the cost of operating the animal facility's procedure rooms, quarantine rooms, and surgery suites. Even the cost of space (which is negotiated with the government) for rooms that house animals involved in research (as long as the animals are not usually removed from the vivarium) can become part of the institution's indirect costs. However, institutions are not *required* to include those costs in the institutional indirect costs. Consequently, if they are not included in the institution's indirect costs, they have to be included as part of the *per diem* rates established by the animal facility. If this were to happen, then the vivarium manager (or whoever is developing the *per diem* rates) would have to find out from the institution's financial office the cost per square foot of space that has been approved by the federal government. It's obvious that the person responsible for calculating *per diem* rates has to have a clear statement from his or her institution about what to include or exclude in the *per diem* calculations.

Of course, your calculated *per diem* rate may be so high that no reasonable researcher could afford to pay it. That's why many institutions subsidize the *per diem* charges to investigators (and they *cannot* do that by using other federal dollars they may have received; they have to use nonfederal money for a subsidy). Even if the institution chooses to subsidize part of the *per diem* rate to make it more affordable to the researchers, your job is still to calculate the *actual* rate. In the following section, we will learn how to calculate the actual *per diem* rate. Keep in mind that your institution may actually charge a *per diem* rate that differs from the one you calculated.

Determining the Number of Animals Housed as Part of the *Per Diem* Calculation

The basic concept behind establishing a *per diem* rate for any species is to have the cost of all of your activities for that species covered by animal facility services that generate income. Over a one-year time period, the basic formula for any species is

$$\left\{ \frac{\begin{array}{c} \text{All annual expenses less} \\ \text{all recharged revenues for} \\ \text{species A} \end{array}}{365 \text{ days}} \right\} \div \left\{ \begin{array}{c} \text{Average daily census of} \\ \text{species A animals } (\text{or cages}) \end{array} \right\} = \begin{array}{c} \text{Species A} \\ per\ diem \end{array}$$

Here's an example of how the formula would be used if there were, on average, a total of 350 rat cages per day in the vivarium over one year and you have total rat expenses of $100,000. I'll explain how to get the total annual expenses later on in this appendix.

$$\frac{\$100,000}{365 \text{ days}} \div 350 \text{ cages } (\text{on average}) = \$0.78 \ per\ diem \text{ rate for one rat cage}$$

This formula is slightly different from the one used in Chapter 4. It recognizes that there can be nonanimal care revenue generated for a particular species (e.g., from a charge to an investigator for administering a drug to a rat). In most institutions, the *per diem* charge is only for animal care and related costs. The charges for the drug and associated labor and supplies are usually billed directly to the investigator. Therefore, if you directly charge the investigator for the drug, and the labor involved in administering the drug to an animal, you cannot try to recoup these charges a second time through the *per diem* charge. That means you must exclude the charges for the additional labor when you calculate the labor portion of your *per diem* rate, and you also must exclude the charges for the drug and supplies (e.g., needle and syringe) you used when you calculate the supplies portion of your *per diem* rate. I'll suggest a way to do this elsewhere in this appendix. However, if you find that these additional revenue-producing activities are really a very, very insignificant dollar amount when compared with the total *per diem* income, you can just forget about them in your calculations because they will have almost no impact on the *per diem* rate.

Let's begin. First, it's quite important to know how to determine the average number of animals housed per day. You can't develop a *per diem* rate without this information. Please take a look at the following example:

Calculating the Average Number of Rat Cages Housed per Day,
over a Month's Time, in a Hypothetical Animal Facility

300 rat cages counted on January 1	=	300 rat cage days
200 rat cages counted on January 2	=	200 rat cage days
100 rat cages counted on January 3	=	100 rat cage days
0 rat cages for remainder of month	=	0 rat cage days
Total	=	600 rat cage days for January

600 rat cage days ÷ 31 days in January = 19.4 rat cages counted/day, on
average, during January

If there were no additional rat cages used for the remainder of the *year*, then

600 rat cage days ÷ 365 days = 1.6 rat cages day, on average, for the entire year

Rather than calculating the number of rat cage days for each day of the year, you may wonder if you can get a "close enough" estimate of the average number of rat cages maintained per day by doing a cage count once or twice a month, and then averaging those numbers at the end of the year. The answer for most laboratory animal facilities is no. If you are always filled to the brim with rat cages, you might be able to do this because the average number of cages per day will not change too much. But if the number of cages varies quite a bit during the month, you will have a poor estimate of the average number of cages per day. Also, performing a census only once or twice a month is dangerous if you are counting individual rats rather than rat cages. That's because the number of rats per cage typically changes over the course of time, unless every animal is individually housed. For this reason, it's easier if you charge your *per diem* rate by the cage rather than by the animal. It's certainly easier to count cages rather than individual animals (at least with rodents), but even here you would have to assume that the number of cages stays fairly constant over the time interval between cage counts. That assumption usually is unrealistic, and for that reason, averaging your cage or animal count is discouraged if you want an accurate reflection of your cage days.

Here's the good news. In a real-world situation, it isn't hard to determine the number of cages (or even individual animals) present per day. Most laboratory animal facilities already calculate their monthly billing for animal care based on rat cage days (or mouse cage days, etc.). These charges originate with the daily animal census form, which is filled out by either the investigator, the animal facility staff, or a combination of both (see the example in Chapter 4). So, it should be a relatively small problem at the end of the year to get the total number of cage days (or animal-days) by species. In larger facilities, this is likely to be computerized information. In smaller facilities, you may just have to do some simple addition. One way or the other, you will be able to get those numbers. Once you have those numbers, you will use them for either prospective or retrospective *per diem* calculations, as will be described soon.

When you are calculating a *per diem* rate, you are usually making a plan for the future. It's a rate you may begin charging in a month, six months, or even more, but it's not something you will be charging tomorrow. Investigators need lots of advance notice, so they can plan their own budgets, but you have to balance that notice against your need to use current information in your calculations. Whenever possible, I like to tell investigators what the new rates will be about six months before the end of the current fiscal year. Some institutions impose a *per diem* increase (usually about 3% or 4%) every year. Therefore, they can let investigators know years in advance what their *per diem* rate will be. If that's what your institution does, then the only reason for calculating the actual *per diem* rate is to remain in compliance with federal regulations about the frequency of calculating *per diem* rates, because no matter what your calculated actual *per diem* rate is, the investigators are going to pay just a few percentage points more than they did the previous year. That's not, in my opinion, the best way to do business, and I'll comment more about this toward the end of this appendix.

You can approach your *per diem* calculations either prospectively (as with a zero-based budget) or retrospectively. The truth is, the calculations are the same, but it's technically easier to do them retrospectively. That's because you already know what your labor and other expenses were for a past time period. Prospectively, you have to do your best to estimate your expenses and the number of animals (or cages) you will house per day. Therefore, I will first describe a method of doing a retrospective *per diem* rate calculation. *If you use a retrospective per diem rate calculation to help you set per diem rates for a future time period, it is assumed that the number and percentage of animals of each species will be roughly the same for the coming year (or other time period), and they will be housed about the same (e.g., on average, three mice per cage). Therefore, if 25% of your animal population was mice this year, it should be about 25% next year.* Why are the above assumptions important? If, for example, you know that next year you will be housing 1000 more mouse cages per day than during the past year, you very well may have to buy more cages, more water bottles, more stoppers, more mops, and so forth. You will have to budget for these increased expenses and recover their cost through next year's *per diem*. Likewise, you will need more labor to take care of those animals. Sure, you'll eventually recover all or part of your increased labor cost via the increased revenues from the *per diem* charges, but you may be fighting an uphill battle when you try to hire people and there is no "up-front" money in your budget. If you foresee these or similar problems, you may as well do a prospective budget and a prospective *per diem* calculation.

Retrospective Calculation of a *Per Diem* Rate

Here are the general procedures to be followed in calculating a *per diem* rate for each species you are housing. *In the example used, I have intentionally not included any indirect costs. This will simplify your understanding of the process needed to calculate the per diem rate.*

Basic Needs for Calculating a Retrospective Per Diem *Rate*

1. **Have your animal facility's financial (budget) reports with you. You probably receive these on a monthly basis, and they provide you with a breakdown of your**

income and expenses for that month, and often for the year-to-date. I am not aware of any institution where these are not available.

2. Know the average number of animals housed per day, by species. The means of getting this information was explained above.

3. Know the distribution of labor, by species, for you and your staff. This is described briefly in Chapter 4, and in more detail later in this appendix. Labor is the largest part of the *per diem* charge, and you cannot perform a *per diem* calculation without knowing the percentage of labor effort to assign to each species you house. Labor has to include all people involved in operating the animal facility, including those not directly working with animals, such as a business manager. The only exception here is that if you know that a person working for the animal facility is paid from your *institution's* indirect cost allotment, you can't include his or her salary in the vivarium's *per diem* rate calculation. That would be charging twice for the same person.

4. Know the cost of all other items used in animal care, such as food, bedding, and cleaning supplies.

General Instructions for Calculating *Per Diem* Rates

Just as a reminder, here's the formula for calculating the *per diem* rate:

$$\left\{\frac{\text{All annual expenses less all recharged revenues for species A}}{365 \text{ days}}\right\} \div \left\{\begin{array}{l}\text{Average daily census of}\\ \text{species A animals or cages}\end{array}\right\} = \begin{array}{l}\text{Species A}\\ \textit{per diem}\end{array}$$

If we have only one species, such as rats, it's going to be easy calculating the *per diem* rate because all annual expenses are only for rats. Let's get specific now about calculating *per diem* charges when there are multiple species. Multiple species make it harder—not a lot harder, but nevertheless harder.

Assume the total expenses to operate your animal facility for fiscal year 2016 were $500,000. The fiscal year was January 1, 2016, to December 30, 2016. Let's also assume that today's date is March 3, 2017. How do you establish your *per diem* rates for fiscal year 2018, which will go into effect on January 1, 2018, which is ninth months from now? The most recent complete data we have is from 2016, so that's what we will work with.

For simplicity, assume that all of your income came from *per diem* charges. If that's not true (and it usually isn't), just delete any income that did not come from *per diem* charges. As I wrote above, there may have been fees for veterinary services that were charged directly to an investigator and paid by the investigator. You can't include the same charge in the *per diem* calculation because that would be including the same charge twice. Later on, I'll describe a simple method you can use to separate fee-for-service charges from *per diem* charges. Remember, the general concept is to assign to each species housed a percentage of your total expenses, so by the end of the 2018 budget, your total *per diem* income will roughly equal your total *per diem* expenses.

The question is, "What percentage of the $500,000 total annual expenses should I assign to each species?" What follows is a suggested way of doing this. It will not be absolutely exact (no *per diem* calculation is), and it doesn't do things the exact way the NIH *Cost Analysis and Rate Setting Manual* [3] does, but it's much easier. It's also much better than guessing wildly or looking at every other animal facility in the neighborhood and setting rates that are someplace in between everybody else's. *It is important to ensure that for your facility, the relative proportion of expenses assigned to each species is correct.* Thus, rat users should not pay an unfairly high percentage of the total cost; otherwise, they might be subsidizing a monkey or rabbit user's research. Because the largest single item in the *per diem* rate is labor, tackle this first. You must have time-motion data to properly track labor effort, at least if you are basing the labor part of your *per diem* rates for 2018 on the labor activity of 2016.

Tracking Labor through Time-Motion Records

A time-motion study, in the context of *per diem* calculations, simply means recording which species people work with and how much time they spend with each species. This information is crucial to calculating *per diem* rates because labor is about 60%–70% of most *per diem* rates.

Time-motion records are typically obtained in one of two ways. The easiest way, as you might have guessed, is the most expensive. If you want to know who is in a particular animal room at a particular time, you can use a computer-linked key card (or a radio frequency identification chip) that records when a person enters and exits that room. It really doesn't matter what the person is doing in the room, since all we want to know is how much time is spent with a particular species. Computer key cards may be appropriate if one species is housed in the room. There will be times when you enter a room but have no intent of working with a species, for example, if you go into a room to look for a person or to see if you left something there the day before. For practical purposes, those instances are relatively quick and minor and will lead to no significant problems when calculating *per diem* rates.

Another way of recording the time spent working on a particular species is manual recording of time-motion on an activity log. This is probably the most common method in use today. A sample form for keeping such records is shown. Everybody fills out their own form. Notice there is no space for a person's name. That's intentional, and it helps drive home the point that an activity log is not used as a punitive device to see who is or isn't doing their job. You may have to work hard to get your staff to understand that the logs are used to help calculate *per diem* charges. You will also notice on the sample log that people are asked to enter the month that the log was filled in. Empirically, I have found that asking people to fill out the logs for one week out of a month is adequate to get a reasonable picture of work activity, as long as you do not use the same week of the month each time. For instance, in January you would record activity in week 1, week 2 in February, week 3 in March, and so on. It's important to be able to identify a person's job title (e.g., animal care supervisor) because we use the total salaries and benefits in a job title to help calculate the *per diem* rate.

Sample Time-Motion Activity Log

University laboratory animal resources					
Animal care tech ☐ Vet tech ☐ Animal care supervisor ☐ Vet tech supervisor ☐ DVM ☐ Office Staff ☐					
Enter the total minutes worked that were not directly billed to the investigator					

Hour of day	Mouse	Rat	Swine	Other Activity	Lunch	Total
7–8 a.m.	45			15		60 minutes
8–9 a.m.	60					60 minutes
9–10 a.m.		30	30			60 minutes
Etc.						60 minutes
		Use only these minutes to calculate a person's effort				
Total						
Month_____Day____ Year _____						

A few explanatory points are in order. First, I ask for the time spent at lunch because I want to be sure that people clearly understand that every hour has to account for 60 minutes (the only exception is when people start or end work on the half-hour). That time does not get entered into any calculations of effort. Next, I make it clear that I am not interested in anything but time that is directly related to working with animals (including cage changing, cleaning the animal room, and environmental enrichment). The "Other Activity" category includes meetings, breaks, talking to investigators, cleaning hallways that serve rooms housing many species, and similar items that cannot be attributed to one species of animal. That time also does not get entered into any calculations of effort. Once again, that column is there just to remind people that there are 60 minutes in each hour that have to be accounted for.

Shortly, I'll give you some information about how to calculate the time allocation (by species) for office staff who never have anything to do with animals, but many office staff have no problem allocating their time by species. For example, if a person spent a day ordering animals, he should be able to estimate the percentage of time he spent ordering each species. If a person orders supplies for animals, she should be able to estimate the time she spent with each species.

Finally, in some facilities, there is a different *per diem* rate charged for working inside or outside of a barrier area. You can set these activity logs up in any manner you choose. If you charge differently for different-size cages, or if there are different charges for cages kept in a biocontainment suite, you can include them as separate columns on the log. One time-saving suggestion that you might consider is to computerize the activity log so that the raw data can be entered electronically rather than manually on cards. This is not for a beginner in programming, but at the University of Massachusetts Medical School, our information technology (IT) department developed a very user-friendly program that allows the data to be entered and collated electronically. It literally saves me many hours of work every month, and because it also automatically populates the spreadsheets that I use for calculating *per diems*, it is a tremendous time-saver when it comes to getting summarized labor data for our annual *per diem* rate calculations.

Using the Data from Time-Motion Logs

The next step in the process is to convert the total minutes of annual effort into the percentage of effort, per year. Either manually or electronically (if you are using a computerized program), you will total the number of minutes spent with each species shown above in the activity log for each person working. Using a hypothetical example, this has been done for you in Table A2.1. We'll make it easy for ourselves and assume that there are three animal care technicians, one animal care supervisor, and one part-time veterinarian working in this vivarium. The animal care technicians, taken as a group, spent 74,880 minutes caring for rats and rat housing rooms, trimming rat nails, and so forth, during 2015. They spent 280,800 minutes caring for mice and 18,720 minutes caring for swine. Some of that time was regular time, and some of it may have been overtime; it doesn't matter. The total time that the animal care technicians spent working directly with animals was 374,400 minutes for the entire year. Now, all we have to do is set up a proportion to change minutes into percent effort. It's easy; let's just look at rats:

$$\frac{74,880 \text{ minutes with rats}}{374,400 \text{ total minutes}} = \begin{array}{l} 20\% \text{ of the time allocated to working with} \\ \text{animals was spent caring with rats} \end{array}$$

Table A2.1   Annual Total Allocation of Labor Effort by Species

	Three Animal Care Techs (minutes)	Percent Effort	One Animal Care Supervisor (minutes)	Percent Effort	One Half-Time Veterinarian (minutes)	Percent Effort
Rats	74,880	20%	12,480	10%	3,100	5%
Mice	280,800	75%	99,840	80%	46,500	75%
Swine	18,720	5%	12,480	10%	12,400	20%
Total	374,400 minutes per year	100%	124,800 minutes per year	100%	62,000 minutes per year	100%

Table A2.2   Allocation of Labor for a Person Not Working Directly or Indirectly with Animals

	Three Animal Care Technicians' Effort	One Supervisor's Effort	0.5 Veterinarian's Effort	Average Effort of 4.5 People	Total Effort to Allocate for an Office Assistant Not Directly or Indirectly Working with Animals
Rat	20% × 3 = 60%	10% × 1 = 10%	5% × 0.5 = 2.5%	16.1%	16.1%
Mouse	75% × 3 = 225%	80% × 1 = 80%	75% × 0.5 = 37.5%	76.1%	76.1%
Swine	5% × 3 = 15%	10% × 1 = 10%	20% × 0.5 = 10%	7.8%	7.8%

Table A2.1 then shows the same calculation for all of the workers and all of the other species.

What happens if you have an office assistant who spends no time at all directly or indirectly working with animals? How is that person's salary allocated to the three animal species we have in the vivarium? We take the needed data from Table A2.1 and enter it into Table A2.2. Then we "weight" the effort of each group of people by the number of people in the group, as shown in Table A2.2. Finally, we make a simple ratio, as shown in Table A2.2, of what we already know about the effort of everybody except, of course, the office assistant. Remember, the veterinarian is only working half-time. All Table A2.2 is saying is that we are estimating that for the office assistant, the percent effort to attribute to each species of animal is the average effort that everybody else recorded for the same species.

Let's assume that during the fiscal year 2015, the total of salary and fringe benefits for everybody in the animal facility was $245,000, and for the three animal care technicians alone, it was $90,000. (We can get the basic salaries from payroll records, and for most institutions, fringe benefits are about 30% of a person's salary. Your business office should be able to give you the exact figure.) We know from Table A2.1 that 20% of the cost of the animal care technicians' salaries and fringe benefits went for work with rats, 75% with mice, and 5% with swine. Therefore, we'll allocate 20% of $90,000% = $18,000 for caring for rats. Let's do that for each species in Table A2.1. It will now look like this for all of the species we house (Table A2.3).

When there is more than one person in a labor group (such as three animal care technicians), we use the *total* salary plus benefits, not the average salary plus benefits. We can duplicate Table A2.3 and its calculations for the supervisor, the part-time veterinarian, and the office assistant, but I'll save you the time and show you in Table A2.4 how it will look when the information is summarized.

Table A2.4 is your best estimate of the labor cost (salaries + benefits) by species. Because, in most budgets, salaries and benefits account for about 60% of all total expenses, you can see why it is critical to correctly allocate labor by species. To put it another way, if you stopped all your *per diem* calculations at this point, you would probably be someplace near 60% accurate. If you continued on and allocated all

Table A2.3   Distribution of Salaries for Animal Care Technicians

	Total Salary + Benefits for Three Animal Care Technicians = $90,000	*Percent Effort by Species*	*Salary Allocation for Animal Care Technicians, by Species*
Rat	$90,000	20%	$18,000
Mouse	$90,000	75%	$67,500
Swine	$90,000	5%	$4,500

Table A2.4   Summary of Salary and Benefits Distribution, by Species

	Animal Care Technicians	*Supervisor*	*Veterinarian*	*Office Assistant*	*Total*
Rat	$18,000	$4,500	$3,750	$5,250	$31,500
Mouse	$67,500	$36,000	$56,250	$26,600	$186,350
Swine	$4,500	$4,500	$15,000	$3,150	$27,150
Total salary + benefits	$90,000	$45,000	$75,000	$35,000	$245,000

other costs at the same percentage that you used for labor, you would probably be even more accurate.

As a secondary issue, if you wanted to calculate just a labor *per diem* rate (that part of the total *per diem* rate that includes only labor), all you have to do is divide the labor cost for a species by 365, and then divide the resulting number by the average daily census. If there is an average of 600 rat cages housed each day, and from Table A2.4 we know that the total annual cost for labor that is attributable to maintaining rats is $31,500, then the labor *per diem* for a single rat cage would be

$$\$31,500 \div 365 \text{ days/year} = \$86.30 \, (\text{labor cost to maintain 600 rat cages for 1 day})$$

$$\$86.30 \div 600 \text{ rat cages day} = \$0.14 \, (\text{labor cost to maintain 1 rat cage for 1 day})$$

In this hypothetical example, it costs $0.14 *in labor only*, to maintain one rat cage for one day.

Distributing Labor Costs for Cage Washing

There always will be unique situations, and sometimes there is no perfect answer to allocating labor costs. For example, if you have a cage wash technician who never handles animals, how is that person's labor effort distributed? We can, of course, estimate that person's effort the same way we did for the administrative assistant in the example used above, but that often doesn't work because rodent cages may have to be filled with bedding, monkey cages may require special handling, and rabbit cages may require scraping. One way to approach this problem is to have the cage wash technician estimate, on his time-motion activity forms, the approximate amount of time he spent working with mouse, rat, gerbil, rabbit, and monkey cages. That is the easiest way to do it. But what if you use the same cage for rats and gerbils? How can the cage wash technician decide the amount of time to record working with cages that housed rats and those that housed gerbils? Fortunately, the cage wash technician usually can tell which is which, but sometimes it's necessary for a supervisor to proportion the rat versus gerbil labor time based on the average daily rat and gerbil cage census. For example, assume that during the time-activity recording period about 90% of the rat/gerbil cages housed rats. The remaining 10% of the rat/gerbil cages housed gerbils. All other cages were for other species. When it comes to reporting the time spent cleaning rat or gerbil cages, the technician reports 90% of that time as being with rats and 10% as being with gerbils.

Food Cost Distribution by Species

Labor is your largest expense. Two other items that you can allocate to a given species, with reasonable ease, are food and bedding. Right now, we'll talk about food.

Your job is to estimate the percentage of your total food expenditures to assign to each species. For species such as monkeys and rabbits, it's easy, since all monkey food goes to monkeys and all rabbit food goes to rabbits. (This is assuming that all monkeys are of the same species and all rabbits are roughly of the same size.) But what about rats and mice? They both eat the same food. How do you figure things out?

In the first two editions of this book, the method I used to get the cost of food that was to be allocated, by species, when different species ate the same food involved "changing" every species to rats. A mouse, for example, eats about one-third of the amount of food a rat eats, so one mouse = 1/3 rat. That method works perfectly well, but in this edition, I show you another method you can use that may be easier for some people. So far, we've only determined the distribution of the cost of labor by the species housed. Now, we are going to move on to the food allocation. So far, here is what we have determined:

	Salaries + Benefits	Food	Bedding
Rat	$31,500	?	?
Mouse	$186,350	?	?
Swine	$27,150	?	?
Total	$245,000		

To make things easy, we'll assume that there were, on average, two rats per cage and five mice per cage. A typical rat eats about 12 g of food a day, so two rats in a cage will eat about 25 g per day. Twenty-five grams is the same as 0.025 kg. A typical mouse eats about 5 g of food a day, so five mice in a cage will eat about 25 g (0.025 kg).

We need two more pieces of information. First is the average daily census of rats and mouse cages. We go to our census records and find that the average daily rat cage census was 600 per day, and for mice, it was 6000 cages per day. The second piece of information we need is the cost of rodent feed. From our financial records, we can easily determine that the total cost of rodent food (i.e., rat and mouse food) during the 2015 fiscal year was, we'll say, $95,000. Let's put this information into Table A2.5.

Table A2.5 tells us that one cage of rats consumed about 0.025 kg of food every day. We had 600 rat cages, so 0.025 × 600 = 15 kg of food consumed daily. Over a year, 15 kg × 365 days = 5475 kg of food was consumed by rats. If we do the same calculation for mice, we find that the mice in 6000 cages consumed 54,750 kg of food during the year. Therefore, in total, 60,225 kg of rodent food was consumed

Table A2.5 Distribution of Costs for Food Consumed by Rats and Mice

	Average Consumption (kg) per Cage per Day	Average Daily Cage Census	Consumption (kg) in 365 Days	Percent of Total Consumption to Allocate	Allocated % of $95,000
Rat (2 rats per cage)	0.025	600	5,475	9%	$8,550
Mouse (5 mice per cage)	0.025	6,000	54,750	91%	$86,450
Total			60,225	100%	$95,000

Table A2.6 Cost of Food for Swine

		Average Daily Pig Census		Total Cost
Swine		2.5		$300

during the year. That amount of food cost $95,000. The next step is to allocate a percentage of $95,000 to rats and mice.

- 5,474 kg ÷ 60,225 kg = 9%. Therefore, allocate 9% of the $95,000 (which is $8,550) to rats.
- 54,750 kg ÷ 60,225 kg = 91%. Therefore, allocate 91% of the $95,000 (which is $86,450) to mice.

Table A2.6 is pretty simple. It shows that there was an average daily census of 2.5 pigs in the vivarium, and the total annual cost of their food was $300. Because only pigs ate the pig food, there is no need for us to worry about distributing the $300 cost among different species.

Here is where we are so far in our *per diem* calculations:

	Salaries + Benefits	Food	Bedding
Rat	$31,500	$8,550	?
Mouse	$186,350	$86,450	?
Swine	$27,150	$300	?
Total	$245,000	$95,300	

Next, we have to consider the distribution of the cost of bedding between rats and mice. This assumes, of course, that you are using the same bedding for rats and mice, and that pigs get either no bedding or some other form of bedding. It also assumes that you will put about the same amount of bedding in the cage, whether there are two mice or five mice in it.

We'll assume that rats and mice use corncob bedding and pigs use wood shavings. The basic calculations are the same as the ones we used for food. I'll make it brief. First, determine from your records the total annual cost of corncob bedding (I'm using $17,000 as the cost). Then estimate the amount of bedding that goes into a rat cage and the amount that goes into a mouse cage. I've estimated 0.3 kg of bedding per rat cage and 0.15 kg per mouse cage. If you change the cages every 10 days, that means there are 36 cage changes per year. All of this is shown in Table A2.7, and the calculations used are basically the same as the ones I described for food. Once again, note that because swine is the only species to use the wood shavings bedding, there is no need to do any calculations; all of the cost of their bedding is allocated only to swine.

So far, we have the cost (by species) of three items (labor, food, and bedding) that are relatively easy to determine. But animal facilities have many more costs. There are always large amounts of supply and equipment items for which there is no simple way to assign expenditures to a given species. Pens, pencils, many cleaning supplies, truck repairs, and a long list of other items may fall into this category. But, to establish a *per diem* rate, we somehow have to transfer the expenses for these items to potential income-producing activities, such as animal care. The income, of course, will be derived from the *per diem* charges for the different species we house. I'm going to simplify things by lumping together all other expenses into one big category, which is "Everything Else," as shown below. Of course, if you can assign

Table A2.7　Allocated Dollars for Animal Bedding (Rodent Bedding Costs $17,000 per year)

	Average Amount (kg) per Cage Every 10 Days	Average Daily Cage Census	Kilograms Used per Year (36 Cage Changes)	Percent of Total Bedding Cost to Be Allocated	Allocated Dollars, Based on % of $17,000
Rat	0.3	600	6,480	17%	$2,890
Mouse	0.15	6,000	32,400	83%	$14,110
Total			38,880	100%	$17,000
		Average Daily Pig Census			Total
Swine		2.5			$200

one or more of these "Everything Else" items to a particular species, you should do so by adding additional columns to the chart shown below.

	Salaries + Benefits	Food	Bedding	Everything Else
Rat	$31,500	$8,550	$2,890	?
Mouse	$186,350	$86,450	$14,110	?
Swine	$27,150	$300	$200	?
Total	$245,000	$95,300	$17,200	

We're going to estimate the allocation (by species) of the costs in the "Everything Else" column. The method to be used is no different than the one that was used in assigning the office assistant's salary to the various species housed (Table A2.2).

The first thing to do is to determine how much money it cost us, in total, to operate the animal facility for one year. This is your total annual expenses, and you can get this information from the accounting information that is typically sent to the vivarium business manager on a monthly basis. There may be some costs you are told not to include for a variety of reasons, but that's something that's not hard to deal with. In the example we've been using, let's assume the total cost to operate the animal facility for one year was $557,500. How much money has to be distributed in the "Everything Else" column? From Table A2.7, we know the following:

$$\begin{aligned} \text{Salaries} + \text{benefits} &= \$245,000 \\ \text{Animal food} &= \$95,300 \\ \text{Animal bedding} &= \underline{\$17,200} \\ &\quad \$357,500 \end{aligned}$$

Since the total cost was $557,500, the remaining amount of money for "Everything Else" is

$$\$557,500 - \$357,500 = \$200,000$$

Now look at Table A2.8, where you can see how the remaining $200,000 was distributed, using the same method of distribution we used in the earlier tables.

You can make another column for major equipment if you want to, but I have not given an example since much of the major equipment you purchase (e.g., a cage washer) is a capital budget expense. The money to purchase that equipment is usually obtained directly from other institutional funds, not from your operational

Table A2.8   Distribution of Remaining Vivarium Expenses

	Salaries + Benefits + Food + Bedding	Percent of $357,500 to be Distributed	Distribution of Remaining $200,000 Based on the % of $357,500
Rat	$42,940	12%	$24,000
Mouse	$286,910	79%	$158,000
Swine	$27,650	9%	$18,000
Total	$357,500	100%	$200,000

Table A2.9   Final Distribution of Animal Facility Expenses

	Salaries + Benefits	Food	Bedding	Everything Else	Grand Total
Rat	$31,500	$8,550	$2,890	$24,000	$66,940
Mouse	$186,350	$86,450	$14,110	$158,000	$444,910
Swine	$27,150	$300	$200	$18,000	$45,650
Total	$245,000	$95,300	$17,200	$200,000	$557,500

budget. If the cage washer was purchased with federal government money, you are not allowed to recover this money through the *per diem* charges. However, if the cage washer was purchased with nonfederal money and its maintenance is not included in your institution's indirect costs that are received from the federal government for grants and contracts, your institution may direct you to include an annual depreciation charge for the cage washer in the *per diem* calculation, as was described at the beginning of this appendix. For now, we'll say the cage washer was purchased with government money and you will not be adding a depreciation charge. Let's put everything together in Table A2.9.

Calculating the Actual Per Diem *Rate*

The actual *per diem* charge is simply the total amount spent for each species for the year divided by 365 days, and the resulting number divided by the average number of animals housed per day, by species. For the three species we house, we have an average daily population of

Rats   600 cages
Mice   6000 cages
Swine   2.5 pigs

The *per diem* formula is

$$\left\{\frac{\begin{array}{c}\text{All annual expenses less}\\\text{all recharged revenues for}\\\text{species A}\end{array}}{365 \text{ days}}\right\} \div \left\{\begin{array}{c}\text{Average daily species cage}\\\text{(or animal) census}\end{array}\right\} = \begin{array}{c}\text{Species A}\\\textit{per diem}\end{array}$$

Using the grand total column from Table A2.9, here are the actual *per diem* rates you have calculated. There were no recharged revenues to deduct.

$$\frac{\$66,940}{365} \div 600 \text{ cages day} = \$0.31/\text{day} = \text{rat cage } \textit{per diem}$$

$$\frac{\$444,910}{365} \div 6,000 \text{ cages day} = \$0.20/\text{day} = \text{mouse cage } \textit{per diem}$$

$$\frac{\$45,650}{365} \div 2.5 \text{ pigs/day} = \$50.03/\text{day} = \textit{per diem} \text{ for each individual pig}$$

What you have calculated is the actual *per diem* cost to you (not to the investigator) for the *past* year. To set the *per diem* rate for a future time period (such as the following year), you can use the rate you just calculated if you do not anticipate any increases in costs (which is unlikely, but possible). However, if you anticipate 4% inflation, you can increase the amount of the grand total cost in Table A2.9 by 4% for each species if your animal facility charges investigators the actual *per diem* cost to the vivarium. Of course, your institution may subsidize part of the *per diem* rate so that an investigator pays less than what it costs you to care for her animals. That's an institutional decision only. Your job is to know the actual *per diem* cost to the animal facility.

Prospective Calculation of a *Per Diem* Rate

We just calculated a *per diem* rate for a future time period, based on figures we generated from a past time period. What do you do if you don't have any past history to use? The answer is, you perform a prospective calculation of a *per diem* rate. The information you need to do a prospective *per diem* calculation is not substantially different from that for the retrospective procedure. The variation is that with the prospective calculation, you must estimate the number of animals you will be housing and what you will need to care for each species, whereas with the retrospective calculation, you do not estimate; you use the actual expenses of the past year.

This was discussed in Chapter 4, under the heading "Working with a Zero-Based Budget."

Why bother to take the time to do a prospective calculation when the retrospective one is inherently easier? First, remember that the retrospective calculation assumes that the number of animals, the percentage distribution of each species, and the method of housing will be about the same for the coming year as it was for the past year. If any of these parameters will be changing significantly in your animal facility, you may have to estimate the number of animals (or animal cages), by species, for the coming year. The second reason is that a prospective *per diem* calculation gives you a much clearer understanding of exactly what is happening financially, because you are forced to examine and question everything you put down on paper.

Labor

The actual labor calculations for a prospective *per diem* rate do not differ from those made with the retrospective one. If your housing method will change, then clearly you are going to have to estimate if there will be any increased or decreased time for animal care and its associated cost for that species.

One thing should be obvious. You will have to do time-motion studies (as described in this appendix), whether you are calculating a retrospective or prospective *per diem*. If you do not have current data on how long it takes to care for animals, you can either do a little pilot study of your own to find out or make a best guess. The pilot study is safer.

Food and Bedding

The cost of food is not necessarily the same on a per-animal (or per-cage) basis if your housing system changes, such as changing from group housing animals to individual animal housing. Very often, animal care technicians will feed one animal the same amount of food they give three animals, and then dump the remaining food in a few days. A lot has to do with the way you care for your animals, but you will have to see if the amount fed per animal will change. The same comments made for food are also true for bedding.

Other Supplies and Equipment

For a prospective *per diem* calculation, there's more work to do. If you know the approximate average number of animals of each species that will be housed each day, you can usually get a very good estimate of what food and bedding will cost. But for office supplies, cleaning equipment, surgical supplies, and so forth, it can be harder to estimate your total costs for the coming year. At this time, you need not record how much they will cost you; just list the items you anticipate having to purchase. You should be able to obtain almost all of this information from your current expense budget. If you are just opening your animal facility, look at a budget sheet and monthly expense reports from another animal facility, just to be sure you aren't forgetting anything.

From this point, the path you have to take is obvious. You will have to estimate all of your costs for the coming year and do the same calculations I described for the retrospective *per diem* rate. If you can possibly put your supplies into distinct categories for a given species, you should do so. For example, you can have a column called "Environmental Enrichment," if you know that certain enrichment objects or methods are only for rats, certain ones only for mice, others only for monkeys, and so forth. Wherever possible, assign items to a given species. That way, there will be far fewer items in that lumped-together category of "Other Supplies and Equipment."

Computer Calculations of *Per Diem* Rates

Computer software programs that calculate your *per diem* rates operate much like what was explained above. They calculate the rate using information you provide about labor, supplies, small-equipment purchases, the size of your laboratory animal facility, cage sizes, and your average daily census of each species. Rather than *you* specifying the percentage of each expense that will be allocated to a species, a typical computer program makes certain assumptions about how the allocations will be made, often based on the amount of total facility space occupied by a particular species. For example, based on your input, the program might determine the amount of space taken up by one mouse cage out of the combined size of all of your animal holding rooms. Then, integrating this with census figures you supply, the computer allocates a portion of all expenses to mice.

Computer generation of *per diem* rates certainly helps make life a little easier. I use a computer program that is based on a spreadsheet I put together myself, using the concepts I described for a retrospective *per diem* calculation. It took me more time than I would like to develop it, but then again, I'm borderline computer illiterate. Fortunately, our IT department was able to integrate my spreadsheet with the electronic time recording program IT developed for our staff, which largely automates all of the labor calculations that are part of the *per diem* calculation process. Remember, though, any computer program used for *per diem* calculations requires data input from you. It will do the calculations for you, but either directly or indirectly, you still have to feed in the raw data.

Fee-for-Service Charges

As indicated earlier in this appendix, it's not unusual for an animal facility employee to provide more than one service. For example, most of the time, a technician may be performing animal care work that is normally included in an investigator's *per*

	Per diem time			Fee-for-service time		
Hour of Day	Mouse	Rat	Swine	Other Activity	Lunch	Total
7–8 a.m.	45		15			60 minutes
8–9 a.m.						60 minutes

Figure A2.1 Section of a time-motion activity log used for recording both *per diem* and fee-for-service activity.

diem charge. However, once in a while, that technician may perform fee-for-service work, such as giving an antibiotic injection to an animal. Let's say that injection took 15 minutes. How is that time recorded? Different animal facilities have developed different methods of recording this time. Where I work, nearly 100% of it is recorded electronically (on a computer program), but it can also be recorded manually. Figure A2.1 is a small section of the sample time-motion activity log (described earlier in this chapter) in which the time recording boxes are divided diagonally. The left (upper) section is used for time recording of typical daily activities that are included in the *per diem* charge, while the right (lower) section is used to record the time (if any) spent on fee-for-service activities.

When you are doing the calculations needed to develop your *per diem* rates, you will subtract the dollar value of the time spent for fee-for-service work from the total salaries and benefits of the labor group they are in. For example, assume that the total salaries and benefits for all of your animal care technicians were $100,000. Also assume your animal facility charges $25 per hour for fee-for-service work performed by animal care technicians, and the animal care technicians recorded 12,000 minutes (200 hours) of fee-for-service work over the course of the year.

$$200 \text{ hours} \times \$25/\text{hour} = \$5,000$$
$$\$100,000 - \$5,000 = \$95,000$$

When calculating your *per diem* charges, you use $95,000 as the true salary plus benefits of your animal care technicians because the $5,000 of fee-for-service work was already charged back to the investigator.

Using Activity Logs

Reality dictates that you cannot expect to receive an activity log from each person on every day of the week that the log is supposed to be used. Some people forget to turn them in; some are on vacation, out sick, at a meeting, and so on. How do you actually calculate the level of effort from these sheets? Here's what I do.

1. Ask all people who work with animals to record their time in minutes. I can assure you that you will have far fewer headaches if people record their time by minutes, rather than by hours and parts of hours. It doesn't matter if the person worked a full day, part of a day, or overtime; just record in minutes the actual time worked with each animal species.
2. Add up the total number of minutes all employees with the same job title (e.g., animal care technician) worked with each individual species (e.g., mice or rats). Then add up the total minutes all employees with the same job title worked with all species (e.g., mice, rats, and rabbits). Divide the first number by the second number. For example, for animal care technicians,

   Total number of minutes/year recorded by the facility's 2 animal care technicians for working with mice = 131,250 minutes
   Total number of minutes/year recorded by 2 animal care technicians for working with *all* animal species = 175,000 minutes

$$\frac{131,250}{175,000} = 75\%$$

This means that working with mice constituted 75% of the work of all the animal care technicians. This is the figure that goes into Table A2.1. Don't worry if somebody was on vacation or out sick. We're looking for a percentage of effort, so a little absenteeism will not have a large effect on the percentage over the course of a year. On the other hand, it is important to try to get everybody to turn in an activity log when they are at work. What would happen, for example, if you had only two people who worked with rabbits and one of them never turned in an activity log? That could make a significant difference in the *per diem* rate calculation.

You can add up the total minutes spent with animals monthly or annually, as you choose, but it makes life easier if it's done monthly, and then you just total everything at the end of the year.

3. If you have two animal care technicians, both of whom work with animals at some time during the day, you would expect a maximum of 14 activity logs in a month (2 people × 7 days in a week × 1 week/month of recording activity = 14 logs). Remember, you only collect logs for one week (seven days) per month. We understand, of course, that most people will not work a seven-day week every week, but that will not throw off your calculations. Similarly, in a year, you can have a maximum of 84 activity logs from any one person (7 days/month × 12 months = 84). Because of vacations, holidays, weekends, and sick days, it will most likely be less than 84 activity logs per person.
4. Disregard any time recorded in the "Other" or "Lunch" columns. Those columns are only there to make sure that the person recording his time

remembers to record 60 minutes for every hour worked (except, as noted earlier, if he doesn't start or stop work exactly on the hour). "Other" not only includes time spent in meetings or breaks but also is used when a person's labor cannot be assigned to a particular animal species. For instance, if a person is mopping a common hallway that is surrounded by rooms housing rats, mice, and rabbits, the easiest solution is to designate that time to "Other" because multiple species are involved. You can, of course, calculate the percentage of cages of each species and proportion the person's time that way, but very few animal facilities will have to go to that extreme.

5. Remember that activity log summaries and calculations are made for each job title, such as animal care technician and veterinary technician. In the example shown above, animal care technicians spent (on average) 75% of their time working with mice. Take this figure and put it into the "Animal Care Techs' Effort" column in Table A2.1. You have now used an activity log to determine the percent of time that your animal care technicians spend working with a particular species of animal.

References

1. National Institutes of Health. 2013. FAQs for costing of NIH-funded core facilities. Notice No. NOT-OD-053. https://grants.nih.gov/grants/guide/notice-files/NOT-OD-13-053.html (accessed March 3, 2016).
2. National Institutes of Health. 2013. Frequently asked questions: Core facilities. April 18. http://grants.nih.gov/grants/policy/core_facilities_faqs.htm#] (accessed March 3, 2016).
3. National Center for Research Resources. 2000. *Cost Analysis and Rate Setting Manual for Animal Research Facilities.* NIH Publication 00-2006. Bethesda, MD: National Institutes of Health.

Appendix 3: Hiring the Right People

If you hire the wrong people, all the fancy management techniques in the world won't bail you out.

Arnold (Red) Auerbach

General Principles

Auerbach was right, of course. Good management is important—very important—but do you really want to spend an inordinate amount of time trying to remedy a person's inappropriate behaviors, inadequate skills, or inability to work with others? Good hiring practices will go a long way toward preventing such problems. In economic terms alone, bad hiring is bad business because of the time and money it takes to train a person. This problem is compounded when we think about hiring senior managers and veterinarians, who are among the higher-paid employees of an animal facility. Then, of course, if you hire someone who fits into your organization like a square peg in a round hole, you start with a second strike against you because you may have to deal with morale problems among your coworkers. Managers are judged by the decisions they make, and as a manager, you will be held accountable for the actions of your team members. Hiring the wrong person will be a reflection on you. Hiring the right person makes your life much easier.

Because managers are judged by the decisions they make, some of the more important decisions you will make concern the people you will hire. In some unionized organizations, you may have little say as to who will be placed in certain positions, but in most instances, whether there is a union or not, you will have a major say about who is hired. By hiring the right person, you may have other potential hires start to knock at your door. Chairpersons of research departments know how important it is to build a strong nucleus of researchers because they can (and do) attract other top-notch researchers. This can work in a laboratory animal facility as well. In fact, it has worked well for my institution, as we have been able to attract some excellent new managers who first "checked us out" with managers they knew at our school.

We hire people for two reasons: to replace a person who has left or to expand our current staff. The first reason is more common in animal facilities, and as managers, we must recognize that very few people stay in the same position or at the same organization for their entire working life. Therefore, you will always have some employee turnover. So breathe easy and evaluate where you are and where you want to be. What do you think the anticipated workload will be in the foreseeable future? Do you think that the existing staff can handle the workload? Can some of the work be outsourced? I know that outsourcing (using an outside company to do work for you) is a controversial topic in the laboratory animal community, but that's the direction in which some organizations are heading.

It's not unusual for new managers to become upset when an employee leaves. They may take it as a personal failure and often want to fill the position right away. The rationale is that if you don't rehire quickly, maybe your boss will think there is something wrong with your ability to attract new people. Or maybe the rest of your staff will start complaining that there's too much work. My advice to you is what I said above: slow down for a second and evaluate where you are and where you want to be. Most administrators would rather see fewer employees doing the same amount of work, so there is rarely a need to impress your boss with how fast you can fill a position. In the long run, by hiring the right person—not just any person—you save money and time, and increase the effectiveness and efficiency of your animal facility.

Only when you are convinced that you need a new person should you start the hiring process. The same general principles for hiring will hold true whether the person to be hired already works for your organization or will be hired from the outside. In this short appendix, I hope to give you some important help while going through the hiring process.

1. Either by yourself or with your supervisor, *determine the optimal and minimal qualifications for the position.* In fact, this may be a good time to reevaluate its requirements. It is critical for every hiring manager to clearly understand the type of person that is needed. Your job announcement should unambiguously state what the responsibilities are and what type of a background is needed.

    The qualifications should fit the position and not incorporate hyperbole that is essentially meaningless. For example, if your opening is for an entry-level supervisor, there's very little need to have 10 years of supervisory experience as a requirement for employment. Under most circumstances, would you really want to hire a very seasoned manager for an entry-level position? If you are hiring a veterinarian who will be primarily responsible for performing large-animal surgery, don't write that "you will be primarily responsible for large animal surgery." Does that mean performing surgery or overseeing surgery? Instead, be specific and write that the candidate must be skilled at performing a large variety of experimental surgical procedures (such as cardiac valve replacements and organ allografts) on swine and other large research animals, and must be experienced in administering large-animal anesthesia and analgesia. Add to the job description, if appropriate, that the successful candidate must have the ability to be diplomatic while interacting with biomedical researchers of differing backgrounds and surgical expertise.

If you cannot find a person who matches your job description, my advice is *not* to give up and simply hire the best person out of those you have already considered. If at all possible, keep the hiring process open until you find the right person. The opposite statement is also true. If you are confident you have found the right person, then don't risk losing him by continuing to interview people until you have met every last applicant. By then, the person you want to hire may have accepted another job. There are times when a human resources policy requires that all applicants with the basic job qualifications be interviewed to help ensure that there are no discriminatory hiring practices. This is ethically appropriate, yet the delay can lead to the loss of the best candidate.

There is one warning note to be made. A person who is a star at one animal facility may not turn out to be such a star at yours. That's because performance is a combination of that person's personal attributes (which may be stellar) and an organization's culture and the quality of supervision that person will receive [1]. One of your goals in hiring a person is to try to make sure the new person will be comfortable in his or her new position. Some of you who have been managers for a few years have likely seen star-quality researchers leave an institution after two or three years because they simply didn't click with the institution's culture or their boss.

2. *Be clear and honest.* Let me be blunt about this. It makes no sense to sugarcoat a job description and make it sound much better than what it really is. I know that it's tempting to do this when there are very few applicants, but you will pay for this deception, in spades, if the new hire is unhappy. You want to start off with trust; otherwise, you're going to have an uphill battle trying to regain it. Under the worst of circumstances, you may have a person who is unhappy enough to sabotage your operations before she quits or is fired.

You should be able to describe the key responsibilities of a job in one or two clear and truthful sentences. Many of us have seen the problems that arise when the advertised position and the actual position are quite different. I don't know if some people sanction this because they think that every ad is supposed to glorify a position, if they actually believe the position is what they advertised, or if they are being intentionally deceitful just to get somebody hired. One way or the other, it's a horrible hiring tactic. It's beyond logic why you would spend the time and money necessary to find, hire, and train a person who has been deceived into taking a position that's significantly different from the one advertised. Let me give you some examples: If you advertise that a position has "unlimited growth potential," but in reality it has almost no growth potential, you start off with a disgruntled employee who can spread her dissatisfaction to others and is likely to leave the position. If you advertise that an employee is expected to be creative, but the job has almost no room for creativity without obtaining up- and down-the-line approvals, then a creative person taking the position is the wrong one for the job. If you advertise that occasional weekend work is required, but the new person is scheduled for every weekend, you have demonstrated that you can't be trusted. As a final but positive example, the New York State College of Veterinary Medicine at Cornell University teaches many of its courses as tutorials, which includes problem-based and case-based learning. This type of educational paradigm may not fit all students (because not all students learn in the same manner), so Cornell makes it a point to let potential students know ahead of time how they will be taught. That way, students can determine if what Cornell offers meshes with what they need.

One can counter that a prospective employee (or student) should sniff around and ask current or former employees about their experiences with their company and the company's culture, but alas, this is not always possible. Indeed, in some companies, current employees who are honest with a prospective employee may become former employees quicker than they intended. Fortunately, in other companies (and schools), prospective students and employees are encouraged to get honest feedback from current students and employees.

Being clear about reporting lines is just as important as being clear about a job's daily responsibilities.

> An animal care supervisor resigned his position to take a managerial position at a different school where there already was a vivarium manager. Before he left his old job, his old boss asked him what his responsibilities were going to be, how they would differ from those of the manager who was already at the new institution, and who his boss would be. The supervisor didn't have a good answer for any of those questions, other than knowing about a few of his responsibilities. He assumed he would report to the new facility director because that was the person with whom he interviewed. Soon after his new job started, the new manager and the manager who was already there began butting heads about job responsibilities and reporting lines.

I don't know how this conflict was resolved, but that's not a concern. Of course, the new manager should have asked more questions about the job, but the problem didn't begin with either the old or the new manager. It began with the interview, and possibly even the advertising that neglected or wasn't clear enough about specific responsibilities and reporting lines. All of this means that the first step toward building trust in the workplace begins with the hiring process. Advertisements should be reasonably correlated to the job, and interviewers should be honest and informative with the job candidate. The person being considered should leave with a factual impression of your animal facility. Honesty does not imply that every positive and negative aspect of the animal facility has to be placed on display. Rather, it means that the person being considered for the position has to have a realistic view of the job, the opportunities in your company for advancement, and equally important, the general organizational culture.

The first step toward building trust in the workplace is to exhibit honesty during the hiring process.

I don't have any hard figures on what it costs to replace a person in a laboratory animal facility, but my best guess is that it's anywhere from 50% to 100% of a person's annual salary, depending on his or her position. Why? As noted elsewhere in this book, we have to consider the "downtime" when a person leaves, paperwork, advertisements, the time cost for the people who have to cover the gap, and the considerable amount of time for training a new person. It adds up quickly, and it's a strong argument for being honest up front.

Here's a final thought: For some positions, I developed a detailed job description that stated what had to be done daily, weekly, and so on. I sent this to applicants I was considering hiring. Most people appreciated this, and although some chose to withdraw their application, on the whole, it worked to everybody's benefit. One downside of this practice is that you can lose some job flexibility and possibly lose a good person. You simply have to decide whether this method will help or hinder you. As a rule of thumb, I would *not* use this technique when recruiting a senior-level manager, but I would consider it for people whose work is more technical in nature.

3. Once you have evaluated the requirements, *make sure the position is advertised where it will do the most good, and make sure that individuals in your organization are aware of the new opening.* Throughout this book, I have emphasized the importance of networking, which is getting to know people through your own contacts and through contacts of other people. When it comes to hiring people, sometimes your best candidates come from networking, not from advertisements. For example, if a senior veterinarian is needed, in addition

to using advertisements or recruiting firms, why not ask colleagues whom they would really like to hire if they had a job opening. Armed with that knowledge, you can contact those veterinarians to determine their interest in moving; if they are not currently available, ask them to suggest others who might fit the position you have open.

Your own staff may wish to apply, or they may know people who fit your requirements. Don't forget, Internet sites such as LinkedIn, Monster, Indeed, or Spoke can be excellent sources for both advertising an opening and finding résumés of potential employees. In fact, they are likely to be the first place many people look. We also have seen video, not just written résumés, on the Internet. This is an interesting yet potentially troublesome phenomenon because it has the potential for leading to discrimination in hiring (e.g., male vs. female, old vs. young, attractive vs. average looking, and so forth). Eventually, this risk will be worked out to everyone's satisfaction, but for now, I have to advise caution with video résumés.

A file of recent applicants who were not hired is always a good source of information. Check through it, as there may be some qualified people still looking for a new job. As a rule of thumb, I suggest keeping all applications for any advertised position for about a year. I also try to keep on good terms with certain key people who have worked for me but have left for another position. Career goals change, and there may be a mutual need to work together once again.

4. I suppose that the easiest (and wisest) thing I can say about hiring is to advise you to *hire the smartest people you can find who are qualified for the position*, even if that means that person is better than you are at parts of the job. There's no doubt in my mind that I would rather have a reasonably intelligent person who needs less care than someone who needs more care. Similarly, I would rather have a really good coworker who stays for three years than a marginal one who stays forever. If you are looking for an average person who will come to work every day, get through their daily work, but do nothing more, that's acceptable if your goal is to have a lot of average people who don't give you any hassles. But if your goal is to have better-than-average employees, then you should have better-than-average people interviewing and making the decision to hire these better-than-average people [2].

Should you consider a person who is likely to be overqualified for a job in your animal facility? The answer is maybe. There is no evidence that I know of that documents overqualified people as caring less, quitting more, quitting faster, or performing worse or better work than people who are probably not overqualified for their job. In fact, one interesting online article provides numerous reasons why you should consider an overqualified candidate [3]. An overqualified person may be looking for a less stressful work environment or have other personal needs that are not immediately obvious. Furthermore, should an opening occur that requires the skills of an overqualified person, and you already have such a person who understands all of the basic operations of an animal facility, your search time and training time may be dramatically lessened.

Of course, intelligence is not, and never will be, the only hallmark for good hiring, since not every person can excel at every job, nor do all people have the self-motivation needed to get them up to speed. I have three college degrees; nevertheless, I couldn't teach a beginning high school math course or build a bookcase if my life depended on it. Therefore, hiring smart people means that you have to find a person who is smart enough for the job you have open, and hopefully, the smartest of all your applicants for that position, but also has the basic qualifications that you consider to be important. If a potential employee seems smart enough to be able to operate your cage washer and ensure that the cages keep flowing, but can probably also take care of the unique demands of your new zebrafish facility, then you're way ahead of the game. That person may have a lot of practical intelligence. For a veterinarian's position, you may need a person who has a good amount of book knowledge and the ability to apply that knowledge. The person's interests and time will determine the best fit.

Should I hire the most experienced person? Not necessarily, although much depends on the needed job skills and the applicant's personality. To put it bluntly, if you can easily train a person to do a particular job, then I would stick with the concept of hiring smart and hiring a person who you feel will integrate well with your existing staff. The potential for a person to learn and grow into a new position is at least as important as a person's experience, and as you may recall from Chapter 3, it can be very hard to change a person's negative behaviors. It's easier to train a compatible new person. More is written about this subject in item 9 below.

When hiring a manager, one of your goals should be to hire people who are more skilled in a particular aspect of laboratory animal science than you are. If, for example, you need to hire a director of animal care, then your goal should be to hire a person whose practical and theoretical knowledge of the subject is greater than yours. If you need to hire a clinical veterinarian and you're a veterinarian who, like me, has drifted away from clinical work, then your goal should be to hire a person who knows more about clinical medicine than you do.

5. *Money helps, but it isn't everything.* One of the biggest problems some managers face is their own company's policies, which often demand hiring a person at a salary that is no greater than the midpoint of the salary range. You should consider the hiring salary range before, not after, you decide on your preferred applicant. What happens if you are forced into hiring average people at an average salary? Most likely, you will get average people doing average work. In some instances, you can hire an exceptional person at an average salary, but in due time, you lose these good people to another department or a

competitor who is willing to pay more for a good employee [4]. Nevertheless, although money is certainly important to most of us, many people look for a position where they not only can make a living, but also can feel comfortable with their coworkers and with the overall corporate climate. This point was underlined for me by a colleague who, during his employment interview, interviewed me and my co-veterinarians as much as we interviewed him. His primary goal in determining where to work was to find a position where his future colleagues were congenial with each other.

6. I suggest that you *ask for a cover letter and a résumé from anybody who is seeking the new position.* This applies to any applicant, no matter what level the position or what level of education it requires. Many human resources departments in larger organizations will require this, and the days of hiring any person who can stand up and breathe are long gone. Cover letters and résumés that are handwritten, on lined paper, have cross-outs, are far outdated, and have other obvious flaws may tell you that this person is not managerial quality. If the person is not applying for a managerial position, less weight is given to style. Do I make exceptions? Sure I do. There are some disadvantaged people who need a helping hand to get started, and this often shows through on a written application. But those are the exceptions, not the rule.

An alternative to having a résumé sent in is to have a brief telephone interview. This tends to work best for hiring people with unique skills, such as a veterinarian or a transgenic facility manager who doesn't reside near your animal facility. An initial telephone interview can be helpful in determining if the applicant truly understands the position's responsibilities, if the applicant's qualifications are appropriate, if the person would potentially integrate well with your current coworkers, if the salary range is acceptable, and any other basic issues. If the preliminary telephone interview seems promising, it can be followed up by an interview at your office or by having the candidate fill out a formal application to be sent to you.

7. *When an applicant is interviewed at your place of business, that person should be expected to be neatly dressed and groomed.* I believe that someone who is serious about obtaining a new job will make the effort to be presentable. This is true for any position—managerial or otherwise. First impressions count (as discussed below), and a bad first impression is hard to repair.

I do practice what I preach. I rejected an applicant for a supervisory position, in part because she arrived for her interview wearing old jeans. As much as I personally like wearing jeans, I would never consider wearing them for an interview. There's a certain amount of tact and common sense needed as a supervisor, and in my opinion, this woman showed she didn't have much of either.

8. *Before the interview, have an idea of what questions you are going to ask.* This advice holds true for both telephone and face-to-face interviews. Also consider who will be the interviewer. Consider using your best people to perform

the interview. If you are hiring a veterinarian, the primary reviewers should be your top veterinarians or others who know what to expect of a laboratory animal veterinarian. Most of these interviewers may not be pros at assessing potential new hires, but they can do a reasonable job. Unfortunately, I've participated in interviews where some of my colleagues asked questions or made statements that were so outlandish that I would have been afraid to take the job had I been the person being interviewed. Every animal facility has people who are good at their jobs but are poor at interviewing. If you know who these people are, don't use them as interviewers. Robert Sutton, the noted management theorist, was using some tongue-in-cheek imagery when he wrote that we should stop our jerks from hiring other jerks [5], and although I wouldn't be that callous, his message is unmistakable.

Interviewers are often influenced by the way the people being interviewed dress and are groomed. Certainly, you want a person to dress appropriately and be reasonably groomed for an interview, but a person's other physical characteristics (e.g., height, weight, and gender) should never enter into the equation unless it prevents the interviewee from performing the responsibilities of the job. There also are many other topics that cannot be broached during an interview (such as how a parent will provide for child care during the workday), and most human resources departments can educate you and your staff on allowable and unallowable interview topics.

Many, perhaps most, interviewers focus on the immediate needs of their department (such as hiring a new manager) and ask questions geared to those needs, such as "What is the most difficult management situation you have faced?" That's a perfectly reasonable question to ask, but don't forget about the goals of your animal facility and the culture of the facility [6]. As an example, the University of Massachusetts Medical School is not far from Southborough, Massachusetts, where the New England National Primate Research Center was located. When we first learned that the primate center was closing, we predicted that demands would be placed on our vivarium to house nonhuman primates. Because of that fact, and because we have very little turnover of animal care supervisors, when a supervisory opening did occur, we only interviewed people having experience with nonhuman primates. Other skills and traits were discussed during the interviews, as were the culture and goals of our department. The people performing the interviews knew the department's goals, and of course, they knew the culture of the department and the school. For the person who was hired and for our department, it was a win–win scenario.

With your department's specific goals in mind, interviewers should encourage applicants to talk by asking questions about their background, and especially why they applied for the job (even if the reason is obvious). What are their plans for the future (perhaps they plan to go back to school in two months), and what do they expect to get out of the job? Although you may want a person with animal-related experience, hiring people for laboratory animal facilities without a formal animal-related background may leave you pleasantly surprised if you are interviewing for a technician's position. Some of the best technicians I've hired liked animals, had a pet when they were kids, but had backgrounds in psychology, literature, and the like. In truth, at the entry level it's unusual to find a person who has a background working in a laboratory animal facility. So, look at the entire person and don't assume that a background working with animals is essential. Ambitious, bright people usually pick up on the daily work routine very fast. I'll add just one word of caution: There are positions that do require some experience, whether from schooling, hands-on, or both. A supervisor of animal care might be a good example of such a position.

Likewise, there are cultural differences to consider. One of the finest people I ever worked with was Korean, and very traditional in his upbringing. He would rarely shake hands and often did not make eye contact when I would talk with him. Nevertheless, it was absolutely clear that he knew his business. It would have been quite a loss not to have had him as a coworker, simply because his actions reflected a different culture.

Should you ask a person if he or she would be able to work in your facility? It makes no sense to do so. You already know what the answer will be. Rather, ask broad questions that don't require a yes or no answer and can potentially show a person's ability to think or act rationally. Some examples are, "How would you approach a dog that is afraid of you?" "What do you think you would do if you found a monkey had escaped from its cage?" I wouldn't even hesitate to ask generic questions about a person's last vacation if it got him to open up a little. I'm more concerned about what a person has to say, in general, than a specific answer to a specific question. In other words, who cares if an applicant who has never worked with monkeys knows nothing about the details of capturing an escaped monkey? Although there are those who disagree with this line of questioning, I'm of the opinion that you want to know how they would approach the problem, not the specifics of how they would solve it.

A former colleague had an excellent knack of getting people to talk so he could evaluate their ability to handle new situations. Rather than asking a question that requires a distinct answer (such as, "How would you capture an escaped monkey?"), he asked much more open-ended questions. For example, when interviewing private practice veterinarians for a clinical position within our department, he said, "I'll bet you came across some unique challenges in your practice. Tell me about one of them and how you handled it." This type of open-ended question allows the respondent to think about both the situation and an answer, not the answer alone. Using open-ended questions is an excellent technique when the situation is applicable. The downside of asking open-ended questions is that some people speak forever, and if that happens, you have to know how to interpret whatever answer you get. If a person gave a reasonable response to the question about a challenge in private practice and another applicant gave a different but nevertheless equally reasonable answer, is there a way of discriminating which response was better? Sometimes, an interview question requiring a more specific response is better; other times, an open-ended question is better. It really depends on the requirements of the position.

You should ask questions that are in line with the job requirements you have established. If initiative is one such requirement, you will not want to hire a person who responds, "I don't know" to the question about the escaped monkey. You would not necessarily expect that person to know precisely what to do, but you should expect an answer that would indicate to you that the potential employee could make a reasonable decision under pressure.

I do have two cautions for you. First, don't assume that an "I don't know" answer is always wrong. Sometimes it's the most honest and best answer you can get from a person, although you should expect some explanation attached to that kind of an answer. Not knowing what to do if a monkey escapes may be a fatal flaw for a prospective animal care supervisor in a primate center, but perhaps not so for a prospective animal care technician who has never before worked with monkeys. The level of initiative required is something you have to think about before the interview.

The second caution is related to the truthfulness of responses. Don't be afraid to probe a person's response to a question to see if they're being honest with you. For example, if you ask a person about the various skills they have, don't be surprised that they can't demonstrate those skills once they are hired. Why? Because they told you what you wanted to hear during the interview process. They may have seen the skill performed by others (e.g., an intravenous injection into a tail vein), and they may have even tried it once or twice, but the needed proficiency isn't there. If certain specific skills are needed, consider digging deeper to determine the extent of a person's capabilities.

Having given you some thoughts about asking appropriate questions, I'll become a little cynical. Many (perhaps most) people really use the interview to evaluate a person's personality and potential fit in their organization. That might take about 90 seconds or so, because that first impression made by the applicant is hard to change, even over the course of an hour's interview. If that first impression is poor, the evaluation is often poor. Some job applicants recognize this and focus more on how they present themselves, rather than on their qualifications. As Mel Kleiman notes, "People who interview the best are often people who have been interviewed the most. They've heard every question and honed their answers over time. They know how to tell you what you want to hear" [7]. Therefore, don't let yourself be fooled into thinking that a pleasant, articulate person is necessarily the right choice. Likewise, don't base your hiring decision on the appropriate or inappropriate answer to one or two questions. Look at the entire person in front of you, and try to determine if he or she is really appropriate for your job and your organization.

9. *Try to determine if the person you are interviewing will mesh with your existing staff.* Even the best of people may present a problem if they just will not fit in. My basic philosophy, which I share with others [8], is that a person's ability to work harmoniously with others is of high importance when lower levels of technical or managerial skill are needed, but as we hire people for more advanced positions, such as high-level managers, that person's personality remains important, but not as important as her skills as a manager and leader. That said, I want to add a caveat and emphasize that a manager or veterinarian who cannot get along with other managers or veterinarians has a high probability of not being able to get along with the research community and can quickly tarnish the reputation of your department. Therefore, although skills are more important than personality when hiring at higher levels, you also must consider what interactions the new person will have with others in your department and with external animal users. The more interactions, the more weight you will want to give to the individual's personality.

Hiring a person who meshes with your existing staff is a reflection of your department's culture. If your goal is to hire a highly motivated and forward-thinking person who will need minimal supervision, first look at the staff you have now. Are any of them self-starters? Do you, as a manager, encourage a culture of providing decision-making discretion, sharing information, minimizing incivility, and offering performance feedback [9]? If you are not setting the stage for a vitalized (or revitalized) departmental culture, why would you consider hiring a person who may quickly leave or not acclimate to the existing culture? Some managers do try to hire a qualified "misfit" when they are trying to make changes in the type of personnel they hire, especially if they are "upgrading" (trying to improve the quality of employee skills or behaviors). If you are doing this, be very careful and make sure that person knows what your intentions are, and make sure that you are willing to do your part to improve the culture. To hire this person and just hope for the best is an invitation to disaster.

It's not unusual for me to ask a potential employee to meet with all or some of the people with whom he will be working (or even supervising). I know that many people disagree with this, but I believe it's helpful. Afterward, I solicit and use the feedback of the animal facility people. Depending on the circumstances, I may even have the entire staff vote on the person they would like to see hired. Staff involvement has a dual benefit: you get feedback and you involve others in making a decision that can directly affect them.

Sometimes, if I *really* want a specific person, I'll ask him or her to meet with upper management. Doing this sends a clear message to the job applicant that "you are really important and wanted." Let's not forget that people's sense of importance in their work is one of the basic requirements of motivation, and if you initiate that process with sincerity while that person is being interviewed, you can enhance your potential to hire her. I'll say a little more about this in just a moment.

In recent years, there has been a trend in interviewing that has become nearly standard for many companies, although I am not aware of it being used (so far) for vivarium hiring purposes. I am referring to personality testing, which is typically a professionally developed questionnaire that is used to uncover a job applicant's psychological profile in order to help make decisions about hiring, not hiring, or where a person might best fit into the company if he or she is hired. While there are advantages to using such a profiling tool, there are certain intangibles that, in my opinion, still require an interview. These intangibles include how a person responds to interview questions (e.g., fast and without much thought vs. slowly and thoughtfully), how a person dresses, and whether the person arrives on time for the interview. Time will tell if standardized personality testing will find a place in vivarium hiring decisions.

10. *You should always check references.* If a person you believe should be contacted is not listed as a reference, get the applicant's permission to contact that person. Although business and academic references are quite important, don't dismiss personal references. Many managers assume they're valueless because the applicant carefully chose those references and they will probably say nothing negative about the applicant. Nevertheless, I have had a number of personal references give an honest opinion of an applicant's suitability for a job. I assume that happens because I point out to the reference that I am confident they would not want a friend to fail in a new position, and therefore an honest evaluation of that person's potential for success was very important.

Don't forget networking. If someone is working at one facility and is looking to come to yours, it's often appropriate to call the manager of the first facility and "get the scoop" on the candidate. There are times, of course, when you can't do this. The most obvious is when a person has not told their current supervisor that they are planning to leave, or has specifically told you not to contact that person. If you call the supervisor, you may be giving a kiss of death to your applicant at his current job, and you can be sure he will not look favorably upon ever working with you. Yet, you can always ask a candidate if she has told her current supervisor of her intent to leave, particularly if the current supervisor is not listed as a reference to be contacted.

Checking references can be an art in itself. Put together a list of questions you want to ask, particularly those that are important for the job. If punctuality is important to you, don't be afraid to ask about the person's arriving late or leaving early. Ask about strengths and weaknesses. If the person you are speaking to is a current employer who is giving a very positive reference, ask if he tried to hold on to the applicant. If not, why not? Some people will not give you much (or any) information over the telephone, and if that's the case, get another reference, if possible. One thing is for sure—always check more than one reference.

11. *Consider the possibility of interviewing an applicant at his or her place of business.* This obviously makes the most sense when the person is working in another laboratory animal facility and it would do no political harm to you or the applicant if you went to that facility. If you are able to do this, you can observe for yourself the applicant's area of responsibility and work habits. Going to a person's place of business is not usual, but if you can do it, much valuable information can be gained.

12. *All individuals who have responded to an advertisement or an announcement should be contacted, one way or another, to inform them of the status of their application.* Most of us have felt the frustration of applying for a job and never hearing anything from the hiring company. It's good business to respond, and in the future, you may have a need for that person's services.

Summary of the Hiring Process

- Determine the optimal and minimal qualifications for the position.
- Be honest with yourself and any applicants about the requirements and opportunities of the position.
- Advertise or network as needed to uncover potential applicants.
- Go after the smartest person who is right for the position.
- Be prepared to offer a salary commensurate with the abilities of the applicant and the needs of your organization.
- Get a cover letter and résumé. Consider a telephone interview.
- Appearances count. An interviewee should be dressed neatly.
- Before the interview, compile a list of open-ended questions.
- Ask yourself if the applicant will mesh with your current staff.
- Always check references.
- Consider the possibility of interviewing an applicant at his or her current place of business.

Hiring the "Hot Prospect"

Before I end this appendix, I'd like to say just a few words about hiring a specific person who has identified herself as being interested in your open position. More important, this is a person that you would very seriously consider for that position. Maybe you know about her personally, through her publications, the strength of her résumé, or other means, but one way or the other, you're happy to see her application. Now what happens? To begin with, perhaps you and your coworkers can give yourselves a pat on the back for establishing the basic culture that would attract such a person. Next, consider yourself on equal footing with her because at this stage of the game, you are willing to seriously consider each other for employment. Third, invite the person in for a chat, not necessarily a formal interview, even if you have to pay travel expenses. If you can't do that much, then maybe the position isn't quite as important as you think, or perhaps you can make do with a less qualified individual. Having that person come to meet you and your colleagues is akin to a salesman getting his foot in the door. A face-to-face meeting is often worth its weight in gold, particularly since you have to consider that there are others out there who may want to hire the same person.

Just as you might want a person who will fit in with your existing group, she wants to fit in as well. She wants, as you already know from Chapter 3, a fair salary, good working conditions, respect, a feeling of accomplishment, and a feeling that her work is going to be appreciated. Are these things you can really provide, or are you going to fool yourself into believing you can provide them? You're going to have to be dead honest. If the salary is fine but the benefits are weak, there's no need to emphasize the weakness of the benefits, but all benefits can be discussed informally so there's no shock factor later on. If you need somebody to help you build cohesiveness within the group, you have to talk about that as an opportunity for growth and accomplishment, not by laying out all of your gripes. But you have to be honest.

You also have to talk about her goals. We'll assume her talents match your job opening, but are her goals aligned with what you can provide? What challenges make her happy? What doesn't she want? If you know, at least in general terms, what she wants (outside of the basics I noted previously), can you provide them? There's no need to talk in detail about a salary when it's early in the hiring process, but do you have to shy away from the subject? I would say no. If the tenor of the early discussion seems positive, then toward the end of the discussion, see if the two of you can agree on a ballpark range, even if it's quite a wide range. If the candidate is adamant that she can't accept less than $100,000, and you know you simply cannot come close to that, then you can part friends early on rather than waste each other's time. I'll grant you that there are times when a candidate's insistence on a high salary is nothing but a negotiating point, but you're not yet in salary negotiations; you're just trying to see if there's even a chance of finding common ground.

Nowadays, people under the age of 30 (the so-called Generation Y or Millennial Generation) look at employment somewhat differently than they did 15 or 20 years ago. The need for positive social interactions holds a great deal of weight, as does the desire to have sufficient free time for personal enjoyment. People are defining themselves by the music they listen to, not by their political or business views. As Nadira Hira wrote, "Nearly every businessperson over 30 has done it: sat in his office after a staff meeting and—reflecting upon the 25-year-old colleague with two tattoos, a piercing, no watch and a shameless propensity for chatting up the boss—wondered, What is with that guy?!" [10]. We now have a consumer generation that has many choices, and our job opening is looked at as if it was just 1 of 10 different cars. In addition to salary and benefits, the new young generation of employees wants honesty, respect, and an opportunity to do good things for people and animals. What does this mean to you if you are recruiting that special younger person? It means if she was born sometime after about 1980, and you're somewhat older than she is, don't expect her to be you. The hiring game has changed, and we have to consider being more flexible with working hours, more concerned about interpersonal relationships at work, and able to provide a choice of benefits, an opportunity to incorporate community service into the workplace, and a basic understanding that salary and benefits remain important, but they are not the keys to successful hiring.

So now that she's at your place of business for an interview, it's important to take her to an informal lunch with some of the core people she's going to interact with in the animal facility. She wants to see if there's a potentially good social network. You want to discuss her interests outside of work and how they might integrate with her work. Perhaps you can talk about your interests and find common ground, but overall, we have to understand that a changing workforce is here, and we have to adapt to it or lose the best and the brightest.

After the Hire Is Made

I'll be brief. After a new person is hired, I strongly suggest you do as much as you can to quickly integrate this person into all aspects of your vivarium's operations. In addition to introducing the new person to those he will work with on a day-to-day basis, take him around and introduce him to the front office staff, other employees, the business manager, the facility director, the veterinarians, and so forth. Make sure he gets a tour through all of your vivariums, if you have more than one. If it's appropriate to do so, show him where the cafeteria is and discuss general expectations, such as arrival time, quitting time, and the overall culture of your facility. If you have descriptive literature about your institution, give it to him so he can see how his work will support the overall institutional mission. During his first few weeks on the job, keep up a friendly dialogue fairly often to help ensure that he has the tools he needs to succeed and is integrating into the vivarium's culture and activities. If he has a problem, be there to support him.

And If a Mistake Is Made

I'll conclude with some information that was developed largely by personnel specialist Jeff Haden [11]. Inevitably, all of us will make a hiring error, and in the long run, that error becomes a problem for the animal facility and for the person who was hired and now may be fired. None of us like to dismiss a person we thought would be an asset to the facility, so Haden offers suggestions on how to quickly spot a person who is likely to be a problem employee. He believes that new employees who are late or absent within the first few weeks of work are prone to have this problem exacerbate over time. Employees who immediately state that they need a new computer or other tools before they are familiar with the details of their work tend to be problems later on. A loud, assertive new hire may always be a handful. And lastly, a new hire who continually suggests that things were better at his old job should have stayed there.

Whatever the reason, there is often a reluctance to end the employment relationship sooner rather than later, particularly if the newly hired person had to relocate his or her family. There is also a feeling that "I made an error, but I don't want to have to admit to it," and likewise, there is often the hope that just given a little more time, everything will work itself out because you don't want to have to go through the whole hiring process again from the very beginning. More often than not, the problem does not work itself out. It just festers and becomes worse over time until it becomes so bad that releasing the employee is the only way out. If your institution is one of those that has a three- or six-month trial period for a new employee, use it wisely to fully evaluate the new employee. In essence, this is a paid tryout period for both the new employee and you, the employer. Don't be afraid to work with your human resources department to release unsatisfactory employees before they do lasting harm to your vivarium. If you allow an unsatisfactory employee to remain past the trial period, the damage will continue to accumulate, and the longer you wait, the harder it will be to rectify the problem.

It's a helpful exercise to ask yourself, once in a while, if you would rehire the people you hired on your team if the opportunity was available to do so [12]. If the answer is yes for most of your hires, you've done a commendable job.

References

1. Cappelli, P. 2013. HR for neophytes. *Harvard Bus. Rev.* 91(10): 25.
2. Girard, B. 2009. *The Google Way: How One Company Is Revolutionizing Management as We Know It.* San Francisco: No Starch Press.
3. Sullivan, J. 2014. Refusing to hire overqualified candidates—A myth that can hurt your firm. ERE Recruiting Intelligence. http://www.eremedia.com/ere/refusing-to-hire-overqualified-candidates-a-myth-that-can-hurt-your-firm/ (accessed February 12, 2016).
4. Dauten, D. 2006. *(Great) Employees Only: How Gifted Bosses Hire and De-Hire Their Way to Success.* Hoboken, NJ: John Wiley & Sons.
5. Sutton, R.I. 2007. *The No Asshole Rule: Building a Civilized Workplace and Surviving One That Isn't.* New York: Grand Central Publishing.
6. Phillips, J.M., and Gully, S.M. 2015. Multilevel and strategic recruiting: Where have we been, where can we go from here? *J. Manage.* 41: 1416.

7. Mel Kleiman Executive Reports. Why stupid people get hired: And what you can do to avoid it. http://api.ning.com/files/MhqovnvrwKq*vZ7YDkEl*xGlD Y4Jop*OWX-NNUeA1Mu06K7Y*EUEz73TjUBqFp36iSMeAN23o1J2JThLs-nczzQwIBpDNBcV/stupidpeopleexecreport.pdf (accessed April 7, 2014).

8. Gary, L. 2005. The high stakes of hiring today. *Harvard Manage. Update* 10(2): 1.

9. Spreitzer, G., and Porath, C. 2012. Creating sustainable performance. *Harvard Bus. Rev.* 90: 93.

10. Hira, N.A. 2007. Attracting the twentysomething worker. CNNMoney.com. http://money.cnn.com/magazines/fortune/fortune_archive/2007/05/28/100033934/index.htm (accessed March 7, 2016).

11. Haden, J. 2011. How to tell a newly hired employee may not cut it. http://www.cbsnews.com/news/how-to-tell-a-newly-hired-employee-may-not-cut-it/ (accessed March 3, 2014).

12. Harnish, V. 2010. *Mastering the Rockefeller Habits: What You Must Do to Increase the Value of Your Growing Firm.* Ashburn, VA: Gazelles.

Appendix 4: Training and Mentoring

> The only thing worse than trained employees leaving the company is untrained employees that stay with the company.

Lean Management Systems Handbook

Kaizen is the Japanese term that means continuous improvement. Along with the elimination of waste and producing value for the customer, kaizen is one of the three building blocks of Lean management (as discussed in Appendix 1). Continuous improvement and training are closely related concepts in managing a laboratory animal facility. For most of us, training connotes a vision of teaching people how to do a task, such as giving an intravenous injection or making basic repairs to robotics. But we also know that Institutional Animal Care and Use Committee (IACUC) members have to be trained in protocol review procedures and pertinent federal regulations, and new office staff may have to be trained in your facility's financial procedures. Managers also have to be trained to be a manager (which is the main purpose of this book), and leaders often have to be taught how to lead. Very often, training includes mentoring. Mentoring and training are closely related, but not quite the same. A mentor is an advisor, a supporter, and a motivator. In this appendix, we'll talk about the basics of training people in the traditional context of task performance, but we'll also touch on mentoring staff members to improve their performance, along with learning new skills.

Training

Even if a person has the basic skills required to perform a job, a training period may be needed to perfect them. It's important for people to clearly understand what is expected of them in terms of job responsibilities and what constitutes satisfactory performance of their jobs. That's basic and has been emphasized in other parts of this book. You probably want to know (or already know) that same information about your own job. Some people will need more training than others to live up to expectations, but almost everybody will need a little guidance. An animal facility manager who can get a new employee oriented, trained, and productive in three months is to be commended [1].

Employees have a right to know what is expected of them and what is considered to be satisfactory performance. Training is when this information is first transmitted.

Orienting the New Employee

Before I describe my thoughts on how to train new or promoted employees, an important question is, who does this training? This is not a simple question. Does the facility manager train a new technician or is that task delegated to a senior technician, or is there a person whose primary responsibility is training people? Who trains (or orients) the new supervisor of animal care, or the new facility director? There are no formal guidelines, but it's fair to say that, although indirectly everybody helps, the primary trainer should initially be a person who is a least one "level" higher than the new person or is the facility's designated trainer.

What's wrong, you may ask, with an operations manager orienting a new veterinarian, even if the veterinarian is to be the operation manager's boss? All she's going to do is give the new person some inside information on how things are done. She won't be teaching him how to be a veterinarian. Isn't that okay? Yes, it is okay to a certain extent. It's expected. But we have to be careful about the limits of that type of training. The new veterinarian will have his or her own opinions on how things should be done, and you don't want to unintentionally engender an adversarial relationship with your new boss by telling her what to do, how to do it, and when to do it. We all try to protect our own way of doing things, and by advancing our own agenda, we can easily offend a new superior. Therefore, with a person who is above you in the organization's hierarchy, tread carefully because we do not work in an egalitarian society.

The remainder of this discussion will concern itself with the more typical situation: training a person who, on the employment hierarchy, will be working for you. In Chapter 2, I discussed the culture of an organization, how it affects your work as a manager, and how you can learn about it. On a new employee's first day, it makes good managerial sense to sit down and discuss the culture of your organization. Describe, in some detail, what your organization does. Does it produce pharmaceuticals, perform biomedical research, or safety-test products? You might be surprised to learn that many employees know their own job, but have only a general idea of what their organization does (i.e., they may know little more than "we do research with animals"). We have to take that to the next level. Describe your organization's mission. I suggest that you do this even if your company puts out a six-color brochure that was given to the person before he was hired, because as I describe below, this information must be emphasized. In my animal facility, we provide new employees with an organizational culture handout that tells them about our philosophy toward treating animals, working hours, what to do if there are problems, and in general, how we do things in our department.

If the new person will report to you, tell her about your style of management and your expectations. Think back to the discussion on the need to communicate clearly. This is an important time to be clear, because what you say, even little jokes, may be misinterpreted. Don't forget to tell the new employee what is expected in terms of lunch hours, breaks, humane treatment of animals, and the like. Let the employee know who to go to if there are questions or problems. If you believe that close supervision will be needed for a certain period of time, say so. It can be nerve racking to a new person to have you constantly looking over her shoulder without her knowing why.

Paint the big picture for your new employees and tell them where they fit in. What is your department's mission? What is the institution's mission? Introduce them to the other people with whom they will be interacting, particularly if the new person is a manager. This will extend your own authority to your new manager. Since we all want to feel important in our jobs, let your new (and established) employees know what their jobs mean to the institution and the animals we care for. Why is our work important? Your personal touch shows that you are an interested and caring manager.

I never thought much about the need to communicate to my own staff about activities outside of the laboratory animal facility until one day, over lunch, some of the people I worked with asked me what kind of work was going on in another part of our building. They all knew we were doing cancer research in the laboratory animal facility, but after a little questioning, I realized that they knew very little about the functioning of the organization as a whole. They knew almost nothing about the different types of cancer we were studying, that we focused on prevention and not treatments, that we also had research in the prevention of cardiac disease, and that there was an entire group in New York City that was involved with epidemiology and bringing our research findings to the public. The staff was fascinated. I was surprised. I had personally told this to many of them when they were being interviewed. I learned two lessons that day. First, people remember what they want to remember during interviews, and second, it's often necessary to repeat important information after a person has settled into a new job.

Training Materials

A manual that describes the specific way things are done on a day-to-day basis is often found in laboratory animal facilities. Part of a page from such a manual is shown below. This manual is usually called the standard operating procedures (SOPs). In some laboratory animal facilities, SOPs are required or have an implied requirement by law or regulation (e.g., the Good Laboratory Practice Act and, for some animal facilities, the Animal Welfare Act regulations). Even if not mandated, it is a good idea for you to have SOPs to help train and orient your staff, and to serve as a reference document. Additionally, many research organizations obtain a large percentage of their income through grants from the federal government. If various government representatives visit your organization to evaluate its research capabilities and support programs, you should not be surprised to have these site visitors ask to tour the animal facility. When that happens, those people who are visiting you are often very familiar with animal facility operations, and it can be helpful to be able to refer to your SOPs. In fact, they may ask to see your SOPs. As one government site visitor said to me, "If it isn't written down, then it isn't so."

SOPs can include procedures for performing intravenous injections, temperature and humidity standards in animal rooms, frequency of cage changing, and anything else that can be formalized. New employees should be given the SOPs when they are hired. The pertinent SOPs should be reviewed and discussed with the new employees. We have to remember that based on new employees' previous work experiences, their understanding of all or part of an SOP may differ from yours. As I noted in Chapter 3, it is often necessary in a laboratory animal facility to do the same task the same way, day after day, year after year, as a means of decreasing research variables. SOPs are one of the information resources you can use to help ensure such uniformity, even though they can stifle creativity. When managing animal research facilities, there is always a need to balance the need for conformity with the need for creativity. This type of challenge is part of what makes management so interesting.

STANDARD OPERATING PROCEDURE A-29
PERSONAL INJURY REPORTING

All employees will report to the supervisor of animal care or his or her designee any accidents causing personal injury. Persons working with macaque monkeys (such as a rhesus monkey) or in any area that has a biohazard sign posted will also report minor accidents, such as scratches.

The supervisor of animal care or his or her designee will record the type of accident (e.g., a cut finger), how it occurred, the specific area of the body affected, where and when the accident occurred, and the date of occurrence. This information will be kept in a log book on the secretary's desk in the laboratory animal facility office. The log will also note if medical treatment was provided (see SOP A-30: Employee Medical Treatment after Accidents).

Approved by _____ Date _____

SOPs that do little more than list skills that are required by a person are not sufficient for a training document. For example, a training SOP may indicate that a veterinary technician is supposed to learn how to restrain a mouse and perform a subcutaneous injection (two skills). But is there any outcome assessment (e.g., does the technician know how tight a mouse should be held, or how much fluid can be administered?)? A well-developed training SOP should always indicate the basic skill to be learned (e.g., restraining a mouse). To this, you might consider adding the details of how a mouse is to be restrained (although adding this amount of detail may be frowned upon by some trainers). But if there is one thing that I believe the SOP has to include, it is appropriate outcome assessment criteria. Therefore, before a person can be certified as being "trained" in mouse restraint and subcutaneous injections, he must do two things. First, he must show the trainer how to restrain a mouse and perform a subcutaneous injection, and second, he must describe to the trainer how he will prevent harming the animal by not using too tight of a grip and how much fluid can typically be given in a subcutaneous injection. The last two items are the outcome criteria.

You might also want to consider two types of training SOPs: one for the training of a new employee and one for continuing training (e.g., an annual training refresher). Long empirical experience tells us that people have to be retrained, or at least reevaluated, in many tasks, especially those that they do not perform every day. But even for those tasks that people do perform routinely, there can be a certain amount of unwanted drift into bad habits. For this reason, many laboratory animal facilities have an annual (or other interval) retraining program for employees. This is strongly recommended, and if you have such a program, you should have an SOP for it that is similar to that used for new employees in that it will describe the skills to be demonstrated and the desired outcome assessment.

Don't look at SOPs as carved-in-granite documents. They should be reviewed annually to make sure they are current. The needs of laboratory animal facilities change, and you cannot expect SOPs that were made 10 years ago to reflect your current operations. Similarly, don't rigidly stick to your SOPs if a particular circumstance demands laying them aside. You have to be a little flexible at times, depending on the situation. If your SOP states that rabbit cages are to be changed daily, but a heavy snowfall leaves you with a skeleton staff, are you going to insist that every rabbit cage be changed? As long as it does not affect the well-being of the animal or the research, you may have to bend for that one day. I firmly believe in sticking to SOPs whenever possible. They help keep the consistency in research and are one way to let people know what is expected of them. But nothing drives me as crazy as when someone blindly insists on following an SOP or another policy simply because it is written on a piece of paper. Even in pharmaceutical companies, where SOPs usually have to be followed very closely because of regulatory requirements, the good manager knows when it's time to back off a little.

Another document that all employees should be given is the *Guide for the Care and Use of Laboratory Animals* [2]. The guide is a standard reference for almost all people working in laboratory animal science, and your employees should have a working knowledge of its contents. Just don't expect most people to read it without prodding. You should consider reviewing sections of the guide at regular staff meetings.

If a large amount of written material is to be used for training purposes, I suggest that you give careful consideration to the ability of people to read and comprehend it. There are still many people in this nation who have difficulty with reading. Whether due to inadequate academic preparation, difficulty with the English language, or other reasons, if somebody cannot understand what is written, then the purpose of your training material has been defeated. Therefore, you may want to add pictures, use films, do on-the-job training, or use other means as teaching devices. If you do use written materials, write short, clear paragraphs, and do not put too much on any one page. The more you fill up a page with words, the less people are prone to read it.

A good trainer will identify a new employee's learning style or simply ask, "How do you learn best? Is it by reading, doing things yourself, watching somebody else do it first, or something else?" Just as different managers have differing managerial styles, different people have different learning styles, and a trainer who is attuned to a person's learning needs can make a big difference in the quality of training.

Other common materials for training employees include formal lectures, slides, and films. Commercial companies will also come to your facility to train personnel. Laboratory animal facilities are fortunate that American Association for Laboratory Animal Science (AALAS) and other organizations have developed training manuals, slide sets (some of which are available on CDs and DVDs), and related materials. For managerial training, AALAS offers advanced training through the Institute for Laboratory Animal Management and the Certified Manager of Animal Resources (CMAR) program. There are also online and printed training programs for IACUC members and other employees, and short meetings that focus on IACUC issues (e.g., IACUC 101) are also held throughout the country. And, during the training period, you can always ask a person about any additional training or practice she would like to receive.

People from different nations and cultures who require training may require special considerations. Although there may be some language difficulties, typically these individuals are reasonably capable of writing, speaking, and understanding English. Still, it is appropriate for the trainer to speak clearly and without using jargon, request frequent feedback from the trainee to ensure that he has grasped a concept, provide written materials that are equally clear (and perhaps in the trainee's native language), and consider the use of multiple means of teaching the same concept (e.g., hands-on, videos, written materials). In Chapter 3, there is an expanded discussion about managing persons from different cultural backgrounds.

Hands-On Training

This section discusses hands-on training for the laboratory animal facility staff, not investigators or their staff. I believe that training requires more time and forethought than most supervisors appreciate. I began thinking about this while watching flagmen at construction sites. It seemed to me that no two of them directed traffic the same way. One would wave the flag when he wanted you to stop, and another did the same thing when he wanted you to go. Still another waved the flag in a figure-eight motion, and I had no idea what that meant. I witnessed an accident when one poor soul (it turned out to be my secretary) could not understand what the wild waving of the flags meant and moved her car forward when she shouldn't have done so. It seems reasonable to assume these flag men should have been better trained. Perhaps someone just assumed that they all knew what they were supposed to do.

People working in the cage wash area require training also, for, like the flagmen's job, this is an important operation. Yet, the requirements to perform the job correctly are not obvious. The functioning of the cage wash area can affect the efficiency of your entire staff. Still, for reasons I will never understand, there are some managers who, in their own minds, stratify the positions under them and give the least training to those areas they view as relatively unimportant. These people think that cage washing is one such area. Cage washing is not unimportant or insignificant. In fact, I have yet to find an insignificant position in an animal facility. If there is one, train a monkey to fill it, not people. All positions in a laboratory animal facility require thorough training. Let's not forget that the pyramid is only as strong as its base.

I believe that most laboratory animal facility employees should be able to fill in for someone who is on vacation, out ill, or otherwise not at work. This is called being cross-trained. There are some obvious exceptions, but in general, most facility managers support the concept of cross-training. Nevertheless, I am not in favor of rotating animal care duties on a regular basis. As far as I can tell, and based on feedback from animal care technicians, animals get used to the person who cares for them. Until I am shown otherwise, I believe that a rotating group of animal care technicians can place an unnecessary stress on the animals. Granted, this is somewhat anthropomorphic and partly based on my own experience when I was hospitalized and the nurses were always being changed, but many animal care technicians have told me that their animals become upset when someone else cares for them. Nevertheless, some colleagues do not agree with this position. It becomes somewhat of a trade-off. I think that our employees can handle the stress of not rotating a lot better than the animals can. There is enough stress on the animals on weekends and holidays when an unfamiliar person takes care of them. Needless to say, if you subscribe to my philosophy, I strongly encourage you to let your staff know your reasons for having the same people routinely care for the same animals.

Your training program should include unambiguous examples of what constitutes adequate efficiency and effectiveness. As I noted previously (and repeatedly), employees have a right to know what is expected of them and what constitutes satisfactory performance. What is satisfactory performance? You can, for example, explain how the cages are monitored for microbiologic contamination. This might be one way of describing the means by which some of an employee's effectiveness can be measured, or even the effectiveness of the cage washing machine.

Explain the consequences to your organization's effort if a poor job is done. Training is more than telling and showing people how to do a job. It must include a description of the rationale behind and the importance of the job.

It will, of course, take more than an SOP manual, the guide, and a pep talk to train most people. A common way of training a new employee is to either train the new person yourself or delegate this responsibility to another employee. If you delegate the responsibility for training a person, choose carefully, for the future performance of a new employee may depend on the indoctrination received during the first few days on the job. Because so much training occurs informally, the last thing you want is a trainer who says something like "I know what he just told you, but let me tell you the real way we do things around here." That's asking for trouble. Use a trusted, even-tempered person who takes nothing for granted and who is willing to repeat the training process until the new person understands and can perform what is expected. If you choose to train a new person yourself, you will find it is an excellent opportunity to reevaluate and sharpen your own skills.

Some training requires many repetitions of the task to be done, and for important tasks, such as restraining a large animal or changing cages in a biocontainment environment, there can be no shortcuts. Yet, there are only 24 hours in a day, so do we have to make sure that every person knows every detail of every aspect of his job? The answer is no, we don't have to know how to flawlessly perform everything we do (sometimes "good enough" is all it takes), but we do have to have very good skills for certain tasks, such as the just mentioned large-animal restraint. This means that a technical person should be practicing and repracticing primarily those tasks that are most important to her work. Changing mouse cages in a conventional environment certainly requires some practice, but changing mouse cages in a biocontainment suite housing animals that have been inoculated with human pathogens may require substantially more skill and practice.

In Appendix 1, I stated that with Lean management, the concept of kaizen dictates that good enough is not acceptable, and there should be a continued effort toward making improvements. But in the above paragraph, I said that sometimes good enough is sufficient. Is this a contradiction? Not really. Good enough can be adequate for noncritical tasks that a person performs infrequently, but it is not good enough for critical tasks (such as surgery) or for the performance of a person's primary responsibilities.

Another question to ask, and oftentimes the most difficult, is, "Are you a trainer?" Do you have the ability to train and teach, or do you just pass on information and assume you have "trained" someone. Who trains the trainers and evaluates their skills? As already noted, there are many resources in our field that can be used to improve and sharpen the skills of the trainer (e.g., AALAS-offered training).

When training new staff members, explain why their job is necessary and how it relates to the overall operations of the laboratory animal facility, and to your entire organization.

I would probably be remiss if I did not remind you that the world is changing. I do not, not for one second, negate the importance of hands-on training. That will always be important. But let's not forget that sitting down and reading SOPs or a training manual may be getting passé. Computer learning is with us to stay. AALAS has moved in that direction, and you too should think about ways of using computer-based training for new and existing employees.

Evaluation of Training

You should evaluate whether the training program was adequate for your employee and if it served its intended purpose. When you do this, you are implementing yet another management control system, and as you know, developing and using control systems is one of the roles of a manager. This evaluation should be ongoing, so that the need to correct, change, or modify an employee's responsibilities can be implemented in a timely manner. The easiest method, and a very effective one, is to ask the new employee if he understands what has to be done and whether he knows how to do it. Your observations, and your discussions with the person who is training the new employee, are also very valuable. Some animal facilities have formal evaluation sheets that are filled out by both the new employee and his supervisor. I like these forms. They help us remain in compliance with certain regulations, such as those of the Animal Welfare Act. Table A4.1 shows part of a form that might be used in a medium-size animal facility. This type of a form can also be used when established employees perfect new skills. For simplicity, I have not included any outcome assessment criteria that I discussed earlier in this appendix.

We all have strengths and weaknesses. A good training program should amplify a person's strengths, generate a positive attitude, and to the extent possible, minimize a person's weaknesses. If a weakness cannot be minimized to the point where it has no significant effect on your operations, you should reevaluate your training or consider moving the person to a more appropriate position.

Table A4.1 Documentation of Training Proficiency

Great Eastern University		
Laboratory Animal Skills Summary		
Employee: Peter J. Taub		
Skill	Date of Proficiency	Trainer
Rat handling	Jan. 4, 2017	K. Jackson
Mouse handling	Jan. 5, 2017	K. Jackson
Change rat cage	Jan. 4, 2017	K. Jackson
Change mouse cage	Jan 5, 2017	K. Jackson
Primate handling		
Cage washer operation	Jan. 25, 2017	R. Smith
Room sanitation	Feb. 15, 2017	K. Jackson
Mouse blood collection		
Rat blood collection	June 18, 2017	G. Rowan

Mentoring

Let's discuss mentoring of up-and-coming managers (mentoring by leaders is discussed in Chapter 7). I suggested earlier that mentoring differs from training in that mentoring is less hands-on, more personal, and more focused on motivating and advising a person. To be more formal, we can say that in the context of animal facility management, mentoring is a mutually agreed-upon professional relationship between an experienced manager and a person desiring to gain additional management skills and knowledge.

Mentoring is a mutually agreed-upon professional relationship in which an experienced manager teaches and advises a less experienced person, with the goal of having the less experienced person gain additional management skills and knowledge.

Mentoring has some obvious benefits. The mentee (the person being mentored is called a mentee or mentoree) finds a safe haven in having a personal relationship with a more experienced person who is willing to help her reach certain career goals. A happy mentee is a happy employee, and a happy employee is a motivated employee who may choose to make a career at the mentor's institution. The mentor receives the satisfaction of helping another person and may gain a supporter when the mentor has to influence others in the organization, and as with any teaching position, the mentor is likely to learn a lot about the mentoring process that will help when other mentees join the institution. There are some potential downsides to mentoring. First, a particular mentor may not be a particularly good manager, and the mentee may learn poor management skills rather than good ones. This can also happen when a mentor suggests that the mentee follow in the mentor's footsteps without considering that the mentee may have a very different personality or management style. Another downside is that mentoring, like managing, requires learning and practice. A mentor who has no experience in being a mentor may not do a good job of mentoring. An additional consideration is that a person who is a good manager but has not shown the drive or the ability to rise to a leadership level may be chosen for mentoring (this type of person can remain on the management ladder but not in a high leadership position). Lastly, a downside of mentoring is that other employees may notice that mentoring is occurring and resent the mentee getting special treatment. There's a certain amount of truth to that perception, but consider this: Would an up-and-coming employee want to stay with you if she didn't see that her skills and ambition were being recognized and utilized? The possible sensitivity of nonmentored employees can be eased, at least a little, by being honest with everybody, letting them know that a person is being mentored and why that mentoring is happening. Of course, consideration for advancement and mentoring must be open to all qualified employees, not just personal favorites.

> **Would an up-and-coming employee want to stay with you if she didn't see that her skills and ambition were being recognized and utilized?**

Like most management skills, mentoring is based on the trust and open communications that are developed between the mentor and the mentee. Mentoring usually does not have any formal teaching sessions (although sometimes that happens); rather, it is more about the mentee being informally guided by the mentor, with the mentor offering constructive feedback and encouragement to the mentee. Although the mentor is often older than the mentee and in a more senior position, this is by no means a requirement. For instance, a new director of an animal facility can be mentored judiciously by an established animal care supervisor about the culture of the organization. Further, the new director can have more than one mentor, either simultaneously or at different times. In addition to the animal care supervisor, the vivarium's business manager may be able to mentor the new director on the institution's current business processes and introduce the director to people he should know. There are many faces to mentoring, and different people with different skills and values can all be effective mentors. Many of us in laboratory animal science have had one or more mentors who have taken us under their wing and coached us, yelled at us, led us, but always supported us in our career journeys. Many of you may have acted as a mentor without realizing that you were mentoring. That's a great starting place, but we want to delve a little deeper into the mentoring practice.

Mechanics of Mentoring

How does mentoring happen? Some animal facilities will routinely match a new employee with an established person who understands ahead of time that she will be the new person's mentor. That can be considered a formal mentoring program, particularly if the new person desires to obtain a management position or advance an already initiated management career. However, if the new person is a technician, what is called "mentoring" is often more akin to a technical training program. There is nothing wrong with technical training, but it's not mentoring. In other facilities, new employees may be encouraged to find their own mentor. Perhaps we can label this as a semiformal program. This semiformal method often leads to a closer bonding between the mentor and mentee, but it has the potential to be a failure if the new person is simply too shy or too overwhelmed to seek out a mentor. However, once a new person is acclimated to his new position, he can often informally bond to a person he interacts with and who will become his mentor, but this may or may not happen, and there may or may not be a clear understanding that a mentoring relationship has been established. Another common way mentoring relationships are established is when a senior manager sees managerial potential in another person and the senior manager actively begins to mentor that person for advancement within the organization.

No matter how the initial mentoring connection is established and whether it is considered to be formal, semiformal, or informal, there are some basic considerations that should always ensue once both parties understand that there is going to be a mentoring relationship. To begin with, mentoring should have a mutually agreeable goal because mentoring is goal oriented; it's not just being friendly to a person. If you are the animal facility's director who is mentoring a young supervisor, one of her goals might be to pass her CMAR examination in one year. That's a clear goal, although you may have two or three additional goals for her that are more oriented toward general management skills than is passing or not passing an examination. You also should have an understanding of how often you will meet (perhaps once a week at first and then monthly), what subjects can or cannot be discussed (e.g., the mentor's relationship with her own boss may be off limits), what information will remain confidential, and anything else of importance to both of you. However, be aware that it's easy to fall into the trap of trying to solve all of a mentee's personal problems. I strongly recommend that you shy away from doing so unless you are a trained counselor.

I will also point out that, in general, it's not the best idea to have a direct report (a person who reports directly to you) as your mentee. You hold too much power over that person and establishing a true mentor–mentee relationship can be difficult, although not impossible.

It is usually best for a mentor not to have a direct report as a mentee or not to try to solve all of a mentee's personal problems.

A mentee's personal goals are every bit as important as the goals the mentor may have for the mentee. A mentee has to ask herself, what do I want to accomplish during my mentoring experience? Do I want to be mentored about personnel management? Do I want to focus on financial management of animal facilities? Do I want to learn more about nonhuman primate medicine? Do I want to enhance my skills in environmental enrichment for research animals? Do I need mentoring in the management skills required by a first-line manager? Do I want more practical experience in leadership so I can potentially direct the vivarium when my boss retires next year?

Once one or more goals are agreed upon between you and your mentor, you will have to establish a general mentoring plan to reach that goal, and as you may have already guessed, you will need to have some specific strategies to accomplish the plan. Along with the strategies, you will need some metrics to see how you are progressing toward the goal. All of this may seem familiar to you because these are all topics covered earlier in this book. In Chapter 1, we discussed the concept of where am I now, where do I want to be, and how am I going to get there. The same chapter reviewed goal setting, planning, and the need for developing specific strategies to fulfill the plan. Then, in Appendix 1, there was an expanded discussion about using metrics to measure productivity and the progress of the plans being used to accomplish a goal. All of these topics are applicable to mentoring, although in a real-world situation, a strategic plan can be more flexible when you are in a mentor–mentee relationship. The reason for this is that most mentoring relationships are somewhat more personal and relaxed.

Let's move ahead. The question is, what do mentors do when they are mentoring? Or, to put it another way, what plans and strategies are often used by mentors? If the goal is to have the senior clinical veterinarian supersede the director in two years, the general mentoring plan may be to increase the visibility and responsibilities of the senior veterinarian during that time period, with feedback being given by the director at monthly mentoring meetings. Here's an example of a plan and strategies for the associate director to succeed the director in two years.

- Plan component: The director will meet monthly with the senior clinical veterinarian.
  - Strategy: A discussion about significant managerial events and problems faced by the director or senior veterinarian. Why did they occur, how were they handled, and what could have been done differently?
  - Note: These discussions are not performance reviews. They are meant to be informal and supportive. Most of what gets said should remain private between the mentor and mentee.
- Plan component: The director will attempt to have the mentee accompany her to meetings with administrators and faculty.
  - Strategy: The mentee will observe the manner in which the director approaches different discussions at meetings.
- Plan component: The director will have the mentee assume some of the tasks performed by the director.
  - Strategy: After one year, and assuming adequate progress, the mentee will represent the director at some institutional meetings.

The details can continue, but I hope you can see the general picture of the director guiding the senior clinical veterinarian and gradually giving the veterinarian more responsibilities.

The mentor has to do more than open the organization's doors for the mentee. The mentor also has to create a safe learning environment because some mentees may be hesitant to ask questions that might be considered "stupid" [3]. To put a somewhat hesitant mentee at ease, a mentor can ask questions that begin with "What did you think about ...?" For example, "What did you think about John's suggestion?" or "Do you think that John's idea has a reasonable chance of working?" Or, the mentor can make an even safer statement to get a conversation going by saying, "I didn't totally understand what John was proposing to resolve the cage washer problem we're having. Were you able to follow his reasoning?"

Along the course of mentoring, the mentor is continually evaluating her mentee to determine if the mentee is living up to expectations, is exceeding expectations, needs some additional coaching, or is struggling to the point of no return. (Coaching and mentoring are not quite the same, but they are not different enough to spend time discussing the differences.) The mentee also has to determine if her mentor's mentoring style and program are appropriate for the goal that was set and, even more important, if the goal is still something that the mentee wants to reach. Sometimes, mistakes are made and the mentor has to either redouble her efforts or suggest an alternate goal for the mentee, or perhaps a different mentor. Likewise, a mentee may have to request additional (or different) support or go so far as to tell her mentor that she very much appreciates the help she has received, but after much consideration, she does not believe that the goal that was mutually set is still the goal she wishes to pursue. These things happen, but as long as there is a good professional relationship between the mentor and mentee, any such discussion will not be viewed as a failure, but as a decision that is likely to be best for both parties.

Caring for and about Our Fellow Creatures

The final comment about training and mentoring is certainly not the least important. We can teach people how to manage, handle animals, care for their daily needs, clean rooms, keep records, and the like. We can be taught to develop budgets, establish *per diem* rates, and mentor a person to adjust his or her management style to meet different people's needs. But it's a little harder to teach some people that animals are our fellow creatures, not four-legged test tubes. It should be axiomatic that we have to be accountable to ourselves and the public for the well-being of the animals in our care. Perhaps this is not training or mentoring in the classical sense, but it is nonetheless important. The need to be sensitive toward animals has to come from the senior management of our laboratory animal facility and, hopefully, from our institution's senior management as well. Remember that we have two clients, animals and investigators. It's easy to come to work every day and think of laboratory animal science as a business and nothing else. But that's not true. We care for animals, and animals can feel pain, fear, hunger, and contentment. As managers, we have to set the example of truly caring about the well-being of our animal charges, and by doing this, help train and mentor all the persons with whom we work to do the same. Yes, laboratory animals are not pets, but they are animals, and as specialists in the field, we must teach others to care for them and to care about them.

References

1. Rice, C., Marlow, F., and Masarech, M. 2012. *The Engagement Equation: Leadership Strategies for an Inspired Workforce.* Hoboken, NJ: John Wiley & Sons.
2. National Research Council. 2011. *Guide for the Care and Use of Laboratory Animals.* 8th ed. Washington, DC: National Academies Press.
3. Rothwell, W.J., and Chee, P. 2013. *Becoming an Effective Mentoring Leader: Proven Strategies for Building Excellence in Your Organization.* New York: McGraw-Hill.

Index